GOLETA BRANCH
LIBRARY

D0455592

SANTA BARBARA PUBLIC LIBRARY
GOLETA BRANCH

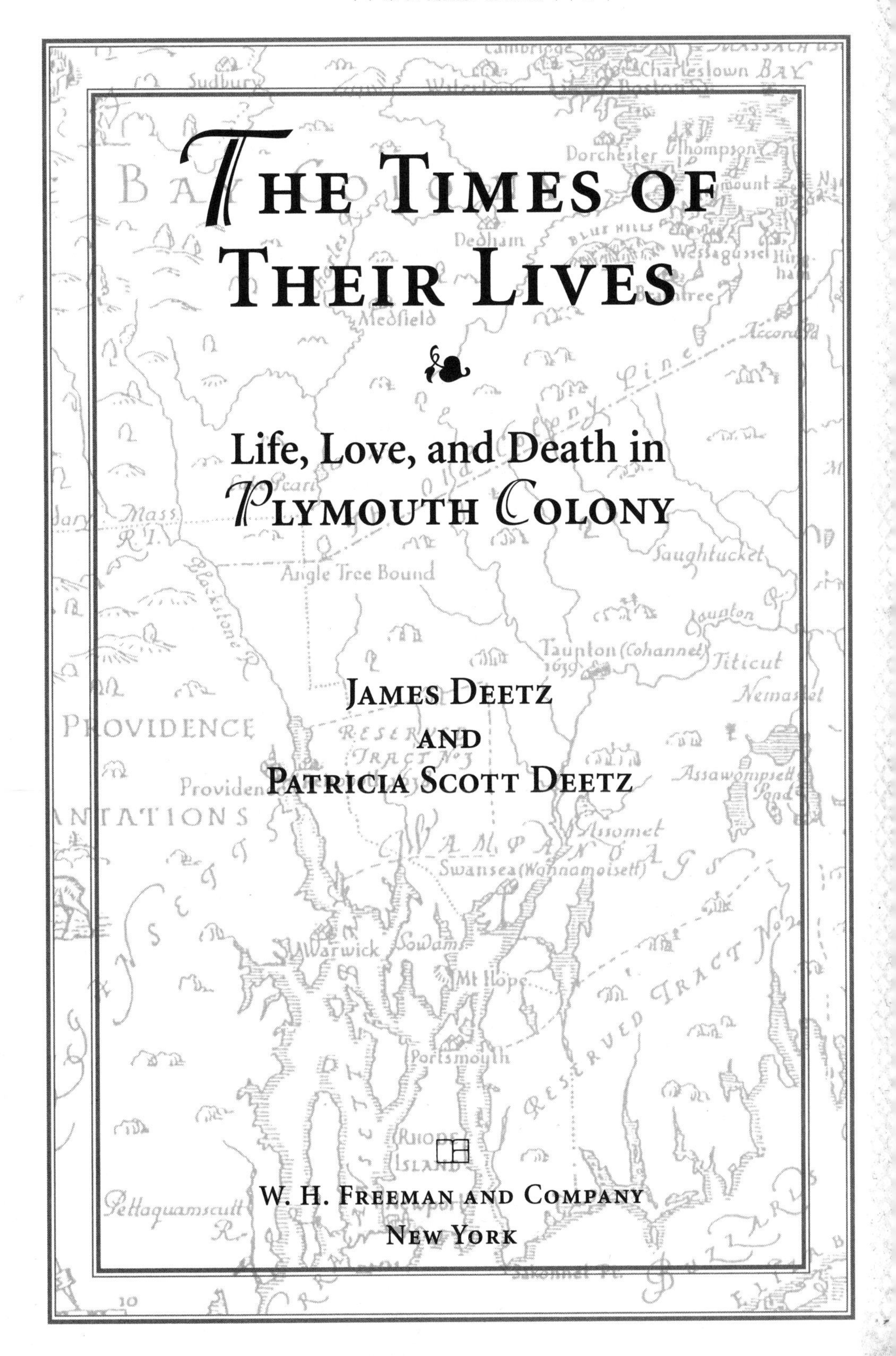

*T*HE TIMES OF THEIR LIVES

Life, Love, and Death in *P*LYMOUTH *C*OLONY

JAMES DEETZ
AND
PATRICIA SCOTT DEETZ

W. H. FREEMAN AND COMPANY
NEW YORK

Text designer: Diana Blume

Library of Congress Cataloging-in-Publication Data

Deetz, James.
The times of their lives : life, love, and death in Plymouth Colony / James Deetz and Patricia Scott Deetz.
p. cm.
Includes bibliographical references and index.
ISBN 0-7167-3830-9
1. Massachusetts—New Plymouth, 1620–1691.
2. Massachusetts—Social conditions—17th century.
3. Massachusetts—Social life and customs—To 1775.
4. Pilgrims (New Plymouth Colony)—Social conditions.
5. Pilgrims (New Plymouth Colony)—Social life and customs.
I. Deetz, Patricia E. Scott. II. Title.

F68 .D4 2000
974.8'202—dc21

00-042947

Printed in the United States of America

Second printing, 2001

W. H. Freeman and Company
41 Madison Avenue, New York, NY 10010
Houndmills, Basingstoke RG21 6XS England

To Harry Hornblower and Ted Avery

Thus out of small beginnings greater things have been produced by His hand . . . and, as one small candle may light a thousand, so the light here kindled hath shone unto many, yea in some sort to our whole nation.

—William Bradford, 1630

Contents

Illustrations

Chapter 1

Chapter 2

Chapter 3

CHAPTER 4

Chapter 5

Chapter 6

CHAPTER 7

Preface

It has only been since the opening years of the twentieth century that Thanksgiving as we know it today has been observed, when it joined a number of other themes that together make up the American origin myth of the Pilgrim Fathers, all deriving in one way or another from the story of the early settlement of Plymouth, Massachusetts. Turkey with stuffing, potatoes and giblet gravy, cranberry sauce, and pumpkin pie—are all familiar icons of one of America's most popular and widely observed holidays. Indeed, if travel statistics are to be believed, America's most popular holiday, because more people travel over that weekend than any other. It is a time for families to gather and offer thanks for their blessings. Not all Americans share this view, for to Native Americans the day has a very different meaning, but in the main, it is a very special day, marked by overindulgence in food, so much so that a cynic might call it the national day of gluttony. Thanksgiving is, in fact, probably the most internationally known American tradition, and for many Americans and immigrants who are being absorbed into American culture it is such an important institution that we felt it would be the natural place to begin our story.

Although the literature on seventeenth-century Plymouth is extensive, with a few exceptions it is very uneven in its coverage of the history and culture of the colony. Much of it was written in the nineteenth century, when the Pilgrim myth was assuming an increasingly robust form. Many of these works are hagiographic, many more genealogical, and sometimes both. There is also a scattering of romantic works, probably owing in some part to the popularity of Henry Wadsworth Longfellow's epic poem *The Courtship of Myles Standish*, published in 1858. There are also a number of town histories. These nineteenth-century works are not without value, but they do need to be used with some caution insofar as historical reliability is concerned, and often require extensive consultation before it is possible to extract information for which one is searching.

With a large number of works on early Plymouth, one could reasonably ask, "Why yet one more?" When we set out to write *The Times of Their Lives* we believed, and still do, that there is no single work that combines between two covers the diverse number of themes that we have included. We cannot stress too much, though, that this work is not of the "debunking" school, but rather an honest attempt to present a realistic, factual accounting based on primary sources. The existence of a rich body of seventeenth-century material that is little known by the public prompted us to begin work on a volume that would focus on life and death in Plymouth Colony and remove some of the widely held misconceptions concerning the "Pilgrims." The enthusiasm of students working for the first time with the primary sources is well summed up in a comment made during the initial seminar we held in 1996 at the University of Virginia. Encountering the "real" Plymouth settlers in the firsthand accounts of the court records, one of them exclaimed, "You told us we would find the seventeenth century come alive, but, man, was I ever there!"

The main sources for the criminal cases that we discuss are the records of the General Court of Plymouth Colony, which provide insightful and remarkable details concerning the lives of at least some of the colonists who settled in Plymouth. The official court records were edited and published for the first time in the 1850s, and were reprinted in 1968 by AMS Press in New York. We have also drawn on probate inventories—the goods and property of individuals listed for tax purposes after their death—and wills for insights into the material world of the settlers. Although one volume of wills and probate records for the colony from 1633 through 1669 has been published, access to those from 1670 onward is less easy. The detailed sources that appear at the end of the book are designed as a guide to those who want to access published and unpublished material. In addition to William Bradford's contemporary history *Of Plymouth Plantation 1620–1647*, we have drawn on *Mourt's Relation* (London, 1622), which is also an eyewitness account of the early settlement. *Three Visitors to Early Plymouth*, first published by Plimoth Plantation in 1963, provides fascinating objective views of the colony during its first years, as do some of the writings of Captain John Smith.

Seventeenth-century spelling was not standardized in the way spelling is today, and to make the texts we have used easier to read and understand we have employed modern orthography, but have preserved the original punctuation and capitalization of the seventeenth-century authors to give something of the flavor of the period. The spelling of names varies considerably, and we have chosen the form most regularly in use by scholars in the field. The term "Pilgrim" is one that we are at pains to avoid using although it has become an apparently inextricable part of the story of Plymouth Colony and the American origin myth, but the people who settled in early Plymouth were not referred to as "Pilgrims" until the end of the eighteenth century. We have also modernized "old style" dates—that is, dates between January 1 and March 25 have been changed to the modern calendar and are considered part of the new year. So what in the seventeenth century would have been January 19, 1675/76, is given as January 19, 1676.

As this book is intended for popular reading and enjoyment, the text is seamless, with no references to distract. We have, however, provided detailed sources and notes for those who are interested in taking aspects of the text further, as we feel strongly that with a subject of such interest and importance it is necessary to underpin statements made and to acknowledge our sources.

While we use the "first Thanksgiving" as our point of departure, and consider the myths, familiar to millions of Americans, that have emerged concerning the "Pilgrims," we then look back to the events that led up to the settlement of Plymouth Colony and, more significantly, the years following that event through 1691, providing glimpses of life in the colony. These years are particularly important because to large numbers of people the early settlers at Plymouth sailed across the Atlantic on the *Mayflower,* had a big dinner the following fall, and disappeared. In truth, Plymouth Colony has an ongoing story that is worth recounting in all its colorful detail, enlivened and expanded by contemporary archaeology, cultural research, and living history.

Acknowledgments

This book could never have been written were it not for 101 students who took the authors' five seminars on the historical ethnography of Plymouth Colony, held at the University of Virginia in the Department of Anthropology from the spring of 1996 through the fall of 1999. They all did an excellent job of mining the primary sources for raw data, and also prepared biographical profiles of 395 people who lived in Plymouth Colony. Of the 101, we want to thank in particular seven students whose research was directed at themes we explore here: Allison Devers, Christopher Fennell, Jillian Galle, Jason Jordan, Lisa Lauria, Anna Neuzil, and Jeffrey Norcross—without them this book would have been years more in the making. Special appreciation goes as well to Andrew Beahrs and Chip Cunningham for their contribution, to Andrew Beahrs and Derek Wheeler for directing the archaeological field school at the Howland site in the summer of 1998, and to Derek for doing so ably on his own in 1999. We express additional warm thanks to Chris Fennell, who is also responsible for a major project in setting up and maintaining the Plymouth Colony Archive Project Web site, which includes archives of historic records and modern analyses, all searchable. The Web address is http://etext.virginia.edu/users/deetz/. We also thank June Webb, of the Department of Anthropology, whose administrative and organizational expertise rescued us time and again.

We deeply appreciate the moral support of Henry Glassie throughout the writing of this book. At one point, when the senior author was seriously disabled, Henry flew from Bloomington, Indiana, to Charlottesville, Virginia, just to cheer him up, and it made all the difference in the world. We also thank Nancy Brennan, director of Plimoth Plantation, and her staff for accommodating our field school sessions of 1998 and 1999 and for always being supportive of our research. James Baker, Webmaster and historian at the Plantation and a walking encyclopedia of Plymouth history, was always available to answer esoteric questions that few if any other people could have. Carolyn Freeman Travers, research manager at Plimoth Plantation, was helpful on numerous occasions, supplying graphic material and information derived from many years of involvement at the plantation. It is significant that her father, David Freeman, was director during the period of radical changes that took place in the plantation's museum program in the 1950s through the 1970s, a program to which he gave his steadfast support.

So many people were involved at Plimoth Plantation and in the archaeological excavations that took place during the years in which James Deetz was assistant director that it is not possible to name them all. Special mention must be made, however, of Ted Avery, to whom this book is dedicated together with Harry Hornblower. Ted, a New Englander from nearby Kingston, is the brilliant photographer whose recording of all aspects of the work was invaluable, and he and his wife, Anna, were a constant and special part of the many parties of personal friends who gathered at the Deetz home in Plymouth. Mention must also be made of Jim and Jody Deetz's nine children, who grew up in the middle of it all—Jamey, Tonia, Joe, Cricket, Eric, Geoff, Josh, Cindy, and Kelley. They left their own legacy.

Our deepest appreciation goes to our editor at W. H. Freeman and Company, Erika Goldman, who had faith in the book before the first word was written, and has been of inestimable value in seeing it through to a successfully completed manuscript. Georgia Lee Hadler, senior project editor, and Jane Elias, our copy editor, have kept us on track with their standard of excellence. We also thank Min Tai for her constant, efficient behind-the-scenes assistance. Our thanks, too, to Sally Brady, our agent, for sorting out a number of logistical problems and for her overall encouragement. With a group like this, how could anything go wrong? Nevertheless, all shortcomings are fully our own. But we are so much the richer for the participation and support of our students and the encouragement of colleagues—our department chair, Richard Handler, stands out for specific thanks and appreciation—as well as family and friends who were there for us under all sorts of circumstances, particularly Eric Deetz, Anna Agbe Davies, Kelley Deetz, Seth Mallios, Anne Roos, Barbara Deetz, Antonia Deetz Rock, Alison Bell, and Derek Wheeler. We salute you all.

The Times of Their Lives

Chapter 1

*P*artakers of *O*ur *P*lenty

The Pilgrim Myth

MR. and Mrs WEST, Viewing the Rock on which our Fore-Father's Landed at Plymouth.

Our harvest being gotten in, our governor sent four men on fowling, that we might after a special manner rejoice together after we had gathered the fruit of our labors. The four in one day killed as much fowl as, with a little help beside, served the company almost a week. At which time, amongst other recreations, we exercised our arms, many of the Indians coming amongst us, and among the rest their greatest king Massasoit, with some ninety men, whom for three days we entertained and feasted, and they went out and killed five deer, which they brought to the plantation and bestowed on our governor, and upon the captain and others. And although it be not always so plentiful as it was at this time with us, yet by the goodness of God, we are so far from want that we often wish you partakers of our plenty.

So wrote *Mayflower* passenger Edward Winslow, in 1621, to a friend in England shortly after the colonists of New Plymouth celebrated their first successful harvest. This brief passage is the only contemporary eyewitness description of the events that were to become the basis of a uniquely American holiday: Thanksgiving. As with many of the facts of those first years in America, this occasion has become so imbued with tradition that it is difficult to place it in the position it occupied in Winslow's eyes. However, a careful critical reading of Winslow's brief text can help restore that perspective.

"Our harvest being gotten in . . ."

Clearly, this occasion took place at harvest time, which in New England, depending on when the crops were planted and how fast they matured, would have been in the early fall, possibly as early as September, and in no case later than early October. And while the observed date of Thanksgiving has been changed over the years, since President Abraham Lincoln's proclamation in 1863 designating the fourth Thursday in November as Thanksgiving, that date has been the one that we observe. A slight change was enacted by Congress in 1941, altering the date from the fourth to the last Thursday in the month, but this only makes a difference in the infrequent November that has five Thursdays.

"Our governor sent four men on fowling . . ."

In all likelihood, the fowl that were killed were ducks and geese, taken with fowling pieces, lightweight firearms that could easily be fired from waist or shoulder, without the musket rest that was needed when using a matchlock musket, a heavy weapon used primarily in military engagements, which required the rest to prop up the barrel. Fowling pieces outnumbered muskets throughout the colony's history, and were the main weapon used in hunting. They were of smaller bore than a matchlock, and fired a single ball. As for turkeys, it is less than likely, though not impossible, that some may have been taken as well. The association of turkeys with Thanksgiving seems to have originated in a passage from William Bradford's history of the colony, written in 1646, many years after the event described by Winslow. A number of historians refer to this passage as a later description of the "First Thanksgiving," including Samuel Eliot Morison in his excellent edition of the Bradford history. But Bradford was writing in more general terms about the fall and winter of 1621:

> They began now to gather in the small harvest they had, and to fit up their houses and dwellings against winter, being all well recovered in health and strength and had all things in good plenty. . . . All the summer there was no want; and now began to come in store of fowl, as winter approached, of which this place did abound when they came first (but afterward decreased by degrees). And besides waterfowl there was a great store of wild turkeys, of which they took many, besides venison, etc. Besides they had about a peck [of] meal a week to a person, or now since harvest, Indian corn to that proportion. Which made many afterwards write so largely of their plenty here to their friends in England, which were not feigned but true reports.

Winslow's letter, written on December 11, 1621, was one of these. But neither Bradford nor Winslow explicitly mentions turkeys as a part of the food taken during the "three days . . . we feasted." Harvest time coincided with the fall migration of ducks and geese, and beyond doubt, the ponds, streams, and ocean were teeming with waterfowl, and it would have been easy to kill as much fowl as would serve the company for almost a week. From the perspective of pure efficiency, it is highly probable that the four men took full advantage of the presence of such a great number of waterfowl, rather than entering the forest in search of turkeys, no matter how plentiful they were at the time. Finally, there is an interesting and possibly important semantic dimension to the quotes from Bradford and Winslow. Bradford distinguished between fowl/waterfowl and turkeys, and while turkeys *are* fowl, the fowl mentioned by Winslow were almost certainly ducks and geese, and therefore fall into Bradford's fowl/waterfowl category.

"Many of the Indians coming amongst us, and among the rest their greatest king Massasoit, with some ninety men . . ."

The winter of 1620–1621 was an extremely difficult time for the people of Plymouth. They managed to cross the Atlantic with the loss of but one life, made up for by the birth of a child while still at sea, appropriately named Oceanus Hopkins. This was a remarkable accomplishment, since shipboard mortality on some transoceanic voyages from England to North America could be as high as 50 percent or more. So the number of people who made landfall on November 9, 1620, at Cape Cod was the same as had departed England, the result of a single death and one birth. But, to quote Bradford,

> That which was so sad and lamentable was, that in two or three months' time half their company died, especially in January and February, being the depth of winter, and wanting houses and other comforts; being infected with the scurvy and other diseases which this long voyage and their inaccommodate condition had brought upon them. So as there died some times two or three of a day in the foresaid time, that of 100 and odd persons, scarce fifty remained.

We are all familiar with paintings, prints, and other graphic depictions of the first Thanksgiving, and without exception, they portray a group of English settlers, many with their eyes turned toward heaven in a most devout manner, and scattered among them, a smaller number of Indians. One watercolor that seems to have found favor among motel owners in modern Plymouth, and that can be found in many of the rooms, depicts eight English*men*, seated at a table with an Indian at either end, and several women standing behind them in a clearly subservient role. The reality could not be more different. The celebration described by Winslow was attended by some fifty English settlers, and over ninety Indians, a ratio of nearly one to two. This difference profoundly transforms the stereotypical image of our national holiday.

"Amongst other recreations, we exercised our arms . . ."

The exercising of arms does not refer to some exotic form of seventeenth-century aerobics, but rather discharging firearms, probably matchlock muskets, and possibly fowling pieces, perhaps in some formal kind of drill, although it may have been rather a random kind of target practice. It is likely that some competition was entered into with the Indians, who may have used bows and arrows in much the same fashion. Since Winslow chose to mention only this one of "other recreations," it must have been engaged in at intervals throughout the three days that the Indians were "entertained and feasted." What might some of the "other recreations" have been?

Various sports were popular with Jacobean Englishmen, and in 1618 King James I even issued a book encouraging "lawful sports" and "honest exercises" on Sundays and holy days, such as "dancing, either men or women; archery for men, leaping, vaulting, or any other such harmless recreation . . . without neglect of divine service," requiring this to be read from every pulpit. Puritans and Separatists were opposed to ball sports and blood sports, but approved hunting, fishing, and martial contests. Any one of the sports listed by King James might have been engaged in, especially considering that this was a mixed community of Separatists and those who had come to the New World without strong religious motives. Only a few months after the events described above, Governor William Bradford objected to a new group of settlers playing stool ball, ancestral to cricket, and to throwing weights on Christmas Day, 1621, a religious holiday not observed by the Separatists. He stated that playing in the streets was not appropriate on a day when others were hard at work, but his specific mention of "recreations" that in this case were actually taking place in Plymouth, allows us to suggest that they featured in the harvest festivities.

Winslow has given us a description of an event that, with only a small amount of interpolation, can be characterized by the smell of gunpowder mingled with the aroma of roasting meat, of a celebration bordering on the rowdy, with the sounds of firearms being discharged accompanying talking and shouting, in two languages, and the consumption of quantities of food. Such an image is entirely at odds with the manner in which Thanksgiving has been portrayed in pictorial form, a solemn group of people seated primly at long tables and partaking of the traditional turkey, among other foods.

"Yet by the goodness of God, we are so far from want . . ."

In a sense, the most remarkable thing about Winslow's brief account is that it makes no mention of giving thanks. Certainly he acknowledges the "goodness of God," but that is as close as he comes, and it is not close enough for us to see the event as the "first Thanksgiving." What was really taking place during those three celebratory days was the observation by the settlers of an old English custom, that of harvest home. It was customary to celebrate the harvest of the year's crop with revelry and feasting. In the sixteenth century this was often done even before the harvest had been completed, leading Henry VIII to berate the farmers for passing up the "opportunity of good and serene weather offered upon the same in time of harvest." But by the time of the initial settlement of Plymouth, the holiday was observed only after all of the harvest had been completed. Thomas Tusser, an Elizabethan farmer-poet (if you can imagine such a combination), wrote:

> In harvest time, harvest folk, servants and all,
> should make all together, good cheer in the hall,
> and fill out the black bowl of blyth to their song
> and let them be merry, all harvest time long.

Winslow in fact has given us a description of harvest home as it was played out at Plymouth in 1621. Whether similar feasts were held at Jamestown is not known, but in view of the difficult and trying times faced by the people there, it would seem highly unlikely. What we recognize today as Thanksgiving was probably the first harvest home festival to have taken place on American soil.

Ye ask, What eat our merry Band
En Route to lovely MARYLAND?

Ye ask, What ate our merry Band
To celebrate our Bounteous Land?

The first of the above couplets was written by Ebenezer Cook, and is taken from his epic poem *The Sotweed Factor*, concerning his adventures in Maryland in the seventeenth century. The poem's title was used by John Barth for his superb novel detailing Cook's New World adventures. The second is simply a paraphrase of the first, but could have well been penned by Cook had he been present at Plymouth's first harvest festival. Winslow tells us that they feasted for three days, but except for the fowl and venison the account is silent on the matter of what foods were consumed. We can be certain, however, that the menu was in no way similar to what we regard today as Thanksgiving dinner, turkey "with all the trimmings" (curiously, in parts of Maryland and Pennsylvania these include sauerkraut cooked with pork), as well as the now traditional cranberry sauce and pumpkin pie. We can say with assurance that the latter two items were not a part of the feast in 1621. Pumpkin pie was not to come along until much later, and although cranberries were abundant, no use was made of them. Emmanuel Altham, who visited Plymouth in 1623, referred to them as alkermes berries, as did Captain John Smith, leader of the Jamestown party, both men believing that they were the same vivid red berries that they knew in Europe. The strange fact is that alkermes berries are actually bright-red pregnant female insects found in the Mediterranean, which were long thought to be a vegetable. Their juice was used both as a dye and a cordial.

Lacking any specific mention of what was feasted upon during those three days, we can only turn to a listing of various foods that Winslow included later in the same letter of December 11, 1621. He states:

> For fish and fowl, we have great abundance; fresh cod in the summer is but coarse meat with us; our bay is full of lobsters all the summer and affordeth variety of other fish; in September we can take a hogshead of eels in a night, with small labor, and can dig them out of their beds all the winter. We have mussels and othus [sic] at our doors. Oysters we have none near, but we can have them brought by the Indians when we will; all the spring-time the earth sendeth forth naturally very good sallet [salad greens] herbs. Here are grapes, white and red, and very sweet and strong also. Strawberries, gooseberries, raspas [raspberries], etc. Plums of three sorts, with black and red, being almost as good as a damson.

Drawing on this listing, supplemented by other references, such as Bradford's mention of the quantities of both wheat and maize in ample supply, it is possible to make at least an informed guess as to what was consumed in those three days of feasting and gaming. Ducks, geese, and venison are the three things of which we are absolutely certain. They were probably spit-roasted, but other possibilities suggest themselves in the light of what we know of English foodways of the time. Pottages, stews of a kind, were a basic component of the Stuart yeoman diet. By cooking corn, and possibly wheat, in a broth obtained by stewing meat, and combining that with pieces of duck, goose, and venison, using a large iron kettle over an open fire, a pottage could well have been a component of the feast. There exists today a recipe for succotash, which bears no resemblance to the familiar combination of corn and lima beans, that resembles the pottage described above. It has been prepared by Plymouth families for at least a century, and probably much longer. Known significantly as

Plymouth succotash, it is a mixture of stewed chicken, corned beef, beans, and hominy. This form of succotash is but a variant of the traditional Virginia dish known as Brunswick stew, and both probably share a common origin in the mother country of England.

Moving to those foods that probably were present, we can include various kinds of fish, eels, and shellfish, including lobster. Lobster, if taken with venison, would constitute the first surf and turf served on American soil. Fall would also have been the time that the various fruits mentioned by Winslow had ripened. The "sallet" herbs that were available in the spring would have matured by the fall to the point where they would have been tough and coarse, and of little use. As for beverages, the likelihood of beer in generous quantities seems quite high. Beer was consumed by all seventeenth-century English people in quantities that today would seem excessive. Almost every household produced its own beer, a custom continued in New England, and brewing it is not a very complex or difficult process. The importance of this beverage to the *Mayflower* passengers is underscored in a "relation," or journal, probably also written by Winslow, possibly with the help of William Bradford. After more than a month of exploring the shore surrounding Cape Cod Bay, the colonists finally settled on a site that seemed to suit their needs. The journal entry reads: "We could not now take time for further search or consideration, our victuals being much spent, *especially our beer*. . . ." Wine may also have been present, and even spirits, since we know from a later entry in the journal that such a drink was offered to Massasoit at the conclusion of a peace treaty drawn up between the English and the local Wampanoag people in late March of 1621. It reads: "After salutations, our governor kissing his hand, the king kissed him, and so they sat down. The governor called for some strong water, and drunk to him, and he drunk a great draught that made him sweat all the while after. . . ." Both wine and "strong water" had to have been a part of the *Mayflower's* cargo, but the beer was probably made from the "indifferent good barley crop" harvested in August 1621. Whatever the precise nature of the harvest home feast was, it was consumed using spoons, knives, which at that time had sharp points expressly intended to spear bits of meat, or with the fingers. Forks would not make their appearance until much later. The earliest fork recovered from an archaeological site in Plymouth dates to the very end of the seventeenth century. Forks originated in Italy, and while the English nobility may have been using them earlier in the century, the common folk considered them a foppish pretension. Few houses had been erected by the fall of 1621, so the entire celebration must have taken place out of doors, and lacking tables and chairs in any great number, most of the settlers and Indians sat on either the ground or any available object, be it a log, sea chest, or a rock. And when one remembers that nearly a hundred and fifty people participated in the feast, in a relatively small area, the final image of our popular concept of the celebration falls away.

So it is that Thanksgiving as we think of it today is largely a myth. But this does not mean that being mythical is bad; to the contrary, historical myths of this kind provide a common shared sense of where we have come from as a people. Such myths are usually referred to as origin myths, and people the world over share in them. And in a nation that is so religiously diverse, it seems most appropriate that a secular event, no matter how transformed over time, serve such a purpose. But the Thanksgiving myth is only a part of a larger national origin myth for Americans, that of the Pilgrims and their supposed role in the making of modern America.

So who *were* the Pilgrims? This question has been a vexing one for modern historians, and depending on the source consulted, different definitions emerge. Were they all of the *Mayflower's* passengers, or were they the minority of religious dissenters among the group? Does the term refer to those who came on four other ships, the *Fortune, Anne, Little James,* and *Charity*, which arrived during the first seven years of the colony? Might the term apply to all of the residents of Plymouth Colony during its existence as a separate colony until 1691? There is no modern consensus regarding this matter, and little wonder, for the people of Plymouth never perceived themselves as a group who would at the end of the eighteenth century come to be known as Pilgrims.

However, if we change the tense of the verb in the question from *were* to *are*, a reasonably concise definition can be offered. The Pilgrims are a quasi-mythic group of people who are looked upon today as the founders of America, and whose dedication to hard work and noble purposes gave rise to our nation as we know it. What most of us know about them we learned as early as grade school, especially around Thanksgiving time. Stern and God-fearing, possessed of the loftiest motives, the women dressed in somber attire with white collars, and the men dressed in gray and black, with buckles on their hats, belts, shoes, and, for all we know, even on their undergarments. Some modern Plymouth residents refer to them as the "Grim Pills." This is the image with which we are all so familiar, but its origins lie more in early-nineteenth-century America than in the reality of a time two hundred years earlier.

With the final stroke of the pen at the signing of the Treaty of Paris in 1783 by representatives of France, England, and her newly independent former American colonies, the American Republic came into being. A decade later the early Plymouth settlers were first referred to as "pilgrims" in a sermon delivered in Plymouth by the Reverend Chandler Robbins, who used a phrase from a copy of Bradford's history, "but they knew they were pilgrims," a quotation from the New Testament. Note the use of the lowercase "p" in the term; Bradford was using it in a generic sense, and in no way singling out the Plymouth party as the sole bearer of the name. In fact, until the early nineteenth century, the term "pilgrim" was generally used to designate any early group of settlers. Those who were adults in 1783 almost certainly retained a strong bond with England, since they were displaced English people. Although separated by an ocean, English colonists still followed the precepts of English law and custom. By 1660, however, a large proportion of the colonial population had never laid eyes on England, and there was a period of about a century during which a different American cultural pattern emerged, but nonetheless, these people were distinctly English. This cultural divergence from the mother country varied regionally, with the southern colonies retaining closer ties with England. By 1760, as shown by both historical and archaeological evidence, America had become more English than it had been since the first three decades of the seventeenth century. But by the time the first generation born in the new republic had come of age, such a bond with the mother country held little if any significance.

By the early nineteenth century, the new nation needed a myth of epic proportion on which to found its history. Who better than the Pilgrims, a term that by that time had narrowed its definition to apply solely to the settlers of Plymouth, whose piety, fortitude, and dedication to hard work embodied a set of ideals that could make every American proud? To be sure, Plymouth was the second-oldest permanent English colony in North America, but Virginia, established at Jamestown in 1607, was hardly a candidate for a national symbol, since it was initially settled by men who were looked upon as a rowdy crowd interested simply in personal gain. Also, relations with the native Powhatan Indians were marked by periods of conflict from the very beginning in Virginia, whereas the Plymouth settlers concluded a peace treaty with the local Wampanoag people, which lasted for over half a century and was honored throughout that time. So it was that Plymouth was chosen to represent the beginnings of the infant nation, but the nineteenth-century construction of the Pilgrims' way of life reflects more of the values of the 1800s than the reality that it was meant to represent. The word "construction" is of particular importance. Although we frequently hear references to *re*constructing the past, this is an impossibility simply because we do not have access to all of the complexities of life in earlier times. What we do is construct the past, and in so doing, decide what is important and what is not. Such constructions invariably reflect, to some extent, the values and biases of the time in which they were written. Our image today of the Pilgrims was strongly influenced by the people of the time in which the term was created, and incorporates as much if not more of how people in the early 1800s saw the world in which they lived.

The Pilgrim myth had matured into a robust tale by 1820, the two hundredth anniversary of the landing of the Plymouth band of settlers. The Pilgrim Society had been established in 1819, and one of the first items on its agenda was the construction of Pilgrim Hall, claimed by many to be America's oldest museum, and which stands today on Main Street in Plymouth. When it first opened, it contained a remarkable assortment of objects, some with genuine "Pilgrim" provenience, but others that had no relationship to Plymouth whatsoever, including Algerian pistols, a pitchfork from Bunker Hill, and assorted seashells. The quantity of "*Mayflower* furniture" that lacked any provenience was so great that a Pilgrim Society member suggested that it was enough to have sunk the ship. Pilgrim Hall today is much different, exhibiting a variety of objects, all of which have authentic proveniences connecting them, if not to the *Mayflower*, to those who sailed aboard her in 1620. It also maintains an excellent

Some early graphic representations of the landing at Plymouth in 1620 and the first Thanksgiving

The Landing of the Pilgrim Fathers, after Charles Lucy, ca. 1850. This engraving is typical of the many graphic portrayals of the landing on Plymouth Rock. Both men and women are present, a portrayal at odds with historical facts in that only males made the first landings and exploratory trips.

The Landing of the Pilgrim Fathers, ca. 1885. A very unusual parody of the Lucy version of the landing, showing that someone at least did not hold the settlers in the reverence that was usual at the time. Could this be the nascent beginning of our modern way of life? The American Express sign prominently displayed suggests that maybe the men paid for their purchase of the famous custom-made Plymouth Rock pants with credit cards. If so, how did they convert $3 into pounds sterling when the dollar had not yet made its appearance?

The Landing of the Pilgrims, by Henry A. Bacon, 1877. Mary Chilton is the central figure in this painting. Family tradition has it that fifteen-year-old Mary was the first to set foot on Plymouth Rock, but the historical facts do not bear this out. Once again there are both men and women shown, as in the case of the Lucy engraving, and only a very small fraction of the passengers and crew of the *Mayflower* are represented.

The First Thanksgiving, by Henry Botkin, ca. 1920. A typical early-twentieth-century depiction of the first Thanksgiving. Note the log cabin in the background; it is now known that such cabins were never built in Plymouth. The table in the foreground, without a tablecloth, is clearly intended for the Indians, although one is seated at the second table with the settlers. The pipe held by the standing Indian offering salutations (Massassoit, Supreme Sachem of the Wampanoag?) is the familiar calumet used by native peoples to the west, but unknown in New England.

library of materials relating to Plymouth Colony and is a valuable resource for anyone conducting research on that subject.

The year 1820 also marked the two hundredth anniversary of the founding of Plymouth, and was celebrated with great enthusiasm. On that occasion the great American orator Daniel Webster gave an address that extolled the virtues of the Pilgrims as they were perceived in those early decades of the nineteenth century. Appropriately, it was delivered while standing by a fragment of Plymouth Rock, which at the time reposed in Town Square. He referred to the rock in his remarks, making both the rock and the Pilgrim myth accessible for the first time to an audience far beyond the confines of the town of Plymouth itself.

The Pilgrim myth did not materialize overnight, but rather was the final and defining episode of an ongoing process that stretches back at least to the later seventeenth century. The Plymouth settlers used several terms to designate themselves. One pair of the terms relates to the makeup of the *Mayflower's* passengers. A minority among them were serious dissenters against the established church, the majority made the crossing in the hope of improving their lot over what it had been in England, where chronic unemployment and increasing shortage of land was making life very difficult for many. The former group was referred to as "Saints," the latter as "Strangers." Two other names were used by the Plymouth settlers to designate themselves, "Old Comers" and "Old Planters," or simply "Planters," but future generations of Plimothians, and later the entire country, would refer to them simply as the "Forefathers." The evolution of the national view of the Forefathers combined with another potentially powerful symbol, Plymouth Rock, form two more strands in the fabric of the Pilgrim myth.

In 1769, a small group of young Plymouth men, all from the more well-to-do families of the town, joined together to form a social organization that they called the Old Colony Club. Among its stated purposes was the establishment of a social environment of a more refined nature than that of the local inns and taverns. One of their first accomplishments was the designation of December 22 as the date to celebrate the landing of the *Mayflower* passengers in Plymouth Harbor. They were even more specific than that, however, and stated that the day would commemorate the landing on Plymouth Rock, which, as we shall see, is more myth than substance itself. The day soon became an annual celebration observed by the people of Plymouth and, as time passed, by people in all parts of America, where it became known as Forefathers' Day and was observed by speeches, parades, and other festive events. It was, in fact, the predecessor of Thanksgiving, but with its emphasis on keeping alive in people's memories both the landing and the rock on which it was supposed to have occurred. While December 22 was the date on which it was usually celebrated, from time to time it would slip back to December 20. The celebration of Forefathers' Day continued into the nineteenth century, and it is still observed in Plymouth, although it had been eclipsed by Thanksgiving in the rest of the country by the opening years of the twentieth century. The Pilgrim myth was given concrete form when in 1859 construction began on an eighty-one-foot-tall monument on a hill overlooking the town of Plymouth. Thirty years in the making, and still standing, when completed it was appropriately named the "National Monument to the Forefathers."

In describing the monument, James Baker writes, "Although it was dedicated to the Pilgrims, they were represented only in the smallest bas-relief elements. Their attributed virtues—Faith, Law, Education, Freedom and Morality—completely overshadowed the human Pilgrim men and women." The monument in fact is an eighty-one-foot-high metaphor, the symbolism of which cannot be missed, for the "virtues" mentioned are represented by very large, full, rounded statues, four seated on pedestals around the base, and the fifth, representing Faith, standing on top. In spite of efforts by a number of writers, some clearly of the "debunking" school, but more objective and serious historians as well, it is this image and the relationships it implies that have come down to us to this day.

The earliest symbol to be associated with the Plymouth settlers is the famous, or perhaps infamous, chunk of granite known as Plymouth Rock. Most Americans know of it, and even a breed of chicken has been named after it. Lacking hard numbers, it is not possible to say that it is the most popular attraction in modern Plymouth, but one has the intuitive sense that such is the case. An apocryphal story in Plymouth has it that two elderly ladies, walking along the waterfront, were overheard when one said to the other, "We must get to bed early tonight, because we have to climb Plymouth Rock in the morning." Such a feat would involve not more than two steps and consume a second or two. It was not always as small as it is today, for over the years it has been repeatedly broken. In his colorful, controversial book *Saints and Strangers*, George Willison gives us an account of the adventures and misadventures of the rock, which cannot be improved upon, so we present it here in its brief entirety:

> For a century and a half after the landing of the Pilgrims the rock lay unmarked, almost unnoticed. Until 1769, on the eve of Independence, it was just another gray granite boulder, one more troublesome bit of glacial debris littering a white arc of beach, impeding development of a busy water front and cursed by many as an obstacle to progress. For another century it was dragged, a broken and mutilated fragment, up and down the streets of Plymouth—first to Town Square, to make a revolutionary holiday; then to Pilgrim Hall, as a museum piece, an altar of filial piety. Restored at length to the beach and enshrined below a box of bones, presumably Pilgrims', embedded in the domed ceiling of an elaborate Victorian stone canopy, the rock enjoyed a half century of comparative quiet and repose before it was snatched up again, in our day, and dumped where it has since remained, at tidewater, sheltered and quite overshadowed by a lustrous Grecian temple of Quincy granite. Here, much like a bear in a pit and not a great deal larger, it lies enclosed within a stout iron paling—dreaming perhaps of other days when it was not gathering moss, and peanuts and odd bits of lunch tossed at it by casual sightseers, and not being apostrophized by every hungry aspirant to public office.

Each time the rock was moved fragments were broken from it. It did not take long for jewelry manufacturers to discover that this granite took a high polish, and would make attractive and "historical" items, such as rings, pendants, and bracelet settings. These items can be found in every souvenir shop in Plymouth today, leading one to suspect that the original rock must have been indeed immense, or, far more likely, that the source of the stones used in the jewelry are in most cases not from fragments of the rock.

Willison also describes the first appearance of the Pilgrims in a dramatic production, presented in Boston in 1808 to an enthusiastic reception. The script of the "New Melo Drama" appears to have been lost to posterity, but a handbill announcing its final performance features the rock in its opening scene:

The Pilgrims
Or the Landing of our Forefathers at
Plymouth Rock
In the course of the Melo Drama, the following scenery, incidents, &c.
A view of the Rock and Plymouth Bay, and the landing of the Pilgrims.
The whole scene represents Winter, with a snow storm.
After returning thanks to Heaven for their safe arrival,
Carver orders one of the Pilgrims to cut on the Rock
December 22nd, 1620
the day of their landing.

We can see here the strong connection between the Rock and Forefathers' Day, which dates to the founding of the Old Colony Club in Plymouth in 1769.

What, if any, factual basis supports the attribution of the Rock as the first spot on which the *Mayflower* passengers set foot? There is one slender thread that, however thin, cannot be entirely dismissed. In 1741, ninety-five-year-old Thomas Faunce asked to be taken for what he thought might be his last look at a certain granite boulder on the beach in Plymouth. Faunce lived two miles south of the town and was brought to the waterfront in a chair. Before a small gathering of people, with tearful eyes, he identified a rock, directly below Cole's Hill, as the very spot "which had received the footsteps of our fathers on their first arrival." He had been told this by his father, who had arrived in Plymouth on the *Anne* in 1623, and who in turn had been told by one of the original party of settlers. This is, of course, a third-hand account, and, as such, not of the greatest reliability, yet it does lend a touch of authenticity to what otherwise would be a story made up of whole cloth. What are the facts? We know from the account in *Mourt's Relation*, published in London in 1622, that a group of passengers and crew left the *Mayflower* in a shallop on Wednesday, December 6, 1620, searching for a suitable harbor and place to settle. Shallops were small craft, primarily propelled by a number of oarsmen, although they did have a mast and a single sail, and featured a leeboard, which allowed the boat to sail into the wind in the same way a centerboard or a fixed keel would, and the usual rudder on the stern used for steering. Two days later, on a stormy Friday night, the group reached Plymouth Harbor, found themselves close to an island, and "fell upon a place of sandy ground, where our shallop did ride safe and secure all that night." On the following Monday, after sounding the depth of the harbor, they "marched also into the land." There is no mention of the rock. William Bradford's account in his history, *Of Plymouth Plantation*, is identical. And if Faunce was referring to the first time the Mayflower docked in Plymouth Harbor on December 16, 1620, *Mourt's Relation* picks up the story only two days later: "Monday the 18th day, we went a-land, [the shallop] manned with the master of the ship and three or four sailors." Bradford simply mentions that after their arrival they "afterwards took better view of the place, and resolved where to pitch their dwelling." These are the only contemporary accounts of the time when the *Mayflower* passengers actually arrived on the mainland at Plymouth.

So the matter stands, but whether or not Plymouth Rock as we know it today was ever trod upon by one or more *Mayflower* passengers is immaterial in the context of the Pilgrim myth. What does matter is that it is possessed of a symbol of great power, as witness the hundreds of thousands of people who pay homage by gazing upon it from above, separated from it by only a sturdy iron railing.

The final component of the myth of the Pilgrims, which made its appearance very early in the nineteenth century, is what is referred to today as the Mayflower Compact. Finding themselves outside the area covered by a patent that gave them rights to settle in Virginia, and, in William Bradford's words, "occasioned partly by the discontented and mutinous speeches that some of the strangers amongst them had let fall from them in the Ship," a covenant was drawn up of a type with which they were very familiar, as covenant agreements were used as a basis for social regulation in England by numerous Puritan and Separatist groups. The document was signed on November 11, 1620, while the *Mayflower* lay at anchor off Cape Cod. The earliest version of this covenant is the one printed in *Mourt's Relation*. It is virtually identical to the version that Bradford gives us in his history, and reads as follows:

> In the name of God, Amen. We whose names are underwritten, the loyal subjects of our dread sovereign lord King James, by the grace of God, of Great Britain, France, and Ireland King, Defender of the Faith, etc.

> Having undertaken, for the glory of God, and advancement of the Christian faith, and honor of our king and country, a voyage to plant the first colony in the northern parts of Virginia, do by these presents solemnly and mutually in the presence of God and one of another, covenant, and combine ourselves together into a civil body politic, for our better ordering and preservation, and furtherance of the ends aforesaid; and by virtue hereof to enact, constitute, and frame such just and equal laws, ordinances, acts, constitutions, offices from time to time, as shall be thought most meet and convenient for the general good of the colony: unto which we promise all due submission and obedience. In witness whereof we have hereunder subscribed our names; Cape Cod, the 11th of November, in the year and reign of our sovereign lord King James, of England, France and Ireland eighteenth and of Scotland fifty-fourth, Anno Domini 1620.

At the same time, the *Mayflower* passengers confirmed thirty-five-year-old John Carver as governor. He would serve in that position for only a few months until his death in April 1621, to be succeeded by William Bradford, by then thirty-one. Carver and Bradford were in fact among the younger members of the Leiden leaders, and below the average age of thirty-four, a figure based on the ages of forty men who sailed on the *Mayflower* whose baptismal or birth dates are known or can be estimated. The settler group has been perceived incorrectly by many as much older. If one assumes that the married women on board (eighteen out of the twenty women) were close in age, or younger, than their husbands, and realizes that almost half of the "children" were in fact young adults, the first Plymouth settlers seem a remarkably young group.

As is befitting a story of mythic proportions, the Mayflower Compact has been endowed with an importance that far transcends reality. In 1802 President John Quincy Adams had this to say:

> One of these remarkable incidents is the execution of that instrument of Government by which they formed themselves into a body-politic. . . . This is perhaps the only instance, in human history, of that positive, original social compact, which speculative philosophers have imagined as the only legitimate source of government. Here was a unanimous and personal assent by all the individuals of the community, to the association by which they became a nation.

Having become an integral part of the Pilgrim myth, this perception of the significance of the Mayflower Compact remains with us to this day. A well-known historian, Henry Steele Commager, commented in a television production in the early 1970s:

> They drew up one of those familiar church or sea compacts, but it was of epic making proportions, the Mayflower Compact that some claim to be the first of all written constitutions. It was drawn up democratically, it was signed by the heads of families and also by some of the servants and hired help. Imagine that, in seventeenth-century England or on the European continent. It was based on the principle that political authority comes from below not from above, and that government derives all of its authority from the consent of the governed. New ideas these in politics, but ideas which were to be the very foundations of American political theory and political practice, and that were to spread throughout the globe.

Many see it as a forerunner to the U.S. Constitution, and it did indeed provide for "political authority [coming] from below not from above" and embody the principle that "government derives all of its authority from the consent of the governed." But a close examination of the list of signatories shows that only four of the ten adult servants aboard the *Mayflower* signed, and none of the women. As for it containing "ideas which were to be the very foundation of American political theory and political practice, and that were to spread throughout the globe," this is, to put it mildly, a bit of an overstatement. In fact, in 1619, Virginia had established the House of Burgesses, which, within limits, provided for a similar type of representative government, although membership in this case was restricted to male landowners. This is not, however, to decry the fact that the basis of Plymouth government was a belief in the rule of law, not nearly as clearly formulated as it is today, but visibly present in what can be seen by 1671 as an embryonic bill of rights.

The myth of the Pilgrims, with its three central themes, the Forefathers, the Rock, and the Compact, became increasingly pervasive in the American collective consciousness during the nineteenth century. It would receive even greater attention, this time on an international scale, when in 1858, just one year before construction began on the

Forefathers' monument, Henry Wadsworth Longfellow wrote an epic poem, *The Courtship of Myles Standish.* Its popularity was almost instantaneous, and more than 10,000 copies sold in London in a single day. Along with *The Midnight Ride of Paul Revere,* it would endow four rather ordinary colonists with heroic and romantic qualities that, although greatly exaggerated, would be perceived as such by those who read the poems. Many of us are familiar with the central story line of *The Courtship,* but far fewer have read the poem in its entirety, which in a way is a mercy. According to Longfellow, Myles Standish, who was in his late twenties or early thirties upon his arrival in Plymouth, became enamored of Priscilla Mullins, the daughter of William Mullins, who died in the sickness of the winter of 1620–1621. Priscilla was seventeen. Standish, however, could not muster the courage to approach Priscilla and make a personal offer of marriage, so he prevailed on his friend John Alden, who was twenty-one, to act on his behalf. When Alden approached Priscilla with Standish's offer, she spoke the now immortal words, "Why don't you speak for yourself, John?" Whether as a result of Standish's request, and it seems highly unlikely that such was the case, or not, John and Priscilla were married in 1623 and produced ten children, six girls and four boys. There was until the 1960s a line of canned goods produced in Massachusetts with the brand name John Alden, which carried a slogan in small print on its label: "It speaks for itself," referring of course to the can's contents, whether peas, corn, beans, or some other vegetable.

Peter Gomes, pastor of Harvard's Memorial Church and an avid student of Plymouth history, comments of Longfellow's poem:

> Had Henry Wadsworth Longfellow devoted himself to the Romance Languages, of which he was Smith Professor at Harvard, rather than to mediocre but memorable verse, the perception of American history may well have been quite different. Paul Revere would have remained an unknown Boston artisan, and the Pilgrims of Plymouth would be little more than aggregate virtue. It was Longfellow's disciplined meters and undisciplined history that launched them both into immortality.

The "aggregate virtue" of the Pilgrims made itself visible in advertising in the later nineteenth century, when graphic portrayals of the

Pilgrim story became widely used. James Baker tells us that of all the products with which these symbols were associated, the commonest by far were liquor and tobacco. Baker reasons quite logically that the use of Pilgrim images to promote such products lent them a legitimacy and respectability that might have been difficult to achieve by other means.

It was not until the opening years of the twentieth century that Thanksgiving was added as a central component of the myth of the Pilgrims, joining the other four: the Pilgrims themselves, the Forefathers, the Rock, and the Compact. Just why it should have taken so long for this to occur is not entirely clear, but James Baker offers an explanation that is both logical and convincing. Nineteenth-century depictions of the event almost always involve conflict between the settlers and the native peoples. A print from *Frank Leslie's Illustrated*, 1869, shows people seated around a table, complete with a turkey at one end, under attack by a group of native people, with an arrow stuck in the door and another in the table that appears to have barely missed the turkey. One man is lifting his musket from its rack on the wall to defend his family, and the others show expressions of alarm, except for the man at the head of the table, who stands with hands folded in prayer, and a woman at the opposite end whose head is bowed. It would appear that the attack took place just as the family was giving thanks.

How this violent image became transformed into a peaceful encounter between colonists and Indians is explained by Baker as follows:

> It was only after the turn of the century, when the western Indian wars were over and the "vanishing red man" was vanishing satisfactorily, that the romantic (and historically correct) idyllic image of the two cultures sitting down to an autumn feast became popular. . . . By the first World War, popular art . . . school books, and literature had linked the Pilgrims and the First Thanksgiving indivisibly together, so much so that the image of the Pilgrim and the familiar fall feast almost ousted the Landing and older patriotic images from the popular consciousness.

This is the image that we carry today, and at holiday time stores are filled with depictions of clean-shaven Pilgrim men, buckled hats and all, equally well scrubbed women with little white caps, Indians, usually with a single feather stuck in a headband, and, of course, turkeys, turkeys, and more turkeys, both in cardboard cutouts and in the frozen food section of supermarkets. Schools the nation over present Thanksgiving plays; most Americans are familiar with such productions, and many have participated in them. By far the most memorable of these is to be seen in the film *Addams Family Values* in which Wednesday Addams and some of the other "misfits" at a summer camp are cast as Indians in a Thanksgiving play. After being greeted by a pretty young blond Pilgrim maiden who tells the Indians that they are not different from themselves, except that the Pilgrims wear shoes and have last names. The Indian members of the cast have revised their part of the script, unbeknownst to the camp counselors, and Wednesday delivers a short speech, which is both funny and, sadly, true:

> I am Pocahontas, a Chippewa maiden. Wait, we cannot break bread with you. You have taken the land which is rightfully ours. Years from now my people will be forced to live in mobile homes on reservations. Your people will wear cardigans and drink highballs. We will sell our bracelets by the roadside. You will play golf and enjoy hot hors d'oeuvres. My people will have pain and degradation. Your people will have stick shifts. The gods of my tribe have spoken. They say, Do not trust the Pilgrims.

The Indians then proceed to tie the Pilgrim maiden who greeted them to a stake and pour gasoline around her feet (we are spared seeing the match applied), then burn the village, and the scene closes with two Pilgrims being spit-roasted together over a fire. Regardless of the mixup between Plymouth and Virginia, and between the Chippewa and the Wampanoag, there is far more than a little truth in Wednesday's, aka Pocahontas's, words. For an important segment of the American people, Thanksgiving is hardly a day to celebrate in a festive way. To Native Americans, Thanksgiving has come to symbolize the beginning of what would eventually become the tragic destruction of their culture.

In 1970, Thanksgiving was declared a National Day of Mourning by Native Americans, and Plymouth was chosen as the location where it would be observed. Native peoples, both local and others from as distant as various western states, converged on Plymouth, assembling on the waterfront adjacent to the Rock and *Mayflower II*, anchored and secured to the wharf nearby. Attendance some years has exceeded five thousand, and while some occasions are marked by more overt protest than others, the Day of Mourning overshadows all other events in town on that day. Drums and singing are a constant part of the event, and speeches, often delivered with great passion, are also a regular feature of the program. On many occasions, Native Americans boarded the ship and climbed into the rigging, and more than once Plymouth Rock has been either painted red or buried in sand, and sometimes both. The participants fast during the day, taking food only after sundown.

There is a significant and understandable irony in the selection of Plymouth and Thanksgiving as the site and date of the Day of Mourning. This selection underlines the power of the Pilgrim myth in the minds of all Americans. Only in the way it is observed is there a dramatic difference. It is historical fact that the Plymouth settlers and their Wampanoag compatriots enjoyed one of the longest periods of peace in colonial history. There were Indian residents within the jurisdiction of the town of Plymouth, and the court records of the colony tell us that they were treated in much the same way as were Europeans for various offenses, and occasionally received a lighter punishment for the same transgression than was meted out to the settlers. In fact, the second execution in the colony, involving three Englishmen, was carried out in 1638 because they had murdered and robbed a Nipmuck messenger from the chief sachem of the Narragansett. However, by the time of King Philip's War (1675–1676), a conflict leading to the greatest loss of life in proportion to population in all of American history, relations had finally deteriorated to the point of open hostility. But in other colonies, particularly Virginia, Indian-European relations were strained from the outset. So one could argue that Jamestown would be a more suitable place for the Day of Mourning to be observed, but the power of the Pilgrim myth is such that Plymouth and Thanksgiving were perceived as the appropriate place and time.

A remarkable event took place on Thanksgiving Day, 1971, which completely escaped media attention. Only those who participated in it, and those to whom they may have mentioned it, were, and still are, aware of it. James Deetz, then a senior staff member of Plimoth Plantation, an outdoor-living history museum in Plymouth that shows what the settlement might have been like in 1627, taught a course on Native

American history during the Harvard Summer School session that year. Among the class members were a number of Native Americans who had been enrolled in Harvard's newly established American Indian Program. In the course of the summer, Deetz developed friendships of varying degrees with a number of these Indian students, and they visited his home, a large Victorian-style house in Plymouth, for socializing at regular intervals, including nearly every weekend. By the time Thanksgiving approached, it was decided that the house, which had been the scene of so many parties, be used to entertain some eighty-odd Native American high school students who were coming to Plymouth to observe and participate in the Day of Mourning.

The high school students, from groups all over America, were part of a program known as "A Better Chance," formed to expose American Indian students to a variety of educational experiences that would be very different from those obtained in mostly reservation classrooms. Plans were accordingly made. A traditional Thanksgiving dinner would be provided after sundown, given by Deetz's circle of Plymouth friends, including a professor of anthropology from Queen's College who attended with his wife and three children. They prepared all the food, enough to feed more than a hundred people. This was a personal invitation, not related to Plimoth Plantation, and a complete and successful press blackout was imposed. Late in the afternoon, some ninety Native Americans appeared at the Deetz home, and when it became apparent that the house, large as it was, could not possibly accommodate all that was planned, last minute arrangements were made to use a nearby church hall. At the beginning, there was a palpable tension in the air, understandable if one puts oneself in the place of the high school students, finding themselves suddenly in an alien environment and perceived as unfriendly. For two hours, conversation was minimal, and it was to the credit of the Harvard American Indian students that they served as mediators between the two parties. Once everyone adjourned to the church hall, however, the atmosphere underwent an immediate change. The Harvard group had brought a drum and there was singing and dancing, this alternating with live bluegrass music, the Plymouth group having among them a number of excellent musicians. Before the evening had ended, around 10 PM when the high school students boarded their bus to return to Boston, it was concluded by all that the festivity had been an outstanding success. But the remarkable, wonderful thing about the entire affair was that it was the first time in three and a half centuries that such a celebration had taken place in Plymouth. It had not been planned in any way to be such an event, but the ethnic makeup of the participants was very close to that of the group who celebrated Harvest Home in Plymouth in the fall of 1621.

In 1976, five years after the event described above took place, Plymouth was the site of what can only be described as a happening. A concert, called the Rolling Thunder Review, was scheduled to tour the northeastern United States, and for the most part it was booked into smaller venues, such as Lowell, Massachusetts, Bangor, Maine, and Durham, New Hampshire, although its final performance was in Madison Square Garden in New York City. The members of the Review included such luminaries as Bob Dylan, Joan Baez, Ronee Blakeley, Sam Shepard, and Allen Ginsberg. The decision to begin the tour at Plymouth, with all of the symbolism it possessed, was tied in with the American bicentennial, and must have been made at some higher level of management, as members of the entourage had a different view of things. Sam Shepard's opinion of Plymouth, published in his *Rolling Thunder Log Book*, was far from complimentary, much in the style reflecting the worldview of the counterculture of the 1960s and 1970s, but probably representative of the group as a whole. Shepard says:

> Plymouth is a donut of a town. The kind of place you aspire to get out of the second you discover you've had the misfortune of having been raised there. Old women dress up in Pilgrim outfits . . . and complain about the damn Pilgrim hat, little white bonnets that keep slipping down their necks.

The inaugural concert was to be given in Plymouth Memorial Hall, a large brick building in the center of town. While paying for some rain boots, Shepard asked the checkout clerk if she is going to the concert:

> "What concert?" "The one in the town here. You know, Bob Dylan, Joan Baez? All those people?" "No, I'm not goin'. Couldn't get tickets. Besides, I don't even know those people."

During their stay in Plymouth, six of the group in a rental dinghy, with Dylan at the helm, attempted successfully to deposit Allen Ginsberg in front of Plymouth Rock. Sam Shepard again:

> Ginsberg is already chanting and chiming on a set of Tibetan bells. . . . The crowds don't seem to recognize Dylan at all. In fact they don't seem to recognize anything that isn't strapped down and carrying a bronze plaque.

Despite this, if outward appearances can be judged, there was at least one time when most of the members took a kindlier view. Plimoth Plantation arranged to entertain the group at a party given at night in the forty-foot square fort that sat atop the hill overlooking the 1627 museum village. A huge bonfire burned outside the door, and inside the fort was a scene straight from Brueghel. Dylan missed it; he got as far as the parking lot, turned and returned to town. Joan Baez made it as far as the door, stayed very briefly, and also departed. Inside, there was orderly chaos—a group of local Narragansett and Wampanoag Indians with a large drum, singing and chanting the entire evening, and opposite a bluegrass band playing with equal volume. In a corner, Allen Ginsberg sat at a little organlike contraption, and chanted "oom" repeatedly for more than an hour. In the center of the floor, some twenty plantation "interpreters," dressed in period attire, performed Playford dances, a kind of dancing commonly engaged in in seventeenth-century England, and related to the more formal Morris dancing that became increasingly popular in the eighteenth century. The party lasted far into the cold, clear November night. Liquid refreshment was available, and Rolling Thunder members and local Plymouth folk mingled freely, engaged in lively conversations, including Scarlet Rivera, who, according to Shepard's account, was "the mysterious dark lady of the fiddle, with whom I never spoke more than three words, not because I didn't want to but because it never happened." After departing Plymouth, the Review moved on to other towns, where they were well received, and the tour was judged a great success.

Plymouth today is a lovely little coastal town, the kind that travel writers describe as "quaint." Even in the absence of the Pilgrim mythos, the town conveys a sense of permanence and history. Gravestones on Burial Hill bear the familiar names of Brewster, Bradford, Doty, Faunce, Bartlett, and many others that go back to the seventeenth century. In the small community of Chiltonville, in the southern end of Plymouth, many of the names on mailboxes can also be found on slate gravestones dating to the eighteenth century. Local businesses still carry familiar names, such as Standish Chevrolet and Bradford Liquors. Three of the most popular motels in town are the Governor Bradford Motor Inn, the John Carver Inn, and, to the south of town, the Pilgrim Sands Motel. Plymouth can boast four seventeenth-century houses, three of which are open to the public and have guided tours; but the fourth, which is possibly the oldest of the group, is a private residence.

Plymouth has its share of monuments, mostly Pilgrim related. As mentioned above, on a hill overlooking the town is the Forefathers' Monument, the tallest granite statue in the country. But the main concentration is to be found in the vicinity of the harbor. In Brewster Gardens, beside Town Brook, which runs through the town center, is the Pilgrim Maiden statue, the only acknowledgment of the women who were members of the settler party. On Cole's Hill, overlooking the harbor and the Rock is a larger-than-life statue of Massasoit. Just below is a sarcophagus containing a small number of bones that many believe are the remains of those who died during the first winter. These were discovered in 1855, when the first sewage system was installed in Plymouth. Apparently some care was taken in exposing and removing the skeletons, since it was noted that they were extended supine burials, and therefore definitely European. This identification was supported by Oliver Wendell Holmes, who among other things was a doctor, when they were sent to him for study in Boston. While they may not have been the graves of the casualties of the first winter, it is virtually certain that they were of the period, although perhaps somewhat later. They were encountered during excavation on Carver Street, which runs along the top of Cole's Hill. They may well have not been the only burials in that general location, for James Thacher, in his 1832 *History of the Town of Plymouth*, comments:

> On this hill, according to common tradition, were deposited the remains of those renowned pilgrims who fell a sacrifice during the perilous winter of 1620–1. About the year 1735, an enormous freshet

> rushed down Middle Street, by which many of the graves of the fathers were laid bare, and their bones washed into the sea.

Apparently none were recovered, so identification of the bones as those of Europeans was impossible. However, in view of the later nineteenth-century discovery in the same area, positively identified as European, it is very likely that Thacher's identification of the remains as "the graves of the fathers" is probably correct, but we cannot rule out the possibility that they were those of Native Americans, since Plymouth was sited on the abandoned Wampanoag village of Patuxet.

The bones that repose in the sarcophagus today were first placed in a canopy that was erected over Plymouth Rock, completed in 1867. By then, the Rock had been moved to its present location, and the bones remained in the ceiling of the canopy until 1919, when it was taken down and replaced by the classical portico that shelters the Rock today.

It is important to remember that the small number of bones that repose in the sarcophagus are the remains of people who experienced life in all of its complexities, its pleasures, as well as its trials and tragedies, and who were instrumental in contributing to the course of Plymouth Colony from 1620 until 1691. In an eloquent passage in his history, *Of Plymouth Plantation*, William Bradford writes:

> As one small candle may light a thousand, so the light here kindled hath shone unto many, yea in some sort to our whole nation. . . .

While it is not clear exactly what Bradford intended when he used the word "nation," it certainly does not refer to the American nation as we know it today. But this in no way diminishes the importance of his words, and although that "light here kindled" has not been extinguished, today it shines with a diminished glow, the result of the gradual but relentless accretions that have attached to the story of the Plymouth settlers as they were transformed into Pilgrims, with all the connotations and multiple meanings that adhere to the term—the Rock, Forefathers' Day, the Compact, and last but perhaps most significant, Thanksgiving. It is time now to strip these layers away, and try as best we can to see these people as they were in the seventeenth century, using a number of avenues of inquiry to accomplish this end.

Chapter 2

I Will Harry Them Out of the Land!

The Early Years, 1606–1627

William Bradford was a member of a Separatist congregation that was formed in 1606 in the tiny hamlet of Scrooby, Nottinghamshire, England. It is important to distinguish between Separatists and Puritans. Although both groups opposed the established Church of England, the Puritans were of the opinion that reform, mostly directed at purifying the church of what they saw as an excess of papist practices, was best accomplished from within, and they remained members of the Church of England. Separatists, as the name implies, had no use whatsoever for the established church, and chose to remove themselves entirely from its membership, forming separate congregations of their own. The Separatist movement had its beginnings in the mid–sixteenth century, as a part of the larger movement of the Reformation. Many different Separatist congregations were formed, each holding a set of doctrines that varied somewhat one from another; but they shared the belief that remaining members of the established church was futile, and they distanced themselves completely from it. During the reign of Elizabeth I (1558–1603), Separatist congregations were not seen as a serious threat to the Church of England. Nonetheless, membership in such a group was considered a high crime against the state, and three well-known Separatist ministers were executed. When King James I ascended to the throne in 1603, however, there was a striking increase in the persecution of Separatist congregations, James I himself declaring, "I will make them conform, or I will harry them out of the land!"

Such was the situation faced by the small Scrooby congregation. They decided to leave England and settle in Holland, where their presence as well as their beliefs would be tolerated. But such a move was far easier to discuss than to put into action. Their first attempt in the fall of 1607 was a disaster. At this stage, James I had tightened his policy toward the Separatists, and the ports were closed to dissenters and Roman Catholics, who were refused licenses to go abroad. A large number of the Scrooby congregation contracted with the master of a ship in Boston, Lincolnshire, to smuggle them in a specially chartered ship across the English Channel to Holland.

> So after long waiting and large expenses . . . he came at length and took them in, in the night. But when he had them and their goods aboard, he betrayed them, having beforehand complotted with the searchers and other officers so to do; who took them, and put them into open boats, and there rifled and ransacked them, searching to their shirts for money, yea even the women further than became modesty; and then carried them back into the town and made them a spectacle and wonder to the multitude which came flocking on all sides to behold them.

Although treated with courtesy by the magistrates, they were imprisoned for a month, after which they, including William Bradford, who was then seventeen, were set free, but seven of their leaders were kept in prison and bound over to the assizes.

The second attempt to depart for Holland in the spring of 1608 was hardly more successful. Some of those from the former group, joined by others, contracted with a Dutchman at Hull who owned a ship and agreed to transport them. The men traveled by land to a deserted area distant from the town to meet the ship, but the women, children, and luggage were transported there by water in a small bark. When the sea became rough, they put into a creek where the bark was grounded when the tide went out. By this time, about half of the men, the first boatload, were already on board, the rest assisting the women and children, when

> . . . master espied a great company, both horse and foot, with bills and guns and other weapons, for the country was raised to take them. The Dutchman, seeing that, swore his country's oath *sacremente*, and having the wind fair, weighed his anchor, hoisted sails, and away.

The trip across the channel took fourteen days, a violent storm having arisen that drove them near the coast of Norway before it abated and they were able to reach their destination in Holland. The men on board were desperate, afraid for their wives, children, and friends, and destitute—as all their money and possessions were in the bark. Those left behind were in an equally terrifying position. The men who had remained managed to escape, except for a few who stayed to assist the women and children.

> Being thus apprehended, they were hurried from one place to another and from one justice to another, till in the end they knew not what to do with them; for to imprison so many women and innocent children for no other cause . . . but that they must go with their husbands, seemed to be unreasonable. . . .

It was also not possible to send them to their homes, for they had disposed of their houses and other property. So, "they were glad to be rid of them in the end upon any terms, for all were weary and tired with them."

The only accounting of these events is to be found in Bradford's history, and he is unclear regarding how the other members of the Scrooby group made their way to Holland, but by 1609, about 125 members of the congregation had arrived in Amsterdam, including their two ministers, Richard Clyfton and John Robinson, as well as William Brewster and Bradford himself. After a year in Amsterdam they moved to Leiden, having been granted permission to live in that city by the Burgomasters upon formal application on February 12, 1609.

After staying in Leiden for almost twelve years, some of the Separatist group felt it better to consider leaving for some new place where their prospects might be improved. While in Leiden, all were gainfully employed, although mostly in menial occupations. Many were engaged in the production of cloth, including Bradford, who worked as a fustian maker. But to many of them, removal to America seemed the most attractive possibility, for in Bradford's words,

> The place they had thoughts on was some of those vast and unpeopled countries of America, which are fruitful and fit for habitation, being devoid of all civil inhabitants, where there are only savage and brutish men which range up and down, little otherwise than the wild beasts of the same.

While Bradford acknowledges the presence of a native population in America, his use of the term "unpeopled" reflects a commonly held view of the time, that land was abundant and there for the taking. A cartoon in *The New Yorker* printed some time in the 1960s shows two Pilgrims, buckled hats and all, conversing aboard the *Mayflower*. One says to the other, "Religious freedom is my first concern, but real estate is my long range goal." A minority of the Scrooby church felt the necessity to remove themselves from Holland, and Bradford states their reasons. First, they foresaw little if any improvement in their employment and thus financial status should they stay. Second, they were concerned that Holland would again be at war with Spain, a truce between the countries being due to end in 1621. Third, they were extremely anxious that their children would become like the Dutch, not only speaking a language not native to them, but influenced in ways that the English community found particularly disturbing. To quote Bradford yet again:

> But that which was more lamentable, and of all the sorrows most heavy to be borne, was that many of their children, by these occasions and the great licentiousness of youth in that country, and the manifold temptations of the place, were drawn away by evil examples into extravagant and dangerous courses, getting the reins off their necks and departing from their parents. . . . so that they saw their posterity would be in danger to degenerate and to be corrupted.

The fourth reason for removal to America is slightly suspect. Bradford states that it would permit an opportunity "for the propagating and advancing the gospel of the Kingdom of Christ in those remote parts of the world. . . ." Propagation of the faith was used repeatedly as a rationalization for pursuing other motives by many European colonizing groups. A good example of this is the founding of the Franciscan missions in Alta California beginning in 1769 and continuing into the early nineteenth century. While these missions did in fact gather a large number of Californian native people in several central locations, where before the population had been more widely distributed, and did convert them to the Catholic faith, the primary motivation was economic, to obtain a large work force to produce the commodities needed by the missions in order to survive. It is for this reason that there is considerable opposition today to the proposed canonization of Father Junipero Serra, who established the missions of Alta California. In any event, efforts by the settlers to convert the Indians in Plymouth Colony did not begin to equal their preoccupation with establishing a colony where they could worship as they pleased, acquire land, and prosper economically.

Having decided to remove themselves to America, the next and very vexing problem was to find the financial backing to make such a venture possible. Through a series of complex negotiations, they managed to form a joint stock company with a group of merchant adventurers in London, representing the Virginia Company. Such contracts between merchant adventurers and planters, those who would cross the Atlantic and plant a colony, were a common arrangement of the time, and formed the basis of much of the early English settlement of North America. The agreement stipulated that the adventurers would provide the capital for the planters in return for various commodities that would be sent from the colony back to England. In the case of Plymouth, these goods included furs, fish, clapboards, and sassafras. The joint stock company would be dissolved at the end of seven years, when presumably the debt would be settled. The terms of the agreement between the Plymouth group and the Virginia Company stipulated that all property would be held in common by all of the planters, and labor would contribute to a shared common goal. The Leiden group bridled at this arrangement, and it had not been resolved in a satisfactory fashion at the time of their departure.

A small ship, the *Speedwell*, was purchased in Holland. It awaited the group at Delftshaven and transported them to England, where a larger ship, the *Mayflower*, awaited them with the rest of the company, most of whom had been recruited in England to bring the total number of planters to a level thought to be sufficient for the successful planting of a colony. On August 5, 1620, both ships departed Southampton, but after sailing only a hundred leagues off Land's End, the master of the *Speedwell* complained that the ship was so leaky that it would sink. Both ships returned, this time to Plymouth, provisions and as many passengers as could be added being transferred to the *Mayflower*, which set sail again on September 6, this time successfully, but leaving some of the group behind.

There were 102 passengers crammed on board the *Mayflower* in addition to the ship's crew, probably around twenty to thirty seamen, including the ship's master, Christopher Jones. When we realize that she had a carrying capacity of 180 tons, and that in the seventeenth century merchant ships similar to the *Mayflower* were only about 113 feet in length with a keel of approximately sixty-four feet and width of twenty-five feet, it is no surprise to learn that it was so crowded that people were even sleeping in their dismantled shallop, which had been stored below the top deck. What most do not realize is that not only were the sixty-nine adult passengers mainly in their thirties, but that there were some fourteen young adults between the ages of thirteen and eighteen on board, and nineteen children twelve and under. The leadership and driving force of the emigration lay, of course, with the adult members of the Scrooby congregation from Leiden. They comprised 42 percent of those on board. Theirs was a close-knit community, the majority being the younger, more vigorous members of the congregation, deliberate policy as they would have the hardest part of establishing the new colony. In addition to being friends, some were related, by both blood and marriage. Surgeon Samuel Fuller had not only his brother Edward and Edward's wife on board but his sister, Susanna, married to William White, and two nephews, Samuel Fuller Junior and Resolved White. Degory Priest, who made the voyage ahead of his family, was married to Isaac Allerton's sister, Sarah. Allerton's wife, Mary, had attended Elizabeth Barker at her marriage to Edward Winslow in 1618. There were also family connections with some of the passengers who joined them in

London, passengers who in fact made up the greater proportion of those who sailed on the *Mayflower.*

Little did anyone setting out to make the Atlantic crossing on that tiny ship imagine that within a year, of the nineteen women on board, fourteen of the eighteen who were married would be dead, as would six of the children, three of the young adults, and over half of the men. In some cases entire family parties died, or almost all. Fifteen-year-old Elizabeth Tilley lost her parents and her uncle and aunt and was left with the care of her toddler cousin Humility Cooper. Her cousin Henry Samson, sixteen at the time of the voyage, survived, and Elizabeth later married fellow passenger John Howland. Seventeen-year-old Priscilla Mullins was also the only survivor of her family group, little dreaming that her name would be forever linked to that of military captain Myles Standish in verse written to celebrate her love for John Alden, who joined the ship in Southampton as a cooper. Mary Chilton was only thirteen when she made the crossing with her parents, and she too was orphaned; her father, at sixty-four the oldest passenger, died on board the *Mayflower* while it lay at anchor off Cape Cod, and her mother and older brother, Joseph, died in April the following year.

Centuries later Mary would live in mythical fame as the first woman to set foot on Plymouth Rock, mythical since the first landing party was all male. She would meet and marry a younger brother of Edward Winslow (John, not Gilbert who was on board with Edward), and have a daughter who would be involved in a murder trial in Plymouth. Then there were the More children, Jasper, Richard, Mary, and Ellen, whose mother was divorced after a scandal in which her husband discovered that the children were sired in an adulterous relationship in which she was involved. He had them sent to America in the care of the leaders of the Leiden group so that they could grow up in an environment unprejudiced by their illegitimacy, but he had no way of knowing that three of them were going to their deaths. Only Richard was to survive and live a full and remarkable life. The reckless, turbulent Billington family all made the passage, but never anticipated that John Billington Sr. would be the first man hanged for murder in the new colony. And William and Dorothy Bradford, who had left their small son in Leiden, had no idea of the dark shadow that would separate them as death claimed Dorothy when she fell off

Jennie Brownscombe's *Landing of the Pilgrims*, ca. 1920, dominated by Mary Chilton.

the *Mayflower* and drowned shortly after the hazards of the voyage were over. All the adults who sailed on the *Mayflower* knew the risks that they were taking, and were as prepared for them as any immigrant can be, always hoping that the final price would not be asked of them, or of those for whom they were responsible, but concluding that the risks were worth the gains of planting a new colony on the shores of the New World.

Neither *Mourt's Relation* nor Bradford's history tells us much about the crossing to America; in fact, neither account even mentions the *Mayflower* by name. At first, the weather was fine and the ship made good progress, but then:

> They were encountered many times with cross winds and met with many fierce storms with which the ship was shroudly [shrewdly] shaken, and her upper works made very leaky; and one of the main beams in the midships was bowed and cracked, which put them in some fear that the ship could not be able to perform the voyage.

But save for the buckled main beam, the master of the ship, Christopher Jones, affirmed that it was "strong and firm under water." The main beam was raised using a "great iron screw the passengers brought out from Holland," and the ship sailed on in fit enough shape to complete the voyage. It has been suggested by a number of writers that the iron screw was a part of William Brewster's printing press, but this is entirely speculative. In one of the storms, "a lusty [lively] young man called John Howland, coming upon some occasion above the gratings was, with a seele [pitch or roll] of the ship, thrown into the sea." Fortunately, he managed to get hold of one of the topsail halyards, and was hoisted by that rope until he could be fished from the water with a boat hook. John Howland, at the time a servant to John Carver, became a highly regarded member of the Plymouth community, and together with his wife, Elizabeth Tilley, fathered ten children, and died in 1673 at an age of "above eighty years."

Only two people died during the crossing. The first, a seaman (the *Mayflower* carried around twenty or more), described by Bradford as a "proud and very profane young man," who made no secret of his hope that many of the passengers would die on the voyage, was afflicted with a serious illness in mid-voyage. In Bradford's words, "but it pleased God before they came half seas over, to smite this young man with a grievous disease, of which he died in a desperate manner, and so was himself the first that was thrown overboard." The second person to die was an apprentice to surgeon Samuel Fuller, William Butten, who passed away shortly before the ship made landfall at Cape Cod. Four more would perish during the time the *Mayflower* lay at anchor in the harbor of what would become Provincetown, at the extreme northern tip of the cape. Land was sighted on November 9, and on November 11 the *Mayflower* entered the harbor, where it would serve as a base of operations until the final decision was made to settle at Plymouth.

As the passengers looked across the frigid water at the land where they were to make their home, it must have filled them with a certain dread, for they had no idea of what they would encounter once ashore. The familiar ordered landscapes of both England and Holland were all that they knew, and that which they now beheld could not have been more different. In Bradford's words,

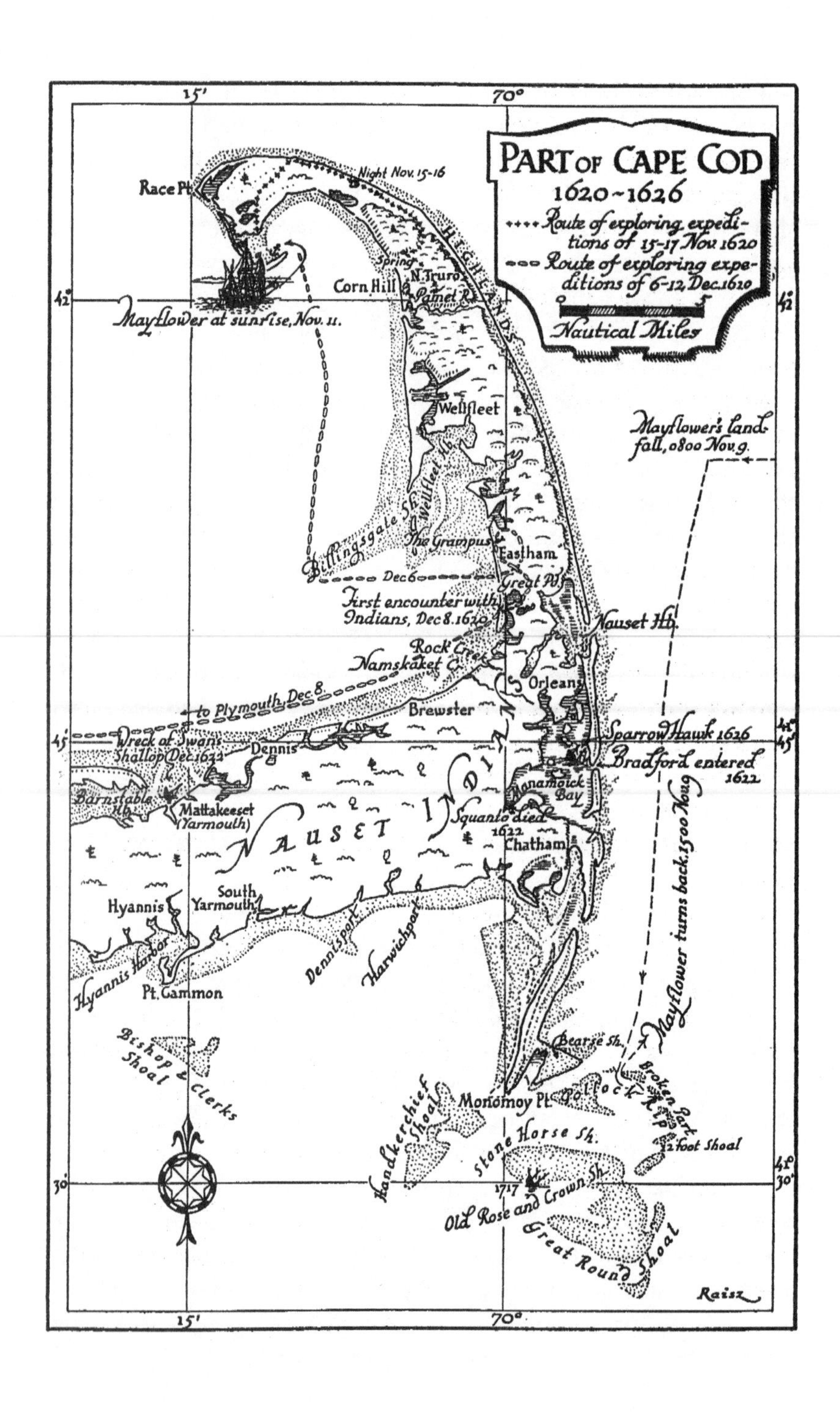

PART OF CAPE COD
1620~1626
Route of exploring expeditions of 15-17 Nov 1620
Route of exploring expeditions of 6-12 Dec. 1620
Nautical Miles
Race Pt.
Night Nov. 15-16
HIGHLANDS
Spring
Corn Hill
N. Truro
Pamet R.
Mayflower at sunrise, Nov. 11.
Wellfleet
Mayflower's landfall, 0800 Nov. 9.
Wellfleet Hb.
Billingsgate Sh.
The Grampus
Eastham
Dec 6
Great Pd.
First encounter with Indians, Dec 8. 1620
Nauset Hb.
Rock Creek
Namskaket C.
Orleans
to Plymouth, Dec. 8.
Brewster
Wreck of Swan's Shallop Dec. 1622
Dennis
Sparrow Hawk 1626
Bradford entered 1622
Monamoick Bay
Barnstable Hb.
Mattakeeset (Yarmouth)
NAUSET INDIANS
Squanto died 1622
Chatham
Hyannis
South Yarmouth
Dennisport
Harwichport
Hyannis Harbor
Pt. Gammon
Mayflower turns back. 1500 Nov. 9
Bearse Sh.
Bishop & Clerks Shoal
Handkerchief Shoal
Monomoy Pt.
Pollock Rip
Broken Part
Stone Horse Sh.
2 foot Shoal
1717
Old Rose and Crown Sh.
Great Round Shoal
Raisz
15'
70°
42°
45'
30'
41°

> And for the season it was winter, and they that know the winters of that country know them to be sharp and violent, and subject to cruel and fierce storms, dangerous to travel to known places, much more to search an unknown coast. Besides, what could they see but a hideous and desolate wilderness, full of wild beasts and wild men—and what multitudes there might be of them they knew not. . . . If they looked behind them, there was the mighty ocean which they had passed and was now as a main bar and gulf to separate them from all the civil parts of the world.

On November 11, the same day that the Mayflower Compact was signed, a party of sixteen armed men went ashore to gather wood, take a close look at the land, and attempt to meet with some of the inhabitants. They transported themselves from the *Mayflower*, which stood at anchor in deep water, in a boat of sufficient size to hold sixteen men. Ships of the time usually carried one or more long boats, which could have been put to such use. The group accomplished its first two goals, but failed to encounter any other persons. That a long boat was used to make this initial one-day trip is suggested by a short passage in *Mourt's Relation*:

> At night our people returned, but found not any person, nor habitation, and laded their boat with juniper, which smelled very sweet and strong and of which we burnt the most part of the time we lay there.

The use of a long boat is confirmed in a later passage from Mourt, "till at length we came near the ship, and then we shot off our pieces, and the long boat came to fetch us." However, extensive exploration of the coast was not possible until the shallop was made seaworthy.

> Monday, the 13th of November, we unshipped our shallop and drew her on land, to mend and repair her, having been forced to cut her down in bestowing her betwixt the decks [this would have been in the space known as the 'tween decks, between the top deck and the one over the hold], and she was much opened with the people's lying in

> her, which kept us long there, for it was sixteen or seventeen days before the carpenter had finished her.

As a result, the party was forced to restrict their exploration to the immediate area, but it was not without several important discoveries.

The most detailed description of the exploration of the cape from November 13 until the final decision to settle at Plymouth on December 22 is to be found in *Mourt's Relation*. In the account that follows, we will present a series of vignettes, with commentary, taken from this work. There are comic moments as well as revealing comments, which, when taken together, convey the strong impression that these Englishmen were like babes in the woods, confronted as they were by an entirely new set of experiences, completely unlike anything they had encountered in their mother country. The entry under Monday, November 13, continues: "Our people went on shore to refresh themselves, and our women to wash, as they had great need." This passage is frequently cited uncritically as the origin of Monday as wash day, which at least until recently has been thought of as a part of everyone's household routine. But it is not even clear if the women were washing clothing or themselves, and Monday is considered wash day in parts of the world that have nothing to do with these women washing on a November day in 1620, South Africa being but one of a number of notable examples. Monday wash day is yet another part, albeit small, of the Pilgrim myth. On the same day, it was decided to send sixteen men out, each with his musket, sword, and corslet (armor), under the direction of Captain Myles Standish, with William Bradford, Stephen Hopkins, and Edward Tilley as advisors. This group was set ashore on the fifteenth, and proceeded single file along the beach. It was not long before they encountered six "savages" with a dog, who fled and whistled the animal after them. The Standish party trailed the group by following their footprints, but night fell before they were able to overtake them, if they could have in any event. They built a fire, set out three sentinels, and spent the night. The next morning they resumed following the "trace" hoping to come upon their dwellings, but with no success:

> We marched through boughs and bushes, and under hills and valleys, which tore our armor in pieces, and yet could meet with none of them, nor their houses, nor find any fresh water, which we greatly desired . . . for we brought neither beer nor water with us, and our victuals was

> only biscuit and Holland cheese, and a little bottle of aquavitae, so as we were sore athirst.

Of particular note in this passage is the reference to "boughs and bushes" tearing their armor in pieces. This would almost certainly indicate that the corslets that they wore were of a type known as a brigandine, quilted cloth with small metal plates sewn inside. Metal armor, which they possessed, could not have been torn to pieces by underbrush.

By this time they had marched over ten miles, and at long last "found springs of fresh water, of which we were heartily glad, and sat us down and drunk our first New England water with as much delight as ever we drunk drink in all our lives." Having refreshed themselves, the party continued on its journey, attempting successfully to reach the shore. Thinking that it would be best to walk along the beach sands, they did so, but it so tired some of the men that they directed their course inland once more. They found,

> A little path to certain heaps of sand, one whereof was covered with old mats, and had a wooden thing like a mortar whelmed [overturned] on the top of it, and an earthen pot laid in a little hole at the end thereof. We, musing what it might be, digged and found a bow, and as we thought, arrows, but they were rotten. We supposed there were many other things, but because we deemed them to be graves, we put in the bow again and made it up as it was, and left the rest untouched, because it would be odious unto them to ransack their sepulchers.

Such sensitivity toward disturbing native graves and removing things from them is somewhat remarkable for the time, and can probably be attributed to the religious beliefs held by the group, but we cannot be certain of their true motivation. In a way, it foreshadows the controversy that began in the 1970s over the treatment of Native American burials and their contents by professional archaeologists. Codified into law by Congress in 1990, the Native American Grave Protection and Repatriation Act sets forth clear and stringent guidelines regarding the excavation of burials, and professional archaeologists and Native American groups have been working together to provide for both reburial of human skeletal remains and the grave goods that accompanied them.

Leaving the burial ground behind, the group continued its journey, passing through two small fields of stubble where corn had been harvested. They came

> . . . to another which had also been new gotten, and there we found where a house had been, and four or five old planks laid together; also we found a great kettle which had been some ship's kettle, and brought out of Europe. There was also a heap of sand, made like the former—but it was newly done, we might see how they had paddled it with their hands—which we digged up, and in it we found a little old basket full of fair Indian corn, and digged further and found a fine great new basket full of very fair corn of this year. . . . The basket was round, and narrow at the top; it held about three or four bushels, which was as much as two of us could lift up from the ground, and was very handsomely and cunningly made . . . after much consultation, we concluded to take the kettle and as much of the corn as we could carry away with us. . . . The rest we buried again, for we were so laden with armor that we could carry no more.
>
> Not far from this place we found the remainder of an old fort, or palisade, which as we conceived had been made by some Christians [Europeans].

This newly harvested corn would form a part of the seed planted the following spring, which produced the bountiful harvest of the fall of 1621, which was celebrated by the settlers as described in Winslow's letter of December of the same year. There is little reason to doubt their identification of the "old fort, or palisade" as being one built by Europeans. Although they had seen native palisades, they could hardly be confused with a European-style fort. And even though they had been in New England for only a period of days, Myles Standish, the group's leader, was very familiar with the manner in which European forts were constructed, having seen service in the Low Countries. This building's remains, and the iron kettles that were found during their explorations, are evidence of the presence of Europeans in the area prior to the arrival of the *Mayflower*. Traders and fishermen from the Old World, mainly England and France, are known to have moved up and down the east coast of North America for at least a century, but they never established any permanent colony in New England as the Plymouth settlers were on the verge of doing. A second exploration party was soon to make another discovery that would add remarkable new evidence of such an earlier European presence.

They spent the night next to a pond in pouring rain, but managed to keep their matches burning. These matches were long cords that were lit on one end and lowered into the firing pan of a matchlock musket, hence the name, but in wet weather they could pose a problem:

> In the morning, we took our kettle and sunk it in the pond and trimmed our muskets, for few of them would go off because of the wet.

For a short time, they lost their way, and were "shrewdly puzzled":

> As we wandered we came to a tree, where a young sprit [sapling] was bowed down over a bow, and some acorns strewed underneath. Stephen Hopkins said it had been to catch some deer. So as we were looking at it, William Bradford being in the rear, when he came looked also upon it, and as he went about, it gave a sudden jerk up, and he was immediately caught by the leg. It was a very pretty device, made with a rope of their own making and having a noose as artificially [skillfully] made as any roper in England can make, and as like ours as can be, which we brought away with us.

They finally regained their bearings and came near the ship, where they fired their muskets, and were brought back to the *Mayflower* in a long boat.

> And thus we came both weary and welcome home, and delivered in our corn into the store, to be kept for seed, for we knew not how to come by any, and therefore very glad, purposing, so soon as we could meet with any of the inhabitants of that place, to make them large satisfaction.

For two or three days, the number is not specified, passengers and crew kept busy at finding wood, hafting tools, and sawing timber to be used in completing the outfitting of the shallop. They could not go from the ship to the shore and back except at high tide, and even then, they often had to wade

> . . . to the middle of the thigh, and oft to the knees, to go and come from land. Some did it necessarily, and some for their own pleasure, but it brought to the most, if not to all, coughs and colds, the weather proving suddenly cold and stormy, which afterwards turned to the scurvy [probably pneumonia], whereof many died.

The deaths referred to here are clearly those that took place later that winter, not while off Cape Cod. Another passage adds to this:

> It blowed and did snow all that night, and froze withal; some of our people that are dead took the original of their death here.

When the shallop was thought fit for use in exploring the coast in search of a suitable place to settle, twenty-four of the passengers and ten members of the crew, all armed, set out on a second, more ambitious and extensive exploration of the northern coast of Cape Cod. Christopher Jones, master of the *Mayflower*, was chosen as the leader of the group. They made little progress the first day, the weather being very bad, but on the next morning, about eleven o'clock, they set sail with a fair wind, explored one harbor that was thought to be unfit for their needs, and went ashore that night. The ground was covered with half a foot of snow, and they made camp under a few pine trees. They "got three fat geese and six ducks for supper, which we ate with soldiers' stomachs, for we had eaten little all that day."

The following morning, they returned to the place where they had dug up the first lot of corn, to which they gave the name "Cornhill," and took the rest of the corn that had been left behind. Further digging in the area produced a total of ten bushels, which "will serve us sufficiently for seed. And sure it was God's good providence that we found this corn, for else we know not how we should have done." At this point, the party divided, with Christopher Jones returning to the shallop taking sixteen of the men who were sick "home" with him, as well as the corn, eighteen men remaining to explore further and await the return of the shallop the next day, with some mattocks and spades, for the weather had worsened and ground had frozen hard. They had used their cutlasses and short swords for digging, but that was now not possible. The word "home" appears in the original manuscript, and it is a revealing reference, for it more than suggests that at this point, home was the *Mayflower*, as it would be for many of the passengers for much of the winter.

The next morning, lighting their matches, the group proceeded along a broad, beaten path, which they thought to have been made to drive deer in. Then, having marched five or six miles into the woods, encountering none of the local inhabitants, they returned by a different route and made a remarkable discovery.

> As we came into the plain ground we found a place like a grave, but it was much bigger and longer than any we had yet seen. It was also covered with boards, so as we mused what it should be, and resolved to dig it up, where we found, first a mat, and under that a board about three quarters [of a yard] long, finely carved and painted, and three tines, or broaches, on the top, like a crown. Also between the mats we found bowls, trays, dishes, and such like trinkets. At length we came to a fair new mat, and under that two bundles, the one bigger, the other less. We opened the greater and found in it a great quantity of fine and perfect red powder, and in it the bones and skull of a man. The skull had fine yellow hair still on it, and some of the flesh unconsumed; there was bound up with it a knife, a packneedle, and two or three old iron things. It was bound up in a sailor's canvas cassock, and a pair of cloth breeches. The red powder was a kind of embalmment, and yielded a strong, but not offensive smell; it was as fine as any flour. We opened the less bundle likewise, and found of the same powder in it, and the bones and head of a little child. About the legs and other parts of it was bound strings and bracelets of fine white beads; there was also by it a little bow, about three quarters long, and some other odd knacks. We brought sundry of the prettiest things away with us, and covered the corpse up again. . . .
>
> There was a variety of opinions amongst us about the embalmed person. Some thought it was an Indian lord and king. Others said the Indians have all black hair, and never any was seen with brown or yellow hair. Some thought it was a Christian of some special note, which had died amongst them, and they thus buried him to honor him. Others thought they had killed him, and did it in triumph over him.

It would seem certain that the adult was a European, not only because of the color of the hair, but also because of the grave goods that accompanied the body. The child, on the other hand, was likely an Indian, in view of the beaded adornments of the body and the "little bow." The powder was almost certainly red ochre, a substance used by the Indians to cover the body before burial. This in turn suggests that the interment was done by the local Indians, but the exact circumstances that surrounded the event will forever remain a mystery. Considering the lavish fashion of the grave furnishings, it is very likely that he had not been killed with malice, but the other opinion expressed by some of the party, that he was "of some special note," and had been buried in a manner to honor him, seems a reasonable explanation.

Thomas Jefferson is generally acclaimed as the father of American archaeology, and rightly so. In his *Notes on the State of Virginia*, written in 1781 and published in 1787, he gives a description of his excavation of an Indian burial mound on the Rivanna River. The work is in the form of a series of answers to queries posed to him by a "Foreigner of Distinction" (a Frenchman, Barbé Marbois), and his description of the excavation is in response to the query, "A description of the Indians established in that state?" It is important to remember that this motivation for digging into the mound was a combination of curiosity and an attempt to determine which of a number of locally held explanations of the true purpose of the mound was the correct one. "For this purpose, I determined to open and examine it thoroughly." This he did, by cutting a trench, a "perpendicular cut through the body of the barrow [mound], that I might examine its internal structure." On the basis of Jefferson's discoveries, a large number of human bones, he concluded that the mound was a repository for the dead added to at intervals, one layer deposited atop another with soil between them.

But over a century and a half earlier, a party of sixteen Englishmen excavated the burial site described above, an excavation that, in its motivation, method, and conclusions, is remarkably similar to that described by Jefferson in his *Notes*. Both were motivated by curiosity, and the less extensive but nevertheless detailed description of the grave's contents, and the sequence of deposition of various artifacts, closely parallels Jefferson's description. True, they were investigating a single grave, not a large mound with multiple burials, but this distinction seems beside the point. The different explanations given by the members of the exploring party were given after the fact, while Jefferson was attempting to resolve which of the explanations offered were correct before he conducted his investigation of the mound. Yet, when one compares the two accounts, they are remarkably similar in a number of ways, so much so that one could argue that the first "archaeology" conducted in America was in fact done in 1620. This is not to detract from Jefferson's accomplishment, for it was remarkable for its time. But in both cases, archaeology as an organized discipline, with research agendas that would change over time, had yet to come into being. So both excavations were conducted to satisfy an understandable curiosity, but with no larger framework in which to place them.

In the course of their explorations, the party came upon ample evidence of a human presence on the cape, but as yet had encountered only the six people with their dog on their first day ashore. Shortly after their excavation of the grave, they came upon two houses that, by all appearances, had to have been inhabited very recently, but were not occupied at the time:

> Two of the sailors . . . by chance espied two houses which had been lately dwelt in, but the people were gone. They . . . entered the houses and took out some things, and durst not stay but came again and told us. So some seven or eight of us went with them . . . The houses were made with long young sapling trees, bended and both ends stuck into the ground. They were made round, like unto an arbor, and covered down to the ground with thick and well wrought mats and the door was not over a yard high, made of a mat to open. The chimney was a wide open hole in the top, for which they had a mat to cover it close when they pleased. One might stand and go upright in them. In the midst of them were four little trunches [stakes] knocked into the ground, and small sticks laid over, on which they hung their pots, and what they had to seethe [boil]. Round about the fire they lay on mats, which are their beds. The houses were double matted, for as they were matted without, so were they within, with newer and fairer mats. In the houses we found wooden bowls, trays and dishes, earthen pots, handbaskets made of crabshells wrought together, also an English pail or bucket; it wanted a bail, but it had two iron ears. There were also baskets of sundry sorts, bigger and some lesser, finer and some coarser; some were curiously wrought with black and white in pretty works. . . .

This description of native houses is one of the most detailed given in early accounts of the region. The exterior mats were probably made by sewing bulrushes arranged in parallel rows; the interior mats were most likely woven in a basic over-and-under checker weave, and could have been made of either bulrushes or cedar bark. Houses of this type ranged in size from fifteen to twenty feet in diameter and housed one or two nuclear families, that is, a husband and wife with their children. Such hemispherical houses are commonly referred to as wigwams, although the Wampanoag term for them was *witu*. Many people confuse the terms wigwam and tipi, but the latter is the conical buffalo hide–covered portable house used by nomadic groups on the plains.

The reason that the exploring party from the *Mayflower* encountered so few of the native peoples is that they followed a seasonal pattern of settlement. During the summer and fall, the population dispersed along the ocean to take advantage of fishing. Small garden plots were cultivated adjacent to the houses, which during these months would have been wigwams. But at the onset of winter, the people moved inland, away from the coast, where it was warmer, and lived in larger houses thirty feet wide and ranging in size from fifty to one hundred feet in length. These larger structures would be occupied by a number of related families who had occupied separate houses along the coast during the summer. These long houses, oblong or oval in floor plan, were covered with slabs of bark, and each family unit had its own hearth, with a smoke hole left in the roof over each.

Since it was December when the exploration of the cape took place, most of the local inhabitants had moved inland. What is somewhat puzzling, however, is the number of objects found within the houses described in *Mourt's Relation*. We lack the necessary information to suggest that leaving houses at least partly equipped was a common practice, but the possibility exists, if such a practice was honored as a matter of mutual trust and agreement.

By now it was December 4, and the party debated the pros and cons of where to establish the new colony. The situation was becoming desperate. They were low on supplies, and

> . . . cold and wet lodging had so tainted our people, for scarce any of us were free from vehement coughs, as if they should continue long in

> that estate it would endanger the lives of many, and breed diseases and infection among us. Again, we had yet some beer, butter, flesh, and other such victuals left, which would quickly be all gone, and then we should have nothing to comfort us in the great labor and toil we were like to undergo at the first.

Some of the party urged that they go as far as "Anguum" or "Angoum" (Agawam, today Ipswich, Massachusetts, north of Boston). It was decided in the end to settle somewhere within the bay:

> Besides, Robert Coppin, our pilot, made relation of a great navigable river and good harbor in the other headland of this bay, almost right over against Cape Cod, being in a right line not much above eight leagues distant, in which he had been once; and because that one of the wild men with whom they had some trucking stole a harping iron [harpoon] from them, they called it Thievish Harbor. And beyond that place they were enjoined not to go, whereupon a company was chosen to go out upon a third discovery. Whilst some were employed in this discovery, it pleased God that Mistress White was brought a-bed of a son, which was called Peregrine.

On the next day, one of the Billington sons nearly blew up the *Mayflower*. He was making squibs with gunpowder in his father's cabin, and having spilled half a small barrel of gunpowder, fired off a fowling piece, setting off a fire. Fortunately there was no explosion, and "by God's mercy, no harm done."

The "third discovery" was much like the earlier two, with the shallop carrying a party of ten passengers, including Standish and Bradford, two of the master's mates, and three sailors. Setting out on Wednesday, December 6, they explored the inner coast of Cape Cod as far south as the point where the cape curves toward the north in the vicinity of present-day Eastham. The better part of the expedition was devoted to searching for an appropriate location to settle, but they were unsuccessful in this attempt. Two incidents are cited in *Mourt's Relation* that are worthy of mention.

> As we drew near to the shore, we espied some ten or twelve Indians very busy about a black thing—what it was we could not tell—till

> afterwards when they saw us, and ran to and fro as if they had been carrying something away. We landed a league or two from them. . . .

These "ten or twelve Indians" were the first they had seen since their initial trip ashore, on November 15. As to the identity of the "black thing," the mystery was resolved on the next day, when

> We found also a great fish, called a grampus, dead on the sands; they in the shallop found two of them also in the bottom of the bay, dead in like sort. They were cast up at high water, and could not get off for the frost and ice. They were some five or six paces long and about two inches thick of fat, and fleshed like a swine. . . .

They then returned to the place where they had seen the Indians, and

> When we were there, we saw it was also a grampus which they were cutting up; they cut it into long rands or pieces, about an ell [45 inches] long, and two handfull broad. We found here and there a piece scattered by the way, as it seemed, in haste.

The animal referred to as a "grampus" by the exploring party was actually a kind of small whale, later to be commonly referred to as blackfish, which could be hunted from small boats just offshore, or driven ashore with great ease, where they would be stranded when the tide went out. As many as four hundred at a time could be obtained in this manner, and each would produce a barrel of oil. The shore whaling industry became a very important and lucrative one during the seventeenth century, with Wellfleet, not far north of the place where the exploring party found the grampuses, becoming an important center for the taking of blackfish.

The party tracked the Indians along the beach to a point where they turned inland, and followed them further, in the course of which they found another burial ground, surrounded by a palisade. The graves "were more sumptuous than those at Cornhill, yet we digged none of them up, but viewed them and went on our way." They made camp that night, and at dawn, after depositing their arms and armor on the beach to be loaded onto the shallop, prepared themselves breakfast.

> Anon, all upon a sudden, we heard a great and strange cry, which we knew to be the same voices, though they varied their notes. One of our company, being abroad, came running in and cried, "They are men! Indians! Indians!" and withal, their arrows came flying amongst us. Our men ran out with all speed to recover their arms, as by the good providence of God they did. In the meantime, Captain Miles Standish, having a snaphance [a type of flintlock musket] ready, made a shot, and after him another. After they two had shot, other two of us were ready, but he wished us not to shoot till we could take aim. . . .

At this time, there were only four men behind a barricade that they had erected the previous night, and they were most concerned to defend it, since it stood between the Indians and the shallop, just offshore. They hoped that those aboard the shallop would defend it, and

> . . . called unto them to know how it was with them, and they answered, "Well! Well!" every one and, "Be of good courage!" . . . The cry of our enemies was dreadful, especially when our men ran out to recover their arms; their note was after this manner, "*Woach woach ha ha hach woach*." Our men were no sooner come to their arms, but the enemy was ready to assault them.

What followed was mercifully a bloodless encounter, with the Indians finally fleeing. The party collected eighteen arrows, with brass, hart's horn, and eagle claw points, which were later sent back to England with Master Jones. They called the place where they and the Indians had their first face-to-face confrontation, and not a happy one, "The First Encounter," a name by which it is still referred to today, with First Encounter Beach marked clearly on modern road maps of Cape Cod. The next face-to-face encounter was to be very different, culminating in the peace treaty concluded between Massasoit, supreme chief of the Wampanoag, and the Plymouth settlers in March 1621. This agreement lasted for half a century, so that from the beginning of the settlement relations between the two groups were on the whole amicable, benefitting both.

Following this incident, they sailed all that day, Friday, December 8, but finding no convenient harbor pushed on to where Coppin believed Thievish Harbor to be, despite the fact that it soon began to snow and rain, and the weather rapidly deteriorated:

> About the midst of the afternoon, the wind increased and the seas began to be very rough, and the hinges of the rudder broke so that we could steer no longer with it, but two men with much ado were fain to serve with a couple of oars. The seas were grown so great that we were much troubled and in great danger, and night grew on. Anon Master Coppin bade us be of good cheer; he saw the harbor. As we drew near, the gale being stiff and we bearing great sail to get in, split our mast in three pieces, and were like to have cast away our shallop. Yet, by God's mercy, recovering ourselves, we had the flood with us, and struck into the harbor.

What seems to be implicit in this portion of *Mourt's Relation* is a sudden sense of urgency. After spending twenty-seven days exploring the cape only as far south as the vicinity of present-day Eastham, they proceeded to traverse the remaining coastline in a single day. No parties were sent ashore to explore, and it seems as though there was a consensus to reach Coppin's "good harbor" as soon as possible. After all, the settlers had been on board the *Mayflower* for sixty-six days at sea, and three weeks in exploration. Winter was closing in, supplies were dwindling, sickness increasing, and the immediate discomfort of three days of sailing and searching, when from the first day "it was very cold, for the water froze on our clothes and made them many times like coats of iron," had to have affected their thinking at this point. While the *Mayflower* was anchored in Cape Cod Bay groups of crew and passengers probably went ashore from time to time, as they did that first occasion on November 13, although there is no record of this happening, but the leaders of the expedition searching for a suitable harbor for their settlement must have been very aware of the distress and discomfort that living on board for so long was causing passengers and crew alike.

After spending the night on an island in the harbor (later named Clark's Island), exploring it, and taking the Sunday as a day of rest, on Monday, December 12, according to *Mourt's Relation*, the members of this final expedition of discovery set foot on the mainland at Plymouth for the first time.

> On Monday we sounded [measured the water for depth] the harbor, and found it a very good harbor for our shipping. We marched also into the land, and found divers cornfields, and little running brooks, a place very good for situation, so we returned to our ship again with good news to the rest of our people, which did much comfort their hearts.

It is worth noting at this point that there was in existence a map of Plymouth Harbor, with soundings, made by a Frenchman, Samuel de Champlain, the founder of Quebec, who visited Plymouth in 1605 during an exploration of the coast of North America in search of a suitable place to establish a French colony. He made a number of excellent charts of the coast, including one of the Port St. Louis (Plymouth Harbor), and published them in 1613 together with an account of his voyages. The chart shows soundings, identifies their anchorage, the channel to the inner harbor, sand dunes and mud flats, and is illustrated with drawings of Indian dwellings surrounded by cornfields. However, since the first thing that the shallop expedition from the *Mayflower* did once they finally reached Plymouth in December 1620 was to take soundings of the harbor's depths to find suitable anchorage for the ship, it is unlikely that they knew of Champlain's work.

The *Mayflower* departed Cape Cod (Provincetown Harbor) on December 15, and arrived at Plymouth the next day, having had some difficulty with the weather, which made it appear at first that they would have to return to the cape, but on Saturday, December 16, 1620, they "came safely into a safe harbor," three months and ten days after leaving "Old Plymouth" on the shores of England.

The "Landing of the Pilgrims," an event celebrated in paintings and even on nineteenth-century English blue and white Staffordshire dinner plates, a scene with which most of us are familiar, actually never took place as it is usually depicted, with all of the ship's passengers coming ashore at one time and stepping onto the famous rock. Rather, it was a gradual process, with small parties of men traveling from ship to shore to explore the area. In fact, there was not an immediate decision to settle in the location where they finally did. Some thought Clark's Island in Plymouth Harbor, which they had already

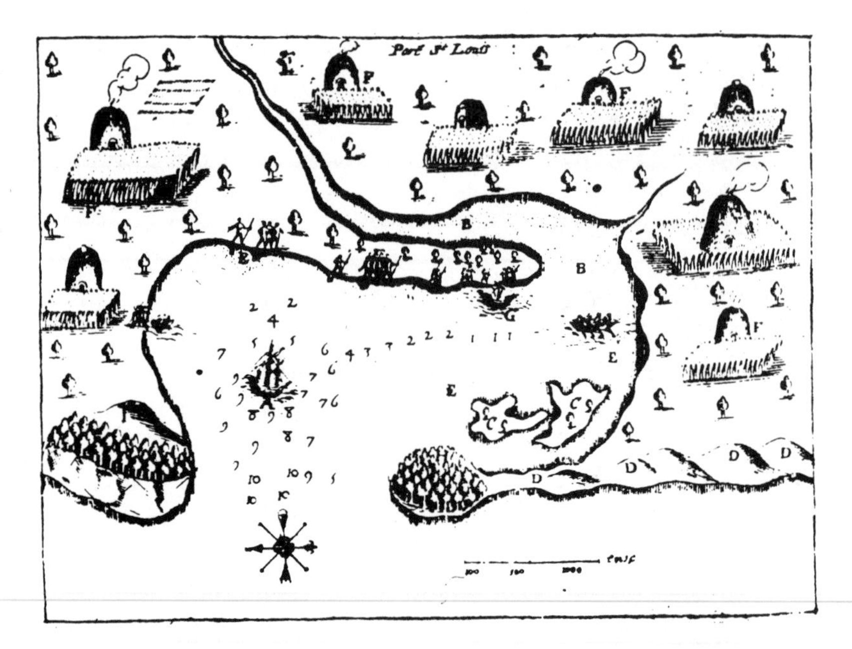

Key: **A.** The place where vessels can anchor, with Champlain's pinnace. **B.** The Channel to the inner harbor. **C.** The two islands—Clark Island, where members of the third exploration party spent their first night ashore in 1620, and Saquish Head. **D.** Sand dunes that stretch along Duxbury Beach to the Gurnet. **E.** Mud flats along Plymouth Harbor and Duxbury Bay, exposed at low tide. **F.** Wampanoag *witu* or wigwams. **G.** "Spot where we ran our pinnace aground" and from which Champlain sketched this map (Brown's Bank, adjacent to Long Beach). **H.** The Gurnet—a conspicuous landmark, visible from four to five leagues out to sea.

Samuel de Champlain's chart of Port St. Louis (Plymouth). From *Les Voyages* (1613).

explored, would be suitable, primarily for defensive reasons. But it lacked a source of fresh water, and was heavily wooded, which would require a great amount of labor to clear for planting corn. Others favored the place where the Jones River enters the bay, located today in Kingston, Massachusetts, some three miles north of Plymouth:

> This place we had a great liking to plant in, but that it was so far from our fishing, our principal profit, and so encompassed with woods that we should be in much danger of the savages, and our number being so little, and so much ground to clear, so as we thought good to quit and clear that place till we were of more strength.

Three days after the *Mayflower* anchored in Plymouth Harbor, on December 19, it was finally decided to settle

> ... on a high ground, where there is a great deal of land cleared, and hath been planted with corn three or four years ago, and there is a very sweet brook runs under the hill side, and many delicate springs of as good water as can be drunk, and where we may harbor our shallops and boats exceeding well, and in this brook much good fish in their seasons; on the further side of the river also much corn-ground cleared. In one field is a great hill on which we point to make a platform and plant our ordnance, which will command all round about.

The site selected for settlement was the location of the Wampanoag village of Patuxet, all of whose inhabitants had died from an epidemic that had swept this part of the coast in 1617. The only disadvantage that the party perceived was the distance they would have to traverse to obtain wood, "which is half a quarter of an English mile, but there is enough so far off." Between December 23 and December 28, various parties went ashore to fell trees and carry them to the site for use in building:

> Monday, the 25th day, we went on shore, some to fell timber, some to saw, some to rive [splitting logs for clapboards and palings] and some to carry, so no man rested that day.

Slowly, a small village began to appear on the hillside. The settlers had a clear notion of how their town would look when completed. In a seventeenth-century version of town planning:

> Thursday, the 28th of December, so many as could went to work on the hill where we purposed to build our platform for our ordnance, and which doth command all the plain and the bay, and from whence we may see far into the sea, and might be easier impaled, having two rows of houses and a fair street. So in the afternoon we went to measure out the grounds, and first we took notice how many families there were, willing all single men that had no wives to join with some family, as they thought fit, that we might build fewer houses, which was done, and we reduced them to nineteen families. To greater families we allotted larger plots, to every person half a pole in breadth

> and three in length [8¼ by 49½ feet], and so lots were cast where every man should lie, which was done, and staked out. We thought this proportion was large enough at the first for houses and gardens, to impale them round, considering the weakness of our people. . . .

On January 4, 1621, Captain Standish and a party of "four or five" went out in the hope of meeting with the local inhabitants, but met with no success. They did see some houses, but they were not inhabited. On their way home, "they shot at an eagle and killed her, which was excellent meat; it was hardly to be discerned from mutton." Today, when one is asked what some exotic meat tastes like, such as alligator, turtle, or rattlesnake, the reply is almost invariably, "It tastes like chicken." Is it possible that in the seventeenth century, mutton was the equivalent standard of comparison? While few if any of us have eaten a bald eagle, which the one in question almost certainly was, one would expect it to have a decidedly fishy flavor. But then, alligator doesn't really taste like chicken either.

The first building to be erected was a common house, about twenty feet square, which was inhabitable though not fully thatched by January 9. Five days dater, on January 14, those still on board ship "spied their great new rendezvous on fire." Apparently a spark had ignited the thatch, but the loss was not as serious as first supposed. At the time of the fire, those onshore were living in the common house, and it was "as full of beds as they could lie one by another, and their muskets charged, but, blessed be God, there was no harm done." House construction continued whenever the weather permitted, and *Mourt's Relation* mentions that the strong wind and rain on February 4 caused much of the daubing [clay plaster] of their houses to fall down. A small house used for the sick members of the party, and there were many, caught fire that afternoon, also from a spark in the thatch, but little damage was done. Just how many houses stood by mid-February is unknown, but by December 11, 1621, Edward Winslow, in the same letter in which he gives his account of the fall harvest celebration, stated, "We have built seven dwelling houses, and four for the use of the plantation [common houses] and have made preparations for diverse others."

But as the little town grew, its inhabitants were reduced in number at an even faster rate. Of the "scarce fifty" who survived that terrible first winter:

> In the time of most distress, there was but six or seven sound persons, who to their great commendations . . . spared no pains night nor day, but with abundance of toil and hazard of their own health, fetched them wood, made them fires, dressed them meat, made their beds, washed their loathsome clothes, clothed and unclothed them. In a word, did all the homely and necessary offices for them which dainty and queasy stomachs cannot endure to hear named, and all this willingly and cheerfully. . . .

Two of the seven were William Brewster, their "reverend Elder," and Myles Standish, who had been chosen as their captain on January 17. Bradford himself was quite ill. Apparently not all of those who tended to the sick were as charitable as those mentioned by Bradford, for at one point, while he lay ill, he asked for a "small can of beer" and was told "that if he were their own father he should have none."

The passengers and the crew of the *Mayflower* were greatly weakened not only from the six-week journey across the Atlantic, but through the "vehement" colds caught through having to wade to and fro from the shore, and from exploring the coast in such bad weather that it must have been difficult to dry out their clothes even once they returned to the ship. Illness would spread quickly among the small shipboard community, and the landing at Plymouth in December 1620 still meant continued living on board until enough housing could be built in the very cold, hazardous weather conditions, which took their own toll. Supplies were running low by the time they arrived at Plymouth, and there was little prospect of replenishing these until the spring. From December 1620 through the end of March 1621, Bradford noted that forty-four passengers died (including the four deaths while the *Mayflower* was at anchor off Cape Cod), and about half of the ship's crew. The passengers included Richard Britteridge, "the first to die in this harbor," on December 21, 1620, and Rose Standish, who died on January 29, after which entry Bradford writes, "This month, eight of our number die." Following an entry noting the death of Mary (Norris) Allerton, wife of Isaac Allerton, on February 25, is a stark entry: "N.B. This month, Seventeen of our number die." Edward Winslow's wife, Elizabeth (Barker) Winslow, died on March 24, two days after the peace treaty with Massasoit was signed, and it requires little imagination to envisage what immense additional stress must have been caused to both husband and wife as he was held hostage through that long day and she lay close to death in the village. Bradford, writing at the end of March 1621, says:

> This month, thirteen of our number die . . . there die sometimes two or three a day. Of one hundred persons scarce fifty remain: The living scarce able to bury the dead; the well not sufficient to tend the sick. . . .

It will be remembered that whole families died, as well as several couples. Of the fifteen men who came out on their own, it seems that only eight survived the winter. There were also some who died in the spring and summer: the governor, John Carver, in April 1621, and his wife, Katherine (White) (Leggatt) Carver, died "five or six weeks after him." Alice Mullins and her adult son, Joseph, were both mentioned in William Mullins's will, which was sent back to England on the *Mayflower*'s return voyage that same April, without any formal endorsement to the effect that two of the chief beneficiaries were dead, so it can be assumed that both were still alive in April 1621 but had died by the end of the summer, as John Pory, visiting Plymouth in August 1622, wrote some five months later that "the Governor told me, for the space of one year of the two wherein they had been there, died not one man, woman or child."

With as few as "six or seven sound persons" available to work during the "time of most distress" it is no small wonder that by the end of the year 1621, when Winslow wrote his letter to his "Loving and Old Friend" in England on December 11, only eleven structures had been erected, seven of which were dwelling houses. Less than two years later, however, a visitor to Plymouth, Emmanuel Altham, wrote to his brother in England: "In this Plantation is about twenty houses," so it is obvious that the surviving members of the community, joined by thirty-five others who arrived on the *Fortune* in November of 1621, had made significant progress in building the village.

By the spring of 1621, however, the worst of the winter seemed to be over, and on the third of March,

> The wind was south, the morning misty, but towards noon, warm and fair weather; the birds sang in the woods most pleasantly. At one of the clock it thundered, which was the first we heard in that country; it was

> strong and great claps, but short, but after an hour it rained very sadly [steadily] till midnight.
>
> The spring now approaching, it pleased God the mortality began to cease amongst them, and the sick and the lame recovered apace, which put as [it] were new life into them, though they had borne their sad affliction with much patience and contentedness as I think any people could do.

On April 5, the *Mayflower* sailed for England, with a depleted but healthy crew. None of the Plymouth residents returned, apparently content to stay and see to the business of constructing their plantation, the Separatists now free from the religious persecution they had left their mother country to avoid.

On March 16, an event took place that would have a profound effect on the relations between the people of Plymouth and their Wampanoag neighbors. Previously, "the Indians came skulking about them, and would sometimes show themselves aloof off . . . and once they stole away their tools where they had been at work and were gone to dinner." But on that March day, "a certain Indian came boldly amongst them and spoke to them in broken English, which they could well understand but marveled at it." His name was Samoset, an Abenaki Indian from Pemaquid Point, Maine. A Captain Thomas Dermer had been sailing along the New England coast in 1619, and stopping at the English fishing station on Monhegan Island, encountered Samoset, and for some unknown reason, transported him to Cape Cod, from where he made his way overland to Plymouth, then the deserted village of Patuxet. He informed the Plymouth people that there was another Indian, Tisquantum, or Squanto, as he came to be called, who spoke even better English than he. Small wonder, for he had actually lived in England for some time, having first been taken captive in 1614 by Thomas Hunt and transported to Spain where he was sold at the slave market at Malaga along with at least twenty-seven other captives. Squanto was one of the few who were acquired by local friars, and apparently was treated well; the fate of the others is unknown. Somehow, he managed to make his way to London, where he lived for several years with a wealthy merchant, John Slanie, treasurer of the Newfoundland Company. Through Slanie he met Thomas Dermer, and accompanied him to New England in 1618, where he jumped ship and traveled to his home at Patuxet, only to find that the village was no longer inhabited. He was in fact the sole survivor, having had the good fortune to have been in England when the epidemic swept through the coastal populations.

Samoset stayed the night in Stephen Hopkins's house, where he was "watched," and departed the following morning for Massasoit's village. He was presented with a knife, bracelet, and a ring, and he promised that he would return in a night or two, and bring some of Massasoit's men with him, as well as some beaver skins, which he hoped to trade with the Plymouth settlers. True to his word, he returned in two days, with five other men, described in *Mourt's Relation* as being tall and proper. Apparently the day's meeting was quite congenial, the Indians leaving their bows and arrows a quarter of a mile from the town, as they had been requested to do. They were well fed by the colonists, and entertained them with dances, accompanied by singing—the Wampanoag did not have musical instruments, so the music that accompanied dances was mainly singing, but may have included beating on the ground with hands or sticks. They had brought three or four skins, but no trading took place; they were asked to bring more when they came to visit again, so that they could be traded for in a single lot. They also returned the tools that had been taken before. The six were presented with other "trifles," and then some of the settlers, in arms, escorted five of the Indians to where they had left their weapons. The sixth, Samoset, stayed behind, claiming sickness, and did not leave for three more days. On the following day, by then Thursday, March 22, 1621, Samoset returned, this time accompanied by Squanto, whom the English now met for the first time. The men told the settlers that Massasoit was close by, accompanied by his brother, Quadequina, and some sixty men. After an hour had passed, the group appeared at the top of a hill "over against us," presumably Watson's Hill, on the south side of Town Brook. Judging from the description of the event in *Mourt's Relation*, the settlers were on another hill, now known as Burial Hill, just north of the brook, which we can assume to be the same hill where they had built the platform for their ordnance. At first, neither group seemed willing to venture further, but Squanto crossed the brook and negotiated for a single person, Edward Winslow, to act as an emissary, which he did. Wearing armor and sword, Winslow informed Massasoit that his people desired to have peace with him and engage in trading. Massasoit was presented with gifts—a pair of knives and a copper chain with a jewel in it; his brother was given a knife and a jewel earring. Both were given a pot of "strong water," a good supply of biscuits and some butter, which were graciously accepted. Winslow made a short speech through his interpreters, telling Massasoit that King James "saluted him with words of love and peace," and that their governor, John Carver, desired to talk with him and make arrangements for trading, probably mainly beaver skins from Massasoit's people in exchange for whatever commodities the English would provide. Finally, holding Winslow hostage in the custody of Quadequina, Massasoit and twenty of his men crossed the brook, leaving their weapons behind. They were greeted by Captain Standish and a companion, and escorted to a partially built house, where he was ensconced on a green rug with some cushions. The governor then made an impressive entrance, "with drum and trumpet," accompanied by some musketeers. After providing Massasoit with "strong water" and a little meat that was eaten by the entire group, they drew up a peace agreement as follows:

> 1. That neither he nor any of his should injure or do hurt to any of our people.
> 2. And if any of his did hurt to any of ours, he should send the offender, that we might punish him.
> 3. That if any of our tools were taken away when our people were at work, he should cause them to be restored, and if ours did any harm to any of his, we would do the like to them.
> 4. If any did unjustly war against him, we would aid him; if any did war against us, he should aid us.
> 5. He should send to his neighbor confederates, to certify them of this, that they might not wrong us, but might be likewise comprised in the conditions of peace.
> 6. That when their men came to us, they should leave their bows and arrows behind them, as we should do our pieces when we came to them.
>
> Lastly, that doing thus, King James would esteem of him as his friend and ally. . . . So after all was done, the governor conducted him to the brook, where they embraced each other and he departed. . . .

Six of Massasoit's men were kept as hostages against the return of Winslow, who was kept with the Wampanoag while Quadequina and another group crossed over the brook, where they were likewise entertained. After they had been escorted back to Massasoit and his followers, Edward Winslow was released. In the light of the hostile relations between the Wampanoag and their neighbors the Narragansetts, the English settlers recognized that the peace accord was obviously made in large part to provide a strong ally for them, "especially because he [Massasoit] hath a potent adversary the Narragansetts, that are at war with him, against whom he thinks we may be some strength to him, for our pieces are terrible unto them." The Narragansetts were a large and powerful group, who had escaped the sickness of 1617 that had so severely decimated their Wampanoag neighbors. The treaty was a remarkable achievement. It held for over fifty years despite some tensions, lasting throughout Massasoit's lifetime (he died in 1660), and would be broken finally by his son, Metacom, known by the English as King Philip, who in 1675 precipitated the war that bears his name.

Having played a key role in the peace negotiations between the English and Wampanoag, Squanto would make another important contribution that would be of great benefit to the colonists. In the spring of 1621, he instructed them in how to plant their corn to ensure a good crop. He told them that unless they put fish into their corn hills, the crop would be of little worth, as the planting grounds were old and not fertile. This was done, and while the "English seed"—wheat and peas—came to little, the corn crop flourished, producing the bountiful harvest that was celebrated the following fall. Squanto remained with the English for the brief time until his death in 1622. He accompanied the colonists on a number of expeditions, including a successful trading trip to a neighboring tribe, the Massachusetts, which produced a large quantity of beaver furs. But in the end, he put himself in a very difficult position. Being able to speak both languages, he played off one side against the other on a number of occasions, for

> . . . they began to see that Squanto sought his own ends and played his own game, by putting the Indians in fear and drawing gifts from them to enrich himself, making them believe he could stir up war against whom he would, and make peace for whom he would. Yea, he made

> them believe they kept the plague buried in the ground, and could send it amongst whom they would, which did much terrify the Indians and made them depend more on him, and seek more to him, than to Massasoit.

The upshot of all of this was predictable; Massasoit demanded that the English hand Squanto over to him for what Bradford called "Indian justice," but the transaction, which had actually begun, was interrupted by the arrival of a shallop from a ship sent to New England by Thomas Weston, who had been one of the leading London merchant adventurers, and Squanto was not handed over to Massasoit's men, who departed in great anger. Nothing more came of it, but apart from journeys for trade and exploration, on one of which he died in the fall of 1622, Squanto spent the remainder of his life within the colonists' village.

Just what did this village look like, and how did it develop in the seven years following the arrival of the *Mayflower* in Plymouth Harbor? While there are no maps as such of the layout of the community, there are excellent descriptions of it written by three men who visited Plymouth during that period, and a fourth, somewhat problematic one, by Captain John Smith. The closest thing to a map is a plan made by Bradford entitled, "The meersteads & garden plots of [those] which came first layd out 1620." It shows seven lots facing "the street," which was bisected by a "highway." These lots are located on the south side of the street; nothing is shown on the north side. East of the highway were the lots of Peter Brown, John Goodman, and William Brewster, and to the west those of John Billington, Isaac Allerton, Francis Cooke, and Edward Winslow. Just when Bradford actually made this plan is unclear, but it does agree with Winslow's mention of seven dwelling houses that were in place before December 11, 1621. Atop the hill was the platform on which their ordnance was placed.

Shortly after the departure of the *Fortune* on December 13, 1621, the Plymouth settlers were openly challenged by the Narragansetts, just to the west. In January 1622, they

> . . . sent a messenger unto them with a bundle of arrows tied about with a great snakeskin, which their interpreters told them was a threatening and a challenge. Upon which the Governor [now Bradford], with the

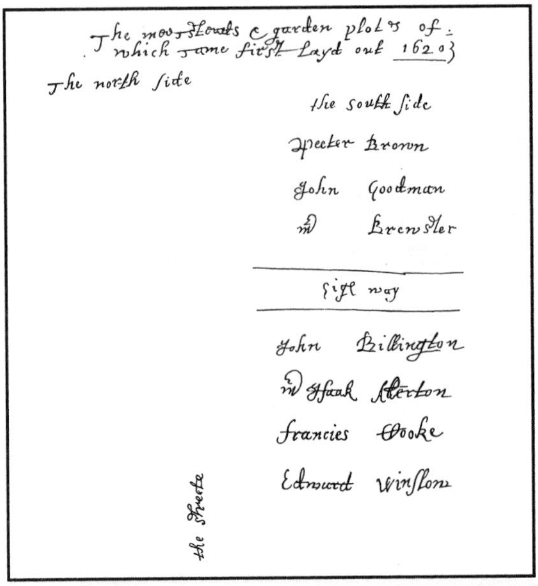

*The meersteads & garden plotes of [those] which came first layd out 1620.

The north side

The south side
Peeter Brown
John Goodman
Mr Wm Brewster

high way

John Billington
Mr Isaak Allerton
Francies Cooke
Edward Winslow

the streete

William Bradford's plan of the original layout of Plymouth.

> advice of others, sent them a round answer that if they had rather have war than peace, they might begin when they would; they had done them no wrong, neither did they fear them or should they find them unprovided. And by another messenger sent the snakeskin back with bullets in it. But they would not receive it, but sent it back again.

However, apparently the challenge was taken quite seriously by the English, for in February 1622, Winslow wrote:

> In the mean time, knowing our own weakness, notwithstanding our high words and lofty looks towards them [the Narragansetts], and still

> lying open to all casualty, having as yet (under God) no other defense than our arms, we thought it most needful to impale our town; which with all expedition we accomplished in the month of February, and some few days, taking in the top of the hill under which our town is seated; making four bulwarks or jetties without the ordinary circuit of the pale, from whence we could defend the whole town; in three whereof are gates, and the fourth in time to be.

Until the village was fortified by this strong palisade, it had been defended only by the ordnance on a platform built on the hilltop in December 1620, with the full complement of ordnance put in place by late February 1621. That the 2,700-foot fortified enclosure was completed in a month was remarkable, considering the fact that it was winter, the wood for the pales (at least ten-and-a-half feet long, to allow eight feet aboveground) would have had to be cut and sharpened, posts and rails prepared, and the whole assembled with three gates and four defensive bastions. With the addition of the new colonists on the *Fortune*, the male population of Plymouth capable of building would have been around fifty-three, including some six young male adults. Women and children also could have assisted, as it is fair to assume that in view of the threat of imminent attack by the powerful Narragansetts, everything was put aside and the efforts of the whole town directed to this one goal.

For the next few months the newly fortified village was defended by the four bulwarks, to which Captain Standish assigned four companies, together with the platform with ordnance on the hill above the houses. At the end of May 1622, word reached Plymouth of the Indian uprising in Virginia that had taken place the previous March 22, one year to the day after the conclusion of the peace treaty with the Wampanoag. Although the main reason for the decision to build a fort within the fortified town has been attributed to alarm at the events in Virginia, Edward Winslow gives more weight to the fact that the colonists were visibly weak for lack of adequate food supplies, to such an extent that

> . . . the Indians began again to cast forth many insulting speeches, glorying in our weakness, and giving out how easy it would be ere long to cut us off. Now also Massassowat [Massasoit] seemed to frown on us, and neither came or sent to us as formerly.

Also, it should be remembered that it was at the end of May, when the shallop from the *Sparrow* arrived with seven of Weston's settlers, that Massasoit's messengers were in the process of demanding that the governor of Plymouth "cut off [Squanto's] head and hands" for them to take to Massasoit, who had sent his own knife for the execution, and that they left "mad with rage" when Bradford delayed his answer due to the shallop's arrival. It was this shallop that also brought news of the Virginian uprising.

In June 1622, work began on further fortification, a project that would take ten months to complete. Bradford writes,

> This summer they built a fort with good timber, both strong and comely, which was of good defence, made with a flat roof and battlements, on which their ordnance were mounted, and where they kept constant watch, especially in time of danger. It served them also for a meeting house and was fitted accordingly for that use. It was a great work for them in this weakness and time of wants, but the danger of the time required it; and both the continual rumors of the fears from the Indians here, especially the Narragansetts, and also the hearing of that great massacre in Virginia, made all hands willing to dispatch the same.

The fort was built on the top of the hill, replacing the ordnance platform. In a section of his *Good Newes from New England* dated June 1622, Winslow commented:

> We have a hill called the Mount, enclosed within our pale, under which our town is seated, we resolved to erect a fort thereon; from whence a few might easily secure the town from any assault the Indians can make, whilst the rest might be employed as occasion served. This work was begun with great eagerness and with the approbation of all men, hoping that this being once finished, and a continual guard there kept, it would utterly discourage the savages from having any hopes or thoughts of rising against us.

By August they had already made excellent progress, for John Pory, stopping at Plymouth *en route* to England from Virginia, at the end of his three-year term as secretary to the governor and council of

Virginia, commented in a letter to the earl of Southampton on the industry of the colonists:

> As well appeareth by their building, as by a substantial palisado about their [town] of 2700 foot in compass, stronger than I have seen any in Virginia, and lastly by a block house which they have erected in the highest place of the town to mount their ordnance upon, from whence they may command all the harbor.

Pory wrote these words in January 1623, and by March the fort had been completed, despite the fact that some of those who had begun work on it with such enthusiasm wanted to give up, "flattering themselves with peace and security, and accounting it rather a work of superfluity and vainglory, than simple necessity."

Less than a year after Pory wrote to the earl of Southampton, his description was corroborated by Emmanuel Altham in a letter to his brother in September 1623. Altham was one of the merchant adventurers who had invested in the New Plymouth Company, and captain of the *Little James,* the pinnace that the company sent to Plymouth for fish and fur trading. He wrote that Plymouth was

> . . . well situated upon a high hill close unto the seaside. . . . In this plantation is about twenty houses, four or five of which are very fair and pleasant, and the rest (as time will serve) shall be made better. And this town is in such manner that it makes a great street between the houses, and at the upper end of the town there is a strong fort . . . with six pieces of reasonably good artillery mounted thereon. . . . This town is paled about with a pale of eight feet long, and in the pale are three great gates.

The third description of the town is that of Captain John Smith, dated 1624. Although best known for his critical role in the development of the English colony at Jamestown, including his rescue by Pocahontas from execution at the hands of Chief Powhatan, John Smith was no stranger to New England. In fact, it was he who gave that name to the region. He first published the result of his 1614 explorations on land and coastal survey in his *Description of New England* (London, 1616). It includes a map of New England that he had presented to Prince Charles, son of James I, "humbly entreating

NEW ENGLAND

The most remarqueable parts thus named. by the high and mighty Prince CHARLES, nowe King of great Britaine.

THE PORTRAICTUER OF CAPTAYNE IOHN SMITH ADMIRALL OF NEW ENGLAND.

These are the Lines that shew thy Face; but those
That shew thy Grace and Glory, brighter bee:
Thy Faire-Discoueries and Fowle-Overthrowes
Of Salvages, much Civilliz'd by thee
Best shew thy Spirit; and to it Glory Wyn;
So, thou art Brasse without, but Golde within.

If so; in Brasse, (too soft smiths Acts to beare)
I fix thy Fame, to make Brasse Steele out weare.

Thine as thou art Virtues,
John Davies. Heref:

Captain John
Smith's map of
New England, 1616.

his Highness he would please to change their barbarous names for such English, as posterity might say Prince Charles was their God-father." Among the twenty-nine places renamed was Accomack, which was given the new name of Plimoth by the prince, later marked on the map as New Plimouth. Smith had offered his services to the Separatists at Leiden who were planning to emigrate to America, but evidently Myles Standish was prepared to charge less for his than the experienced Smith, who commented wryly in his *True Travels, Adventures and Observations*, published in London in 1630, that the "Brownists of England, Amsterdam and Leyden, [who] went to New Plimouth, whose humorous [fanatical] ignorances, caused them for more than a year, to endure a wonderful deal of misery, with an infinite patience; saying my books and maps were much better cheap to teach them, than myself." So it is evident that there was a copy of Smith's map of New England, showing the exact location of Plymouth, in the possession of those on the *Mayflower*.

It is generally agreed that Smith's accounts of his time at Jamestown are reliable, so what about the one he published concerning Plymouth? The colorful nature of his description of his military and other exploits in Eastern Europe in the early years of the seventeenth century, before he sailed to Jamestown in December 1606, has in the past given rise to considerable skepticism as to just how reliable his work is. In particular, there is an echo of the Pocahontas episode in Smith's account of his capture by Turks, being sold as a slave, given to a wealthy young woman in Istanbul who fell in love with him and sent him to her brother to be trained for high office as a Turk, evidently with a view to marrying him. Harshly disciplined and mistreated, Smith finally murdered her brother and escaped, eventually making his way back to England. However, more recent scholarship has demonstrated that his detailing of his European adventures is probably dependable. One little-known connection between his Turkish experiences and New England explorations is that Smith named Cape Trabigzanda, now north of Boston, after Charatza Trabigzanda, his mistress in Istanbul. Prince Charles renamed it Cape Anne, the name it still retains.

In his *Generall Historie of Virginia, New England & The Summer Isles*, published in London in late 1624, Smith gives a description of New Plymouth as it was in that year. Earlier in the volume he provides a lengthy paraphrase of *Mourt's Relation*, with which he was familiar, but, of course, he was not personally involved in the events described in that work. Although his description of New Plymouth appears to be based on his actually having been there, he only made two voyages to New England, the first in 1614, and a second, abortive expedition, in 1615. He evidently had access to some recent communications from Plymouth, and as he retained a vigorous interest in promoting the welfare of the country that he had explored and that bore the name he had given it, it is reasonable to infer that he kept in touch with travelers to the New World. Smith's account reads:

> At New-Plimoth there is about 180 persons, some cattle and goats, but many swine and poultry, 32 dwelling houses, whereof 7 were burnt the last winter, and the value of five hundred pounds in other goods; the Town is impaled about half a mile in compass. In the town upon a high Mount they have a fort well built with wood, loam and stone, where is planted their Ordnance: Also a fair Watch-tower, partly framed, for the Sentinel . . . they have made a saltwork, and with that salt preserve the fish they take, and this year hath fraughted [filled] a ship of 180 tons.

One detail of particular importance in this account is its contribution to our knowledge of the fortified town, where Smith gives the length of the palisade: "the Towne is impailed about halfe a mile in compasse." His dimensions virtually match those given by John Pory in January 1623, but as far as we know, he did not have access to Pory's letter, although it is quite possible that he discussed New Plymouth with him after Pory's return to London. "About half a mile in compass" would be about 2,640 feet, as opposed to Pory's 2,700 feet.

When one description matches the other, there is good reason to accept the accounts as accurate in their details, and the three quoted above are given further veracity by a fourth, written in 1628. The last description of Plymouth in its early years comes to us from a letter written by Isaack de Rasieres, chief trading agent for the Dutch West India Company as well as secretary to the director-general of New Netherland, who visited Plymouth in 1627. His is the most detailed description of the four:

> New Plymouth lies in a large bay to the north of Cape Cod . . . which can be easily seen in clear weather. Directly before the commenced

town lies a sand-bank, about twenty paces broad, whereon the sea breaks violently with an easterly and east-northeasterly wind. On the north side there lies a small island where one must run close along, in order to come before the town; then the ships run behind that bank and lie in a very good roadstead. . . .

At the south side of the town there flows down a small river of fresh water, very rapid, but shallow, which takes its rise from several lakes in the land above, and there empties into the sea. . . .

New Plymouth lies on the slope of a hill stretching east towards the sea-coast, with a broad street about a cannon shot of 800 feet long, leading down the hill; with a [street] crossing in the middle, northwards to the rivulet and southwards to the land. The houses are constructed of clapboards, with gardens also enclosed behind and at the sides with clapboards, so that their houses and courtyards are arranged in very good order, with a stockade against sudden attack; and at the ends of the streets there are three wooden gates. In the center, on the cross street, stands the Governor's house [Bradford], before which is a square stockade upon which four patereros are mounted, so as to enfilade the streets. Upon the hill they have a large square house, with a flat roof, built of thick sawn planks stayed with oak beams, upon the top of which they have six cannon, which shoot iron balls of four and five pounds, and command the surrounding country. The lower part they use for their church, where they preach on Sundays and the usual holidays. They assemble by beat of drum, each with his musket or firelock, in front of the captain's door; they have their cloaks on, and place themselves in order, three abreast, and are led by a sergeant without beat of drum. Behind comes the Governor, in a long robe; beside him on the right hand, comes the preacher with his cloak on, and on the left hand, the captain with his side-arms and cloak on, and with a small cane in his hand; so they march in good order, and each sets his arms down near him. Thus they are constantly on their guard night and day.

Edward Winslow is the only writer to give a description of the shape of the palisade which enclosed the fortified town. His account shows it to be a four-sided enclosure, with four bulwarks "without the ordinary circuit of the pale, from whence we could defend the whole town." To have the clear sight lines required down each side of the palisade, the bulwarks, or flankers, as Bradford calls them, most probably would have been at the corners, forming an enclosed defensive structure, much as had been constructed in Jamestown in 1607, although there the fort was within a triangular palisade. Two of the early visitors to Plymouth mention three gates: Emmanuel Altham in 1623 notes that there were "three great gates," and Isaack de Rasieres refers to three wooden gates at the ends of the streets. When a square enclosure is drawn to scale using Pory's and Smith's dimensions and superimposed on a modern map of Plymouth, placing the fort at the highest point on Burial Hill (Winslow's "Mount"), a remarkable fit is achieved. Two flankers would face inland behind the fort, and the other two would overlook the bay. In a footnote by J. F. Jameson, who translated de Rasieres's letter from the Dutch, he says, "the street first mentioned was longer [than 800 feet], 1,150 feet." If we take the longer dimension, and begin the street at the fort, the other end just reaches the shore, having passed through one of the three gates. The highway shown on Bradford's plan, which crossed the street, would go through the other two. No mention of the fourth gate ever being added has been traced.

De Rasieres's description furnishes us with two other insights regarding the layout of the town. The key sentence reads: "In the center, on the cross street, stands the Governor's house, before which is a square stockade upon which four patereros are mounted, so as to *enfilade the streets* [italics added]." Since Bradford's sketch plan shows only the lots of named individuals on the south side of the street, this sentence places Bradford's house on the north side, although we do not know whether it was west or east of the highway. The square stockade could not have been large, since it was placed in the center of the intersection of the highway and the street, the only location that would permit the patereros to "enfilade the streets." Patereros are small breech-loaded cannons that can fire balls, rocks, or any other suitable objects placed in them. Such a structure would not suffice in any way as a defense against attack from without, being far too small. But based on similar stockades that were commonly erected in the English Ulster plantations in the seventeenth century, they appear to have been riot control devices used in the event of insurgency from within. Firing along the highway and street would certainly deter any who might rebel against the established government. Such a situation never occurred at Plymouth, but apparently a stockade of this type was a standard part of town planning on the English frontier at the time.

Between 1620 and 1627, the year of the cattle division, the population of Plymouth slowly grew. Precise figures are almost impossible to obtain, but we can arrive at some close approximations. The *Fortune* arrived in November 1621, bringing thirty-five more colonists, "most of them lusty young men," led by Robert Cushman, who went back permanently to England on the ship's return voyage. With the deaths that took place in the spring and summer bringing the total deaths to around fifty, just over half of the *Mayflower* passengers had survived, so the tiny colony's population was now increased to some eighty-nine. While there is no doubt that these new people would provide a valuable addition to the colony's number, they were virtually without supplies. As Bradford states:

> So they were all landed, but there was not so much as biscuit-cake or any other victuals for them, neither had they any bedding but some sorry things they had in their cabins; nor pot, or pan to dress any meat in; nor overmany clothes. . . . But there was sent over some Birching Lane suits in the ship, out of which they were supplied [Birching Lane was a street in London where cheap, ready-to-wear clothing could be obtained].

Obviously the *Fortune*'s passengers placed a great strain on the tiny community, and rations for all were cut in half. Early in June of 1622, Edward Winslow traveled to Maine to purchase food from fishing ships. He met with some success, but the captains of the ships were low on provisions and could spare only a small amount. To make matters even worse, in late June two other ships arrived, the *Charity* and the *Swan*, temporarily adding sixty more "lusty men" to the ninety-odd colonists living in the village. They stayed for the months of July and August, having been sent by Thomas Weston to plant a new settlement. Weston had been one of the leaders of the London merchant adventurers who sponsored the establishment of Plymouth Colony, but who was now independently setting up his own. The site eventually chosen was at Wessagusset (modern Weymouth, some thirty miles north of Plymouth). The settlement lasted until the following March, its inhabitants suffering from the winter cold, Indian hostilities, and a shortage of food. The venture was called off, and most of the Wessagusset community returned to England, but an unspecified few settled in Plymouth.

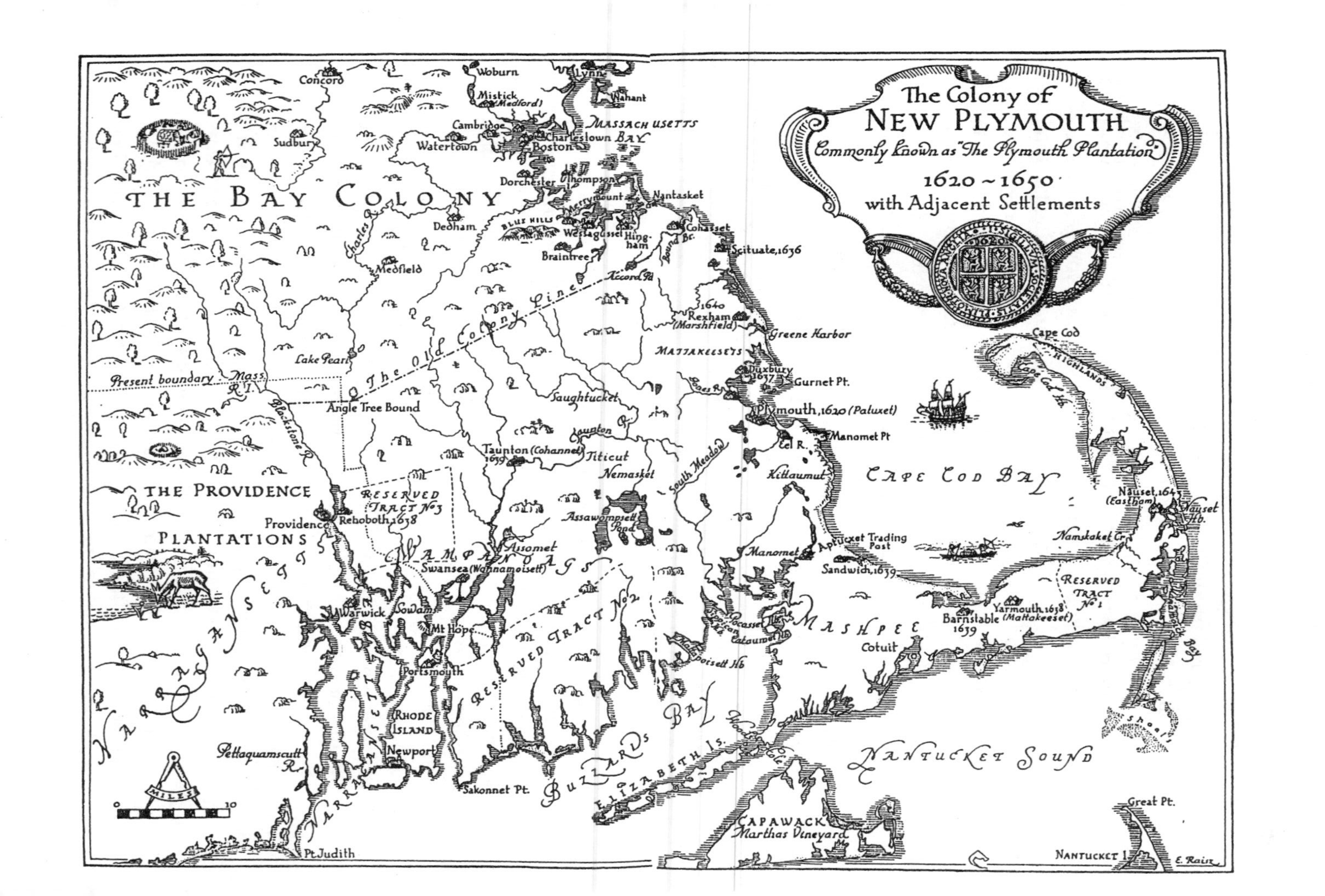
The Colony of
NEW PLYMOUTH
Commonly known as "The Plymouth Plantation"
1620 ~ 1650
with Adjacent Settlements
THE BAY COLONY
THE PROVIDENCE
PLANTATIONS
MASSACHUSETTS BAY
CAPE COD BAY
NANTUCKET SOUND
BUZZARDS BAY
ELIZABETH IS.
NARRAGANSETT
MASHPEE
WAMPANOAGS
MATTAKEESETS
CAPAWACK
Marthas Vineyard
The Old Colony Line
Present boundary
Mass.
R.I.
RESERVED TRACT No 1
RESERVED TRACT No 2
RESERVED TRACT No 3
RHODE ISLAND
Concord
Sudbury
Woburn
Mistick (Medford)
Lynn
Nahant
Cambridge
Watertown
Charlestown
Boston
Dorchester
Thompson
Merrymount
Nantasket
Blue Hills
Wessagussel
Hingham
Cohasset
Dedham
Braintree
Scituate, 1636
Medfield
Charles R.
Accord Pd
Bound Br.
1640 Rexham (Marshfield)
Greene Harbor
Duxbury 1637
Gurnet Pt.
Plymouth, 1620 (Paluxet)
Manomet Pt
Eel R.
Kitlaumut
Lake Pearl
Angle Tree Bound
Blackstone R.
Saughtucket
Taunton R.
Taunton (Cohannet) 1639
Titicut
Nemasket
South Meadow
Assawompsett Pond
Assomet
Swansea (Wannamoisett)
Providence
Rehoboth, 1638
Warwick
Sowams
Mt Hope
Portsmouth
Newport
Pettaquamscutt R.
Pt. Judith
Sakonnet Pt.
Mattapoisett Hb
Cataumet Hb
Pocasset Hb
Manomet
Aptucxet Trading Post
Sandwich, 1639
Cotuit
Barnstable 1639
Yarmouth, 1638 (Mattakeeset)
Namskaket Cr.
Nauset, 1643 (Eastham)
Nauset Hb.
Cape Cod
Highlands
Cape Cod Hb.
Shoals
Great Pt.
Nantucket I.
Woods Hole
Miles
0
10
E. Raisz

In 1623, two more ships, the *Anne* and the *Little James*, arrived carrying some ninety new settlers, including Alice Southworth, whom William Bradford married soon after. They had both been members of the Leiden Church, and she had been widowed in that city. This new group posed a different problem. Sixty of them were sponsored by the joint stock company, and therefore were obligated to work for the "common good" of the colony. But thirty others were under no such obligation, having paid their own expenses needed to make the voyage. They were referred to as "the particulars," having come "on their particular." They were provided with home lots by the government, and could work on their own. Citizenship was not granted to them, however, and they were not permitted to engage in trade with the Indians, understandable in light of the fur trade's importance in paying off the debt incurred with the adventurers. The company from the two ships increased the population of the colony to about 180 people. The "common course," in which planters would hold property in common, and work would all contribute to toward a common shared goal, already causing difficulties, was further strained by this new development. The Leiden group had resisted this policy from the beginning, and had not resolved the question when they departed England. From the start it began to break down, as evidenced by the initial laying out of lots before construction of the town had begun, each lot designated as the property of the aggregate family groups who were to occupy them. In 1623, land outside the palisade was allotted to individuals for their private use, each man, woman, and child receiving a one-acre parcel. In 1624, the *Charity* arrived again with "some passengers and goods," including Edward Winslow, who had been in England for six months, but the most important part of the cargo may well have been the "three heifers and a bull," which, when bred, helped bring the communal herd's total size to seventeen just three years later. No more ships arrived with new settlers for the colony until some five years later.

Livestock were critical to the survival of the settlement. In his description of Plymouth in 1623, Emmanuel Altham added, "Furthermore, there is belonging to the town six goats, about fifty hogs and pigs, also divers hens." Altham's account is the first to mention livestock. There is no way to determine on what ship they were transported to Plymouth by the fall of 1623. At least four vessels had put in at Plymouth, including the *Mayflower*, but no mention is made of any animals being transported there. The *Mayflower* herself could have had livestock aboard, but again, the records are mute on the matter. Of particular note is the number of swine; for there to have been fifty of them in 1623, either a large number were brought across the Atlantic or a smaller number were brought and bred after arrival. Of course, pigs give birth to litters of eight or more, so a small number of breeding stock could have increased to fifty in a relatively short time, but even allowing three years for such an increase, it is quite possible that the *Mayflower* passengers had porcine companions on their voyage. Cattle are another matter, since they usually give birth to a single calf. As mentioned above, the first cattle to arrive at Plymouth came aboard the *Charity* in March 1624, which brought three heifers and a bull, sufficient breeding stock that, when combined with others that came later on the *Jacob*, raised the total number of cattle in 1627 to seventeen, including one unborn calf. The cattle had been held in common until that year, when they were evenly divided between twelve "companies" of colonists, which totaled 156 people by that date. Each company also received two goats apiece. So there were at least twenty-four goats by that time, a fourfold increase in number from four years previous.

John Smith's 1624 estimate of 180 people living at Plymouth, according to the evidence available, appears to be accurate. We know that a few of Weston's settlers joined the community in Plymouth after his settlement's disastrous end. The next population estimate available is that of the cattle division of 1627. Although only 156 people were involved in it, this number would not include the thirty particulars. Adding them to those who received shares in the cattle produces a number not all that different from Smith's.

In 1627, after considerable debate as to how the debt owed the adventurers could be liquidated, it was decided that Bradford and eleven associates would contract to pay off the remaining £1,800 and any other debts owed the company. In turn it was further agreed that this group of twelve, known as the "undertakers," would hold a monopoly on the Indian trade for six years. The plan seemed a good one, but in fact the final settlement of all debts was not accomplished until 1642, helped in large measure by a rapidly developing trade with Massachusetts Bay Colony just to the north, particularly in cattle, but also in corn. However, to produce such commodities in the quantity needed, it was no longer possible for the colonists to remain together within the palisaded village, cultivating the adjacent land. By 1628, people were moving permanently across the bay, first to Duxbury, whose residents complained that they were so distant from the town that "they could not long bring their wives and children to the public worship and church meetings here. . . . And so they were dismissed about this time, though very unwillingly." This initial move began a process known to historians as declension, and after 1627, between 1636 and 1687, twenty towns were established, first along the coast, and then inland as far as Little Compton, which abutted the Rhode Island Colony, and later became part of it. Bradford's description of the circumstances that led to this dispersal has a poignant ring to it:

> For now as their stocks increased, and the increase vendible [marketable], there was no longer any holding them together, but now they must of necessity go to their own great lots. They could not otherwise keep their cattle, and having oxen grown they must have land for plowing and tillage. And no man now thought he could live except he had cattle and a great deal of ground to keep them, all striving to increase their stocks. By which means they were scattered all over the Bay quickly and the town in which they had lived compactly till now was left very thin and in a short time almost desolate.

Chapter 3

There Be Witches Too Many

Glimpses of the Social World

The twenty-one towns that made up Plymouth Colony by 1687 were all rural, farming communities, relying on crops and livestock for their support, and so did not vary much in character one from another, although those on the cape had a somewhat different economy, based in part on exploiting marine resources, particularly fish and whales. At this point we depart from the historiographic approach used thus far and present selected aspects of the historical ethnography of the colony as a whole. In anthropology, ethnography is the systematic description of a culture in as many of its facets as possible. Historical ethnography attempts to achieve this end through the use of primary sources, particularly, insofar as Plymouth Colony is concerned, wills, probate inventories, and the records of the court, as well as through archaeology. We are not suggesting that no change occurred during the period in question, but our emphasis will be on stasis rather than change. After all, Darrett Rutman, in his excellent study of farms and villages in the Old Colony from 1620 through 1692, contends that change in the colony was slow and relatively inconsequential. "The colony's controversies were relatively minor and readily resolved; basic conflicts were none at all. The difference between older and newer villages was small, merely a different position in the same agricultural progression through which all Plymouth's villages were going." Change can come quietly, through the backwaters of daily life and routines, challenged at times by the ripple of greater political events, and by the contentions of different attitudes and demands on a local level. Rutman suggests that for those who approach the history of Plymouth Colony not looking for large-scale changes brought about by conflict, but who are "more interested in ordinary people," a study of the Old Colony "will yield stories great in implication." The material that follows is not comprehensive in scope, it is not intended to be so, but what we do hope to achieve is some sense of what the reality of life in the colony was like during the brief period of its existence. Rutman's "stories great in implication" should emerge in the telling.

It was cool that day, September 16, the year of Our Lord 1633. William Wright lay high on the bolster and pillows of the curtained bed that he and Priscilla had shared since their marriage four years earlier. His sickness was to death. He knew that. Its shadow was cold on him, and it was only a matter of time. He could hear Priscilla now, working in the room next to their bedchamber, the clatter of pewter platters and then the rhythmic pounding of the pestle in the small brass mortar as she prepared herbs to add to the pottage she was making for their main meal. At least he had the house to leave her, standing well in the center of the little town of Plymouth, abutting New Street. And he had been able to complete the fence, enclosing in a palisado their barns, stables, and beast houses as well as the garden plot that gave them fresh vegetables and herbs in the spring and summer, enough to put up some for the long hard winter that was coming again. And perhaps it would bring them an heir. He would never see the child, but his will would provide for her and her heirs—be they theirs or only hers. Will's mind drifted back in time. He had been in New England in America close on twelve years now since

"My passing-bell is tolling, Tolling sweetly,
I lie dying, and my life is from me flying."

he arrived on the Fortune *in November 1621. A bachelor, always in demand as a carpenter and joiner, it seemed right that he had married a woman who bore the name "Carpenter," Priscilla Carpenter—a fine wife and a good trade for a man, carpenter, joiner. He had made their bedstead, and the little chair table in their bedchamber, and the paneled cupboard in the first room for Priscilla as a wedding gift. He pulled the green rug on the bed more snugly around him as he waited for his close friend and brother-in-law, Will Bradford, married to Priscilla's sister Alice, to arrive to complete the will that they had begun some days ago. It would be signed today, sealed in the presence of Will Bradford and Christopher Wadsworth, with whom he had shared mowing ground for two years now. They had been fortunate. Christopher was a kind friend and neighbor, and had promised to help Priscilla by doing Will's share for her, but he was talking about moving to Greene's Harbor. Priscilla did have Alice and Will though, and they would see that she would be alright until she married again, as she must. Plymouth was no place for a woman on her own, and she needed children, heirs who would inherit what Will and then both of them had worked so hard for. Sick and weak though he was, his last will and testament had been made with a clear mind, and in confidence of his own redemption. Priscilla was his great concern. Life with her had been such a joy and comfort for them both. "God give me patience to bear your hand until my change shall come, and I have to leave her and this frail life, so hard now to sustain."*

And he did. By November 6, 1633, before the hard frosts had come and winter finally closed in, Will Wright's body had been committed to the ground and his spirit to God who gave it, an inventory taken of his goods and chattels, and on January 2, the General Court appointed Priscilla Wright Executrix and Administrator of her deceased husband's estate. Less than a year later, on November 27, 1634, Priscilla married John Cooper. As far as is known she bore no child to either man. It is recorded that she died on December 28, 1689 in her ninety-second year. The cycle of life and death had revolved once more.

What might it really have been like to live in Plymouth close to four hundred years ago? Will and Priscilla Wright were ordinary people, living out their lives in a new world, and we would never have known of them had it not been for the last will and testament that Will left, the official records that detailed his possessions, and also those that recorded brief life events for him and Priscilla, as well as her death. It is not possible to know what life was really like, but the records that exist do enable us to obtain a glimpse of what did and could have happened.

By now we have seen something of the reasons for the English settlers having determined to leave the life they had known in the old world for the hazards of the new. And that by the end of the first winter half of those who emigrated had died, families were wiped out, or almost so, a few surviving intact, like those of William Brewster, Stephen Hopkins, and John Billington. New marriages took place; widowed Susanna White married the widower Edward Winslow, William Bradford and Captain Myles Standish, their wives having died by the first spring in New England, married passengers on the *Anne*, which arrived in Plymouth in August 1623, so that a new generation of children born in New England began to increase the number of settlers, although the majority of new arrivals at this time were immigrants from England. It was very different from the modern immigrant experience in America, where the individual rapidly becomes integrated into an immigrant community, part of a national culture of generations of Americans. There were none of the twenty-first-century amenities now taken for granted, so life was cut to the basic need for shelter and survival, a close-knit, face-to-face community, where you walked to your neighbor's house even if it was two or three miles away. All the early settlers had was a scattering of houses where there were no shops such as they had been used to in England, where people were dependent on their neighbors' skills as a blacksmith or house carpenter, joiner, thatcher, weaver, tanner, tailor, fisherman, trader. And always present was the constant fear of the unknown, of bears and wolves, of attacks by Indians, and when the thaw came, new terrors from rattlesnakes, insects, spiders, and other small creatures not like those they were used to in the fields and towns of England and Holland. And for everyone, to a greater or lesser degree depending on personality, there was the shock of transplanting to a new world, where they no longer had the continuities and familiarities of the old. Problematic and difficult as these familiarities may have been, they were at least known, and grief when it came from death was assuaged by the presence of a wider circle of family and friends.

One of the problems of trying to understand belief systems and the cycle of life and death in an earlier time period is that we tend to extrapolate from our own experiences and circumstances, tempered by the knowledge that hundreds of years of life and death have passed, and all that that embraces in changing perspectives, values, and circumstances, as well as by stabilization in place and time. We see through the more familiar framework of the twentieth century, images from the nineteenth: the Plymouth settlers caught up in a web of time, in their Pilgrim suits and hats, working industriously, acting courageously in the face of danger, attending church regularly, bringing up their children with strict ethical standards, founding an exemplary society that became part of the greater nation of America. But we already know that the reality was far different, although aspects of this image remain true. There were indeed many individuals whose lives showed courage, hard work, devotion to God, and who worked this out in their family life and the community as far as they were able. But interwoven into this is a complex pattern of folk customs and beliefs that tend to be lost in the images that have become a common part of the national myth concerning the people of early Plymouth and the world in which they lived.

Magic, Witches, and Folk Customs

Underlying and permeating the approach of seventeenth-century English colonists to the political and social worlds they lived in was a deeply rooted folk tradition of superstition and belief in the supernatural, which existed alongside their religious faith. For centuries the English had looked first for a supernatural explanation of natural or man-made disasters, searching the natural world for explanations, consulting wizards or cunning men and women for cures, potions, charms, and other magical devices to protect them against a world in which tempest, lightning, sudden injury, and death prevailed. Witches were also a real part of the belief system that was transplanted to the New World from Old England as indicated in this 1674 passage by John Josselyn, traveler and naturalist in New England, "There are none that beg in this Country, but there be Witches too many. . . ." From the time that Christianity was brought to England in the sixth century by St. Augustine, the Christian Church used every possible means to replace pagan beliefs, the use of charms and magical potions, consultations with soothsayers, sorcerers, witches, and wizards, teaching on the simplest level that the only origin of good is God, and of evil, the devil, that these are the only two sources of supernatural action. The church did, however, have its own mystic rites to control these forces—the mass, holy water, the sign of the cross—as well as worship of the saints, and taught that the church was the only mediator through which God could be approached.

By the sixteenth century, the theology of the medieval church had become confused at a popular level in the minds of the majority of its lay members, as well as many of the parish priests, with a belief in magic that had been part of their thinking for hundreds of years before England was Christianized. It was in the sixteenth century that the religious movement known as the Reformation took place, a new movement from within the Catholic Church that had for centuries been the supreme force for Christianity. The reformers protested against the corruption of the church's teaching and practices, particularly the rituals that they believed had made it impossible for people to worship God directly. Despite the efforts of the Protestant Church in the later sixteenth and seventeenth centuries, there was such a close relationship between religion, superstition, and magic that the existence of the three flourishing together in rural England could not be easily eradicated. Well into the seventeenth century there was still a large majority of ordinary people, including some of the more educated, whose reliance on magic and a complex cosmology of spirits remained an integral part of their thinking. Patterns of belief, passed down from generation to generation, do not die easily, and this was an age when many still believed that the earth was flat, that people could be endowed with supernatural powers, that there were witches whose malevolent purposes against prospective victims had continually to be guarded against. Wizards were plentiful, often known as "cunning" or wise men or women, to be consulted for healing, love potions, fortune-telling, the magical discovery of missing persons, or the name of a thief. It was a world in which customs of protecting the home and hearth, the center of family life, against witches or other malevolent influences hovering around, were an essential part of folk belief. Disease was more often than not attributed to some evil person whose supernatural powers were able to cause the illness, and many cures were linked to this belief. Due to the influence of the church, numerous non-Christian customs had become laced with terminology and practices reminiscent of those of the priests—prayers, masses, consecrations, exorcisms, invocation of saints and angels coexisted with popular magical practices of inquiry such as conjuration, divination, and necromancy.

It is not possible that the men, women, and children who settled in Plymouth Colony would have been free of such influences. In 1644, Governor John Winthrop of Massachusetts Bay Colony to the north

A conjurer with his familiar (a black dog), practicing popular magic. Witches or fairies appear at the top right, and flames are shown suggestively below.

wrote of the thirty "monstrous births or thereabouts, at once," delivered from a pregnant Anne Hutchinson in 1638, accused on thirty counts for her anti-Puritan schismatic religious beliefs and practices. Winthrop described the births as "some of them bigger, some lesser, some of one shape, some of another, few of any perfect shape, none at all of them (as far as I could ever learn) of human shape. . . ." Deformities were always associated with the devil, and the fact that a well-educated religious and political leader could believe in a physically impossible number of symbolic monstrous births, describe them in such terms, and associate them with parallel heretical beliefs makes it certain that at lower educational levels such connections were the norm. Richard Greenham, a Puritan theologian whose works were owned by Will Wright of Plymouth (listed in the inventory of Wright's possessions taken in 1633), was outspoken against cunning men who used Christian prayers but claimed to achieve supernatural results through their own power, not God's, and warned of the dangers of looking on ministers as magicians or wise men (wizards), who could through "an incantation of words" give relief. And when Greenham was asked for help by a woman who believed she was bewitched, he advised her under no circumstances to consult wizards. It is of particular interest, too, that Richard Greenham retained a belief in good fairies, although he was probably in a minority in so doing among the Puritan leaders. The "little people," hobgoblins, elves, and fairies, were mischievous, sometimes malevolent, malicious spirits, more bad than good, who needed to be propitiated with food and drink to bring good fortune, obtain help in cleaning the house, and to avert their turning the milk, pinching a careless housewife, overturning milk pails and buckets, stealing milk from cows, or even substituting a changeling for an unguarded baby.

It is not surprising that there are few open references in the records of Plymouth Colony to the popular beliefs in magic that undoubtedly persisted there. In the first place, such beliefs would be taken for granted, part of a popular culture that did not need to be detailed. This can also be seen in the scarcity of descriptions of houses, homestead layouts, of farming methods and equipment, crops sown and harvested, meals cooked and consumed, children at work and play; in short, the substance and routines of everyday life. Travelers and diarists wrote of the exotic, the unusual, court records detail more of the aberrant than the norm, wills and probate inventories are tantalizingly laconic, and it is left to later generations to construct their own understanding of how people in the past might have thought and lived. In the second place, popular beliefs in magic would have been opposed by the religious leaders of the colony. The principles that underlay the institution of government in Plymouth Colony, as opposed to Massachusetts Bay, seem to have been remarkably liberal in that there was no effort to force religious conformity, even Quakers being treated with more tolerance despite the harsher laws introduced against them in the 1650s. Records from Massachusetts Bay Colony to the north are another matter. They are far more prolific, and the published works of Puritan leaders, particularly Increase Mather, president of Harvard College, and his son Cotton Mather are good examples. At the height of the outbreak of witchcraft in Salem Village in 1692, Increase Mather stated, "I am abundantly satisfied that there have been, and are still most cursed Witches in the Land." Preaching included graphic descriptions of the devil, and teaching that a witch was someone directly allied with Satan. Linked to this was the association of natural phenomena of thunder, lightning, wind, and tempest with the direct acts of God. God's providential control of the natural world was interpreted as a direct sign of his protecting mercies or of his anger and judgment. With teaching of this nature taking place within the framework of a society already deeply superstitious, in which each individual and every home was under constant threat from an unstable world where familiar spirits could invade at will, fire strike, and violent death come suddenly, anything could be interpreted as a sign or wonder and attributed to witchcraft as much as to God's hand. It is possible that the accusations of witchcraft and subsequent trials in Salem can be seen on one level as a logical culmination of the content and style of the preaching of Cotton Mather and other Puritan leaders, sustaining and developing the complex web of beliefs in the supernatural that was already deeply embedded in popular society.

What did people in the seventeenth century understand by witchcraft? In essence, it was doing evil, the exercise of malevolent magical arts against someone by a person in league with the devil. The law made this clear in that it was a capital offense to make a compact or covenant with the devil, and it was often very difficult for a

A graphic depiction of the world as it appeared to the seventeenth-century mind, nurtured by popular beliefs in spirits that could invade homes, striking at will with fire, brimstone, and sudden death.

jury and magistrates to decide whether the evidence against someone accused of being a witch actually involved direct liaison with the devil, as this connection was not necessarily mentioned by the accuser. Accusations tended to focus more on the signs that appeared to indicate that the accused had malevolent supernatural powers. People lived in a small, face-to-face society where they could not easily move away from neighbors with whom they found themselves in competition, or whom they disliked, and who became natural scapegoats for violent emotions that could not be expressed within the close family circle for fear of damaging it. Such a close-knit society was a fertile breeding ground for accusations of witchcraft, and people who believed in magic and countermagic readily interpreted the strange as bizarre, and had imaginations already primed to see weird beings in flight, people transformed into animals, apparitions, abnormal bodily afflictions associated with someone suspected of being a witch, or simply spilt milk or a dropped kettle as an omen for which someone was to blame. When witch-hunting in Salem and its environs was at its height, Governor William Phips, after he first arrived in Boston in May 1692, described the eruption of

> . . . a most Horrible witchcraft or Possession of Devils which had broke in upon several Towns, some scores of poor people were taken with preternatural torments some scalded with brimstone some had pins stuck in their flesh others hurried into the fire and water and some dragged out of their houses and carried over the tops of trees and hill for many Miles together. . . .

Reading this account in the twenty-first century, the startling dimension is that this was a sober report from the highest level of government.

Given studies of the remarkable resilience of popular beliefs in rural England and what occurred not only in the Bay Colony, but in Connecticut and New Haven in the seventeenth century, it is certain that such were to be found in Plymouth. A number of the settlers came from East Anglia, and customs that survived there into the nineteenth and even twentieth centuries would certainly have been familiar practices. Protection of the house against forces of evil by burying a horse or ox bone in the foundations was one such custom, as was the perceived necessity of protecting the hearth and threshold from entrance by witches and evil spirits through the use of salt-glazed bricks in the chimney, as salt was believed to be a powerful deterrent against evil. An iron horseshoe above the entrance to a house was also common, as iron was considered a source of supernatural protective power. Witch bottles, buried under the hearth, have been found frequently in East Anglia. It was believed that witches could harm prospective victims and their house by possessing or simply touching something that belonged to them, since things that had been in contact were forever linked, no matter what the distance between them might be afterward. Such "contagious" or "touching" magic had a very powerful hold, and could only be counteracted by taking something of the potential victims, normally their urine, mixed with pins and nails (the all-powerful iron), and placing it in a corked stoneware bottle, buried upside down. This was seen as a powerful countermeasure against being cursed, designed to draw the witch to the house by inflicting violent stomach pains and cramps that would compel him or her to come and ask for the curse to be removed before they could damage the intended victim. The bottles, known as Bellarmines or Bartmanns, were made in Germany in the sixteenth and seventeenth centuries, are salt-glazed (another important protective association), and have a bearded face—originally a satyr, sprigged onto the neck of the jar—and a rounded body bearing a town seal. The rounded belly of the jar was seen as symbolic of the bladder of the witch. Witch bottles could also be buried under the threshold, or even outside the house. Fragments of Bellarmines have been excavated in Plymouth, but are not associated with pins or nails; however, as house excavations have been few, that does not necessarily mean that the custom was not in use in the colony, given its persistence in various parts of England in slightly different forms from those found in East Anglia.

There are three explicit references to witches and witchcraft in the Plymouth Colony records. The first is in the 1636 edition of the laws, which were posted and applied in Plymouth well before their first codification. In the list of capital crimes, the third, following treason or rebellion and willful murder, is "Solemn Compaction or conversing with the devil by way of witchcraft, conjuration or the like." Unlike the Bay Colony, however, no one in Plymouth Colony was ever executed as a witch. The second reference is to someone being accused of being a witch, and the third is to the indictment of a woman as one.

The reference to someone accused of being a witch took place in 1661. In March of that year, Dinah Sylvester, nineteen-year-old daughter of Richard and Naomi Sylvester of Marshfield, was summoned to appear before the general court in Plymouth in response to a complaint of defamation. The suit was brought by William Holmes and his wife, also living in Marshfield. Holmes's wife was reported to have been a witch, Dinah Sylvester having claimed to have seen her transform herself into a bear, but under examination Dinah said she could not see what manner of tail it had as it was facing her, a detail evidently considered by the magistrates to be significant in exoneration of Goodwife Holmes. A supernatural event such as shape-changing was associated with the power of the devil to transform himself into an animal, and as a witch was someone who had made a compact with the devil, such a charge, if substantiated, could be sufficient to bring an indictment. Witchcraft, if proven, was a capital offense, and the law required at least two witnesses, but here there was only one. No record of the examination and subsequent debate is given, but the accusation did not stand up to scrutiny by the court, and in May Sylvester was sentenced to pay a fine of five pounds sterling to William Holmes, or be publicly whipped. Alternatively, if she made open acknowledgment of her fault, she would only have to pay the costs of the suit. She chose the latter, and what emerges from her confession is that Goodwife Holmes was a neighbor, that Dinah Sylvester had harbored hard thoughts about her, and her accusation was directly rooted in this animosity.

In some cases of people accused of witchcraft in New England, it has been possible to gain a deeper understanding of the motives that prompted the accuser by examining all the information available about them, and discerning patterns of belief and behavior that could account for them. John Putnam Demos, in his *Entertaining Satan*, has demonstrated this in a scholarly and compelling way. In this instance, the colony records show that there is a pattern of tensions within the immediate Sylvester family that they litigated, and of conflicts with outsiders. Dinah and her eldest brother, John, faced criminal charges after they and a younger sister molested and abused one of the Marshfield constables when he arrested their mother, Naomi, in 1666. They forcibly removed Naomi from him. Naomi's arrest followed a court case against her in which her son-in-law, Edward Wright, and his wife, Lydia, her and Richard Sylvester's eldest daughter, accused Naomi of unjustly keeping back cattle and other goods that belonged to Lydia. Then there was the family's relationship with

John Palmer Jr., who had a long history of problems over a land deal with Richard Sylvester and even had him arrested. It was complicated by Palmer having broken his engagement to Dinah Sylvester shortly after her accusation against Goodwife Holmes, for which he had to pay £20 in damages to her. He was clearly harassed by the Sylvesters over this, for three years later the court publicly recorded the agreement drawn up at the time, in which Palmer had been fully acquitted, so it was evident that bad blood had continued over the affair. Dinah later appeared in court on a charge of fornication (sex between a man and a single woman), and her final mention in the records is when Elkanah Johnson of Marshfield was bound to appear before the July 1669 meeting of the court in Plymouth due to a complaint against him "in reference unto a child layed unto him by Dinah Sylvester."

The question that arises, however, is, to what extent was this a pattern any different from that of the Sylvesters' neighbors, and anyone else living in Plymouth Colony for that matter? Given the numerous records of litigation between neighbors, the way in which people thought, the reading of portents into every action and event, the ebb and flow of hostilities in close-knit communities, it is remarkable that there are only two instances of accusations of witchcraft that have been recorded for the colony. The reason may partly lie in a greater fund of goodwill and tolerance among neighbors than would, of course, be reflected in court records. An indication of this can be found in Naomi Sylvester's situation after the death of her husband in 1663. After the will was read, Naomi petitioned the court for a greater portion of her husband's estate than he had left her, and her neighbors testified to her having been "a frugal and laborious woman in the procuring of the said estate," and the court agreed to increase her share. Whatever relationship the Sylvesters had with others and problems within the family, there were sufficient neighbors to support Naomi Sylvester's request that her husband's will be amended in her favor. The court did not lightly overturn the intent of a will, and Naomi's neighbors' support and goodwill provided important evidence.

The only information traced concerning William Holmes's wife, other than the Holmes vs. Sylvester case, is an early reference in the court records. In 1638, Francis Bauer of Scituate was presented at the general court in Plymouth "for offering to lie with the wife of William Holmes, & to abuse her body with uncleanness." Propositions such as this, if proved, would result in a fine or public whipping. No further mention of the case is made, however, so it is probable that, whatever his suspicions, Holmes could not produce enough evidence for an accusation against Bauer to hold up in court. In 1641 Gowen White of Scituate was fined for assaulting William Holmes, but no reason is given. By 1658 Holmes and his wife had moved from Scituate to Marshfield, where Holmes had one term of office as a constable (1663–1664). They had two sons, Israel and Isaac, and William Holmes died in 1678, aged eighty-six years. At the time he drew up his will in March 1678, his "beloved wife," whom he does not name, was still living. His estate was only worth £70, without house, lands, or his widow's portion, so at this stage of their lives the Holmeses do not appear to have been as wealthy as the Sylvesters (Richard Sylvester's estate amounted to £244, not including lands and housing). The facts are too brief to give a sense of what might have provoked the accusation of "witch" against Goodwife Holmes by a nineteen-year-old girl, but it is significant that in Dinah's case her later accusation against Elkanah Johnson apparently did not stand up to further investigation either, as there is no further mention of it in the court records. It does underline, though, that in a society where belief in supernatural causes of events, however minor, was sufficient to lay blame elsewhere and litigation was a pattern, everyone was at potential risk.

The only instance of someone being *indicted* for witchcraft in Plymouth, as opposed to accused, is the case of Mary Ingham of Scituate, sixteen years after the Holmes vs. Sylvester case.

We can imagine the scene.

March 6, 1677. There were more people than usual gathered at the Meeting House in New Plymouth where the Court convened that day. Meetings of the Court were always events that caused talk and gossip to fly around the town. Before the day was out someone would like as not be in the stocks, whipped at the post, or there would be violent outbursts of bitter arguments as neighbors harboring grudges accused one another before the Governor and his assistant magistrates. There could be physical attacks after a verdict, when the accused was acquitted, and the plaintiff disagreed with the judgement given. Dogs barked, children crowded with adults to see the action, and were cuffed for their temerity when they jeered with the crowd. There was an air of festivity on court days. Ordinaries were open to the public coming in to town, beer and cider easily available, and despite the fact that a couple of years earlier the laws had been tightened and both townspeople and strangers visiting Plymouth had to be out of the ordinary by the shutting in of the daylight, on court days the inn keeper turned a blind eye. There were always sly folks wanting to earn a bit of money on the side as informers, of course, but the inn keeper reckoned it worth the risk of a fine for the extra takings.

This time, though, in addition to the usual settlements of estates, provision for the widows Thacher, Sarah Bobbett, and the Widow Bourne and issuing of letters of administration, fines for committing fornication before marriage, the theft of a kettle and issuing of licences to keep ordinaries, there was the indictment of Mary Ingham for witchcraft. Mary had been brought down from Scituate for examination by a jury of twelve men. If she was found guilty, as the charge would be a capital one, she would be tried at the next session of the General Court, and have the opportunity to defend herself there.

A cold wind was blowing in from the ocean, carrying the plaintive mew of sea gulls, but the icy fear in Mary's heart was colder still as she listened to the indictment being read. If sent to trial she could expect physical attacks from the townspeople, and forced searches of her body by the authorities to find evidence of her crime. Like Margaret Jones of Charlestown, on whom they said they found a teat in her secret parts, freshly sucked, but it was afterward withered, and grew again on the opposite side. Her they hanged, and the same day and hour she was executed there was a great storm and many trees crashed down. Mary shifted her feet wearily. She was splay footed, and standing was exhausting. The indictment was about to be read, the trial had begun.

> Mary Ingham: thou art indited by the name of Mary Ingham, the wife of Thomas Ingham, of the town of Scituate, in the jurisdiction of New Plymouth, for that thou, having not the fear of God before thine eyes, hast, by the help of the devil, in a way of witchcraft or sorcery, maliciously procured much hurt, mischief, and pain to the body of Mehitable Woodworth, the daughter of Walter Woodworth, of Scituate aforesaid, and some others, and particularly causing her, the said Mehitable, to fall into violent fits, and causing great pain to several parts of her body at several times, so as she, the said Mehitable

> Woodworth, hath been almost bereaved of her senses, and hath greatly languished, to her much suffering thereby, and the procuring of great grief, sorrow, and charge to her parents; all which you have procured and done against the law of God, and to his great dishonor, and contrary to our sovereign Lord the King, his crown and dignity.

There is no record of the reasons given by the jury for acquitting Mary Ingham, but she was cleared of the indictment and permitted to return to her home, neither is there any further mention of her name in the colony records in this connection. Reasonable guesses as to why she was accused could include the possibility that she might have been hunchbacked, splayfooted, as is suggested in the construction above, or suffered from some other visible abnormality, since to have any physical deformity was sufficient for the cry of "witch" to go out when misfortune, sudden malady, or other disruption of nature occurred. It was also quite possible to use accusations of witchcraft to pay off private scores, another possible interpretation, and the jury might have had knowledge of a feud between the two families, although there is mention of "some others" whom she was said to have harmed through witchcraft or sorcery. The point, though, is that supernatural explanations for misfortunes were never far from the surface, and neighbors were often targets. It might not be witchcraft, but the actions of malicious fairies, or an ill-fated congruence of certain planets and stars. There was often a reluctance to accept misfortune as the consequence of one's own actions. It was easier to look elsewhere for blame, and it is also possible that the way in which Puritan preachers associated misfortunes with the hand of God in punishment strengthened traditional beliefs in the supernatural and its associated customs.

In New England, women were in general the ones accused of witchcraft, as in Old England, although in Europe it seems that the gender bias was not as clear. Nevertheless, in the *Malleus Maleficarum*, a fifteenth-century treatise commissioned by Pope Innocent VIII to define witchcraft, the way to destroy and cure it, and the judicial proceedings to be followed against all witches and heretics, witches are associated with women almost without exception. This detailed work, essentially a witch-hunting manual, written by two inquisitors, stresses the credulity of women, their being more impressionable than men, morally weak, with "slippery tongues," intellectually weak "like children," and "the natural reason, that she is more carnal than a man . . . since she was formed from a bent rib . . . an imperfect animal," and is "by her nature quicker to waver in her faith, and consequently quicker to abjure the faith, which is the root of all witchcraft." Such views of women were not new in the fifteenth century, and continued through the seventeenth. In New England, where four out of five accusations of witchcraft were against women, one explanation is that in a society under male control men needed to reassert their authority when it was being questioned publicly by some women, the antinomian Anne Hutchinson being a case in point. It also seems clear that accusations of witchcraft against women by men was a way of resolving conflicts between the sexes. So women were vulnerable; midwives in particular were at risk, as infant deaths and breast-feeding difficulties could be blamed on witchcraft, although such instances do not appear to have been as frequent as implied by the extensive coverage given in the *Malleus Maleficarum*. What is of particular interest in regard to Plymouth Colony, however, is the lack of public accusations of witchcraft as compared with Puritan New England, and the greater understanding of the position of women that is found in expansions of the legal system as well as in certain court cases.

There are at least two other Plymouth Colony records that appear to have involved supernatural beliefs. They are both coroner's inquests where the verdict seems to have been based on the ancient medieval practice of trial by ordeal. The first was the indictment of three Indians, Tobias, Wampapaquan (son of Tobias), and Mattushanannamo for the murder of another Indian, John Sassamon, that took place on January 29, 1674. The three men were among the chief counselors of Metacom, better known as King Philip, leader of the Wampanoag Indians, and Sassamon was a literate Indian who had converted to Christianity, attended Harvard, and played an important role in Indian and English relations. At the time he was minister of an Indian "praying town" near Mount Hope, Philip's home. The indictment was on the basis of the testimony of a Christian Indian, Patuckson, who claimed to be an eyewitness, but although there was supporting testimony from another Christian Indian, William Nahauton, an indictment for murder upon the evidence of a single witness was a departure from the law.

On a cold winter's day in January 1675, Sassamon had made a dangerous journey to the Governor of Plymouth, Josiah Winslow. Knowing that he was risking his life, and advising the governor of this, he brought news that Philip was preparing for war, news that the governor ignored. On his return home, at Assowamset Pond, Sassamon met his death, apparently falling through the ice. Five months later, the three counselors were on trial for their life, the jury prudently expanded to include six Indians in addition to the twelve Englishmen, although whether or not they would have ventured to disagree with the English majority is open to question. The initial verdict of accidental death had been revised by exhuming Sassamon's body and examining it for any indications of violence, which there were, and there was evidence that the body was dead before it entered the water. But, according to an account by Cotton Mather, the final evidence was supernatural. Tobias was ordered by the court to approach Sassamon's body, and as he did, the body fell "*a bleeding afresh*, as if it had newly been slain." This test was repeated, and his guilt was considered to be unquestionable as the same fresh bleeding occurred. The jury, English and Indians alike, returned a unanimous verdict. The men were guilty of murder. The general court accepted the verdict, the death sentence was pronounced against them, and Tobias and Mattashunannamo were hanged in Plymouth on June 8, 1675. Some extenuating circumstances were found to delay the death sentence of Tobias's son, but he was shot to death later the same month. The execution of Philip's counselors was the signal for war, and by June 24, what came to be known as King Philip's war had commenced, with ominous signs and portents in the heavens (an eclipse of the moon, but the earth's shadow across it appeared to resemble the scalp of an Indian) seen by soldiers and citizenry alike. The court records do not refer to the supernatural element in the final decision made by the jury, and Mather does not give any source for his account, but its very existence points to the fact that such beliefs were alive and well in the colony, even if not in this instance recorded by the clerk of the court.

The second case took place two years after Mary Ingham's indictment and release. On June 5, 1678, a jury that had been required to search out the cause of death of Anne Batson's child, returned its verdict to the general court in New Plymouth. The composition of the jury is remarkable in that it comprised five women and seven men. The role of women in judicial affairs in the seventeenth century was virtually nonexistent, and it took exceptional circumstances for the authorities to require them to form part of a coroner's jury. The women were Abigail Snow, Faith Winslow, Martha Powell, Mary Williamson, and Mary Branch, whose husband, John, was also on the jury. The sex of the child is not given, but the lower part of the belly and private parts were red and black and swollen. The jury required Anne Batson and several members of the family to touch the dead child, "but there was nothing thereby did appear respecting its death." The fact that the mother herself and family members were suspected of sexual molestation of the child, or worse, was clear, but no one refused to approach the body, and no supernatural occurrence took place to indicate incontrovertible guilt, and so a verdict of death by cause unknown was admitted.

Then, in a different context, but still a window into the seventeenth-century mind, is the case of the guilty canoe. In February 1660, on an icy New England day, when the waters were gray and choppy, Jeremiah Burroughs of Marshfield went to fetch some goods he had left in John Bourne's boat. He approached the boat in a small canoe, and as he stretched out to grasp it, he reached short, fell into the water, and was drowned. Following the verdict of the coroner's jury is a memorandum:

> That some course be thought on and ordered about small and naughty canoes, and in special about this canoe in the which Jeremiah Burroughs went to the boat in which he came by his death.

The sense of "naughty" is used here to mean "blameworthy," and although it could also be read in the later eighteenth-century sense of "inferior in quality," given the deeply rooted pattern of attaching blame to inanimate objects, it does appear that Burroughs's death was blamed on the small and naughty canoe, rather than on his reaching short of his goal.

Wives, Widows, and Spinsters

When images of women in seventeenth-century Plymouth come to mind, they are colored by Victorian sentimentality (soon to die Rose

Standish being carried onshore in the arms of her husband, Myles, in Charles Lucy's 1850 engraving), the iconic quality of Jennie Brownscombe's 1920 painting with a blond Mary Chilton dominating the landing on the Rock, or stereotypes of Pilgrim maids and matrons, depicted in third-rate pictures on motel walls. It is not surprising that stereotypical images have been created and perpetuated, particularly given Victorian sentimentality and the power of the Pilgrim myth, since the lives of women, particularly nonelite women, are far less visible than those of men as they emerge from past records. However, when the court records of Plymouth Colony are examined, we find numerous references to women, and although they tend to be skewed in favor of the unusual rather than daily occurrences and so often read like excerpts from *People* magazine, they give us facts to shape our thinking concerning what women experienced in seventeenth-century Plymouth, and some insight into the changes that took place.

One of the reasons for the greater freedoms given to women in the colony must surely have developed as part of the colonial experience itself, strongly influenced by the nature of its Separatist leaders. Whatever the situation in England, the women who accompanied their husbands to the New World experienced with them the suffering and rigors of establishing a completely new way of life. Of course there was continuity; they were English men and women, transplanted, bringing with them different regional cultures, but as immigrants know, patterns change with circumstances, and the men and women forced to work together to survive would develop new relationships that gradually worked out in changes to the law. And not to be overlooked was the fact that this was a second emigration for the Separatist leaders and their wives; both sexes had already experienced a loosening of the traditional gender roles when they established new lives in the Netherlands. It was a complex process, though, as there was also the whole law-reform debate in England, of which the colony's leaders were well aware, having experienced themselves some of England's unjust and capricious laws. So the freedom of establishing a new society made possible changes within the general framework of English law that could not have taken place in Old England and that included but went far beyond those associated with changing gender roles.

Women in seventeenth-century England had fewer rights and less power than men. They could not vote or hold positions of authority, and married women could not own land, houses, or goods apart from their husbands. It was generally believed that women were not competent to handle their own economic affairs and the majority of women probably agreed with that view. The move to a new world in America, however, brought some important changes, not to positions of leadership in the colony, but a greater recognition of women's rights in regard to land and entitlement to independent, full ownership of a third of their husband's movable property. The laws governing inheritance and deceased estates were based on those of England, but it appears that in Plymouth the marriage laws were more generous to women. The widow was to receive a third of her husband's land during her life, as she would have in England, as well as a third of his goods that she could dispose of as she liked. In the event of her husband dying intestate (without a will), she was also entitled to a third of his land and goods. The fact that she was at liberty to dispose of her third of goods as she pleased was a development of the rights held by women. This evidently was not the case at first, as the original 1636 record of the law stated that when a husband made his will he was entitled to dispose of all his goods in whatever manner he pleased. This was later amended, and the wording altered to read that "if any man die without a will his wife shall have a third part of his Land during her life and a third part of his estate forever." No date is given for the change, but by 1662 only the altered words appear in the laws.

The will and testament of those who had died, where there was one, had to be proved before the governor and assistants within one month of the death of the testator. It was also mandatory that a full inventory, including values, be presented at the same time, and in the case of widowers, the inventory does not appear to have included the widow's "thirds," providing a gap in the description of the contents of a home that can be misleading. A widow's clothing also was always omitted from the inventory taken of her husband's possessions. Only once the estate had been appraised could letters of administration be granted, usually to the widow in the event of the death of a married man, although male "overseers" were almost invariably appointed to assist the widow, often in an ongoing capacity, not just in the administration of the estate. If someone died without a will, "his wife or others nearest to him had to draw up an inventory, duly valued, to be presented to the Governor and his Assistants." The law also stated that in the case of persons dying with more debt than their estate of goods and chattels amounted to, any land that had been bought by the deceased would have to be sold to pay creditors. But in the event that a portion of the land was required for subsistence of the family of the deceased, the land could not be seized by the creditors, an extremely important provision for widows.

In 1646 women obtained a further recognition of their rights concerning the disposal of their husband's property. From July 6, 1646, onward, when the husband sold houses and land, the law required the wife's presence in court so that she could "consent unto and acknowledge the sale thereof also. But all bargains and sales of houses and lands made before this day to remain firm to the buyer notwithstanding that the wife did not acknowledge the same."

An examination of inheritance provisions for women in seventeenth-century Plymouth has shown that there are instances of women inheriting all or part of their father's or husband's housing and lands, and so ownership of land by women regardless of their marital status was quite possible. In 1669 Captain Thomas Southworth died at the age of fifty-two. He was William Bradford's stepson, and he and his wife, Elizabeth Reynor, had one child, a daughter, Elizabeth, who married Joseph Howland, son of *Mayflower* passengers John and Elizabeth (Tilley) Howland. Although Thomas Southworth's wife was still living when he made his will, he nevertheless left all his housing and land to their daughter and her heirs "lawfully begotten upon her own body." The rest of his estate was left in the hands of his son-in-law, daughter, and brother, Constant Southworth, "to be disposed of as they shall see reason for the supply of my wife in her poor condition," Elizabeth having an equal voice in any decisions taken. The court did not question the will, so it presumably considered that Thomas had provided adequately for his wife. She was entitled to a third of his movable goods, and she appears to have outlived her husband by at least ten years.

A common pattern was for husbands to leave widows part or the whole of their house and lands for their lifetime, or until children came of age, and often not just the third of their movable estate but the whole. There was sometimes the specified provision that the widow had the power to dispose of her outright inheritance "according to her discretion as she shall see occasion," an indication perhaps that although colonial law now permitted this, it still needed reinforcing in the light of the earlier English pattern. Normally, instructions were left for the division of all the immovable property among surviving children after the death of the wife. John Howland, however, went further in leaving his wife, Elizabeth Tilley, apart from specified legacies of land, money, and debts that had to be settled, "lands houses goods Chattles; or any thing else that belongeth or appertaineth to me, undisposed of be it either in Plymouth, Duxbury or Middlebury or any other place whatsoever." This was all given "freely and absolutely," which meant that Elizabeth could dispose of it exactly as she wished. She was also made his sole executrix, following the common pattern in the colony, despite the fact that they had mature sons.

Wives who were left as widows frequently remarried, and the usual pattern in the seventeenth century was that on remarriage the widow's possessions would become those of her husband. Wills in Plymouth show that in general if a widow remarried she kept her widow's "thirds," the movable estate that was hers, but that her lifetime interest in any housing and land went to her and her deceased husband's children. Family life in Plymouth Colony was relatively stable in comparison to that in Virginia, where the high death rate and frequent remarriage rate meant complex family relationships among siblings and half- and step-siblings; the situation faced by Agatha Vause of Middlesex County in Virginia, who had lost her father, two stepfathers, mother, and guardian uncle by the age of ten, was not uncommon. While there are not many records of second-marriage contracts concerning provision for stepchildren, as their rights were usually protected by wills, Joanna and Dolar Davis of Barnstable agreed at the time of their marriage that:

> Notwithstanding their marriage and improvement of their stock together for the support and comfort of their family, yet it should not give right to or entitle either of them to the estate of each other, which he or she had before their marriage, but each of them might dispose of his or her estate as he or she should see cause, to his or her own relations and not either of them dispose of the other's estate, but to remain his or her own.

John Demos, in *A Little Commonwealth: Family Life in Plymouth Colony*, states that women in seventeenth-century England did not have the right to make contracts except under exceptional circumstances. The contract between Joanna and Dolar Davis is one of several premarital contracts known to have been made in Plymouth, yet another instance of the greater freedom and respect accorded women before the law.

After the death of her husband, the role of a woman in Plymouth society changed radically. She could no longer depend on her husband for support and had to take responsibility for the raising of their children. She would more often than not, however, have the assistance of one or more "overseers," friends of her husband's to whom he entrusted oversight of the welfare of his family. Some widows were well provided for, and took control of their own affairs, managing land and livestock, and at times selling land that they had inherited. Widows, unlike married women, could be taxed, and make their own wills to provide for the well-being of their children, particularly any unmarried daughters, as well as enter into contracts. Widows filled a completely distinct gender role from their married female counterparts, and while the majority do appear to have remarried, in the earlier years of the colony at least, when the ratio of women to men was higher, there were those who remained single, usually ending their days with a married son or daughter.

A discussion of women in Plymouth Colony would not be complete without reference to unmarried women. They were a very small minority, but one such spinster, Elizabeth Poole of Taunton, has left us valuable information concerning their legal position. She died in 1656, aged about sixty-five, a wealthy woman who had built her own house with adjacent orchard, which was occupied by her brother, Captain William Poole, at the time of her death. She had bought another house from Robert Thornton in which she was living when she died. Mistress Poole also owned a good amount of land in Taunton, and it seems probable that the Mistress Poole granted "competent meadow and uplands for farm land" in Taunton in March 1640 was Elizabeth. In addition to the land adjoining her two houses, she mentions in her will "my forty acres of meadowing at Littleworth," and a hundred acres on the further side of the "great River," as well as another fifty acres of land. In addition to livestock that she owned, Elizabeth Poole held a share in the ironworks at Taunton, worth £25. What is noteworthy is that a single, unmarried woman in seventeenth-century Plymouth had the freedom to own and trade land, employ house carpenters to build a house for her, and own shares in a business enterprise.

Women Before the Court

Although in a minority compared with men, there are numerous instances of women coming before the courts in Plymouth. They appear in cases involving domestic violence, sexual misconduct, and occasionally public drunkenness. Disorderly living, slander, and defamation were also crimes of which they were accused. It was an age when women and men were quick to take offense, and had few scruples about giving it, so the level of verbal and physical abuse tended to be high. Women made frequent use of litigation to get redress for attacks on their reputation, probably more so than men who would use physical violence more readily. In the Plymouth court records from 1633 to 1668 there are thirty references to a woman accusing another person of a crime. The accusations include blasphemous speech, slander, abuse, lewd behavior, beating, molesting, witchcraft, and stealing. In most cases the accusation was taken seriously, and the accused person, male or female, was brought to trial and if found guilty, punished. Widows, married women, and single women all made accusations equally. There is no indication in the records that accusations made by a widow were taken more seriously than those made by a married or single woman, or vice versa. There are far more instances of women being accused of a crime, fifty-three in the same thirty-five year-period. The range is wide, including physical abuse, disorderly living, molestation of other women, lying, slander and defamation, stealing, witchcraft, at least one instance of lesbianism, murder, cruelty to a servant, not attending church, working or traveling on a Sunday, being drunk, mixed dancing, selling "strong waters" without a license and at an excessive profit, exchanging a gun with an Indian, stealing a boat.

Women were also accused of sexual crimes, the most frequent being fornication, but adultery certainly occurred as well, though even if adultery was proved, divorce was not often granted as a solution to the problem as in today's society; and there were also occasional instances of divorce due to desertion. In 1668 William Tubbs was granted a divorce from his wife, Marcy, on these grounds, when she was in her fifties. She appeared to be having an affair with Joseph

Seventeenth-century woodcut of a rogue before the court to answer charges of cozening unsuspecting women.

Rogers, who lived near Duxbury and was single at the time. No one could say how far it had gone, but in 1663 they were presented to the court three times, accused twice of being seen together lying under a blanket, and once of "obscene and lascivious behavior," a standard phrase used by the court to cover lustful actions. They were fined heavily and had to find sureties for their good behavior, and Rogers had to move away from the area. By 1664 Tubbs gave notice that he would no longer be responsible for Marcy's debts. In 1665 she was in trouble again for being entertained by one John Arthur, who had to give account to the court for this and his abusive speech. Finally, in 1668, thirty-one years after their marriage, William Tubbs was granted a divorce from Marcy as she had deserted him and was living in Rhode Island.

Not all desertion ended in divorce, and it was rare for a wife to be cut off in her husband's will, but when Captain William Hedges of Yarmouth drew up his will in 1670, he left his second wife, Blanche Hull, widow of Tristram Hull, only a token twelve pence, "and also what I have received of hers . . . shall be returned to her again," because she had deserted him. John Barnes of Plymouth, knowing well the contentious character of his second wife, Joan, nevertheless left her half of all his housing and lands for her lifetime, but she was only to inherit one-third of his movable estate "provided she does not molest any person to whom he had formerly sold land." If she did, it was to go to his only son, Jonathan. He hoped that his will would be "an Instrument of peace." Despite behavior that landed her in court regularly for slander and defamation and being sentenced "to sit in the stocks during the Court's pleasure, and a paper whereon her fact written in capital letters, to be made fast unto her hat," Joan was well provided for under her husband's will.

What of the provision made for wives whose behavior was outrageous enough to find them in court on a fairly frequent basis? One of Plymouth's most turbulent wives was Hester Rickard, wife of John Rickard and a niece of John Barnes. John Rickard and his brother, Giles, were married on the same fall day in October 1651, John to Hester Barnes, and Giles to Hannah Dunham. Both men were sons of Giles Rickard Sr., a close friend of John Barnes. John and Hester appear to have had only one child, John, born six years after their marriage. Hester was in many ways a regular seventeenth-century honky-tonk angel, living on the wild side of life. In 1660 she was brought before the court for slander and defamation by Joseph Dunham (the brother of her sister-in-law, Hannah), who claimed that she had said that he had offered her money "to be naught [immoral] with her." The case was withdrawn at the "earnest desire" of Hester's father-in-law, Giles Rickard Sr., on the ground that her husband was away from home when this happened. However, only a few months later, the following March, Hester was back before the court, this time convicted of "lascivious and unnatural practices," and sentenced to sit in the stocks during the pleasure of the court, wearing a paper pinned to her hat with the details of her crime written on it in capital letters so that it could be easily read by anyone who came past her. The use of the stocks for the public display of a miscreant was a frequent punishment, and one of the first things required of towns was that they build stocks and a whipping post.

It is not hard to imagine the scene three years later, in March 1664, when the Rickards's son, John, was only seven, and Hester was again before the court, this time with Ann Hoskins, accused of verbal abuse of each other. As most slander attacks were very public in character, and both women lived in Plymouth, they had probably chosen a place where their accusations against each other would have the greatest effect, and where there would be the maximum number of witnesses. Janet A. Thompson, in *Wives, Widows, Witches and Bitches: Women in Seventeenth-Century Devon*, comments that slander suits were often a reflection of social and sexual tensions among neighbors, and with no soap operas, paparazzi, or media coverage, verbal and physical fights were a source of entertainment, as were the punishments meted out by the court. Certainly there would have been plenty of people in and around the court when their case came up. The record reads:

> Ann, the wife of William Hoskins, for speaking most lascivious and filthy language to Hester Rickard, fined twenty shillings.
> Hester, the wife of John Rickard, for most obscene and filthy speeches, fined twenty shillings.
> *Marginal note:*
> These two women were sentenced either to sit in the stocks during the pleasure of the Court or to pay the fines here mentioned, and they chose to pay the fine.

In December 1665, Hester Rickard was back in court, this time in an attempt to clear her name, the case arising from ongoing bad blood between her and Ann Hoskins. As in the case of Marcy Tubbs, it is unlikely that Hester's behavior pattern changed completely, but after this she managed to keep out of the court records for seven years. The last time that she came before the court in Plymouth was in March 1672, when she had to answer for her "uncivil and beastly carriages and speeches" to her husband. Her sentence was to be publicly whipped at the post, but was suspended "at the earnest entreaty of herself and others." It is notable that John Rickard is not mentioned as being among those who pleaded for Hester's sentence to be suspended, and when he died in July 1678, it is perhaps significant that he did not leave a will. His estate, debts deducted, amounted to £73 5s 3d (a shilling, s, is one-twentieth of a pound, and a penny, d, one-twelfth of a shilling), and the court appointed Hester Rickard and her son, John, now twenty-one, joint administrators of the estate because "there is none to enjoy it but they two. . . ." Hester does not appear in the court records after this, and it seems evident that whatever her relationship to her husband in their last years together, colonial law ensured that she was adequately provided for, as John Rickard would have been well aware.

A final note, though, that must be reiterated. Society in Plymouth Colony was not dominated by women whose behavior brought them into regular contact with the court, although their lives certainly had an impact on their neighbors. The majority of women would have been married, living out their lives in the context of the recognized authority of their husbands, and working extremely hard to feed and clothe families that could include foster children as well as indentured servants. Only a small proportion of women left wills, all widows with the exception of two spinsters, Elizabeth Poole, discussed above, and Elizabeth Bacon, who was twenty-eight when she died with a small estate inherited from her father that she willed to different members of her family. So information concerning married women is largely restricted to court records, where, apart from a few agreements recorded concerning children, the average woman is largely invisible. There are a few references to women being licensed to sell liquor, or assisting their husbands to do so, instances of mutual responsibility in business. But largely the records are silent. However, John Demos shows that one can use the deviant cases found in the court records to infer at least three basic obligations in marriage that the law sought to protect. A "good" marriage was between two people only, who were expected to live together on a regular basis and have an exclusive sexual relationship. It should be without fighting between the partners, and the court reserved the right to interfere if they thought it necessary. There are numerous accounts in the court records of men and women complaining of "abusive carriages" one to another; nevertheless, what is noteworthy is that in Plymouth, while life was still strongly influenced by men who held positions of authority politically and socially, the rights of women were upheld, and although they were expected to remain subservient to their husbands, they did have successful recourse to the law and enough influence to shape it in some respects. The only way that we can infer anything concerning the actual relationship between a husband and wife comes through the occasional wills, where the wording is not simply conventional—"my loving wife" as used by Richard Sylvester certainly belied the lack of provision made for Naomi in his will. But Captain Thomas Southworth's concern for his wife's "poor condition," or Henry Howland's sensitivity to his wife's need for privacy in stipulating that she be able to "enjoy the new Room to herself for her own use," and Josiah Cooke's gift to his wife, Elizabeth, of his servant and apprentice, Judah, an Indian, to be "at her Dispose and Improvement," all indicate a loving concern and care for the marriage partner that extended beyond death.

Bound by Indentures—Servitude in Plymouth Colony

In all the American colonies, economic success depended on hard work and cheap labor, and for the colonists in Plymouth cheap labor came in the form of indentured servants and apprentices. At least twelve of the *Mayflower* passengers were servants, but two-thirds had died by the spring of 1621. The surviving colonists, hampered by a severe labor shortage, had to find a labor supply as soon as possible. There is not a great deal to tell us how servants were recruited, but we do know that in 1629 servants were sent to the colony on the *Talbot*, and that in 1637 that redoubtable spinster Elizabeth Poole arrived on the *Speedwell* with fourteen servants. Also on board was Walter Deane with six servants, who, like Poole, settled at Taunton. And there is a contract drawn up the same year, in July, between Edward Doty and Richard Derby, in which they agreed that Derby would:

> Procure one able man servant to be brought over to serve the said Edward Doty for the term of five six or seven years for whose passage the said Edward Doty shall pay five pounds to the said Richard Derby & perform such other covenants to the said servant as the said Edward shall agree upon with twelve bushels of Indian grain at the end of his term.

A year later, the court recorded that William Snow had been "lately brought over out of Old England by Mr. Richard Derby," who assigned him to Doty to serve him for a period of seven years. Edward Doty also agreed to more generous provision for Snow at the end of his indenture, in that he was to be paid "one lively cow calf of two months old, and eight bushels of Indian corn, and a sow pig of 2 or 3 months old, with two suits of apparel, and find him meat, drink, & apparel during his term."

There are some interesting details that emerge from considering the Doty-Derby-Snow agreements. Edward Doty had himself emigrated to Plymouth in 1620 as one of Stephen Hopkins's two servants, possibly close to the end of his servitude. He realized what must have been the ambition of all the indentured servants who came to the New World, in that by 1633 he had become a freeman. His initial land grant as a passenger on the *Mayflower* was the start of his becoming a landowner; he had one apprentice, possibly more, employed at least two servants, and by the time of his death in 1655 his estate was worth £137 19s 6d, of which £60 was real estate. He was a colorful personality, often in court for fighting, slander, debt, and trespass, and he and his fellow servant Edward Lester fought the first duel on New England soil "upon a challenge at single combat with sword and dagger," but Edward Doty demonstrated that it was possible for servants to be successful and upwardly mobile in Plymouth Colony.

Another point that emerges from the contracts made between Doty, Derby, and Snow is that servants did have the power to drive their own bargains for what they would receive at the end of their term of indenture. What William Snow obtained was very different from the minimum that Doty had at first been prepared to offer him. There is no record of Snow having become a freeman, but he owned land and married the daughter of Peter Brown, one of the *Mayflower* passengers. Although Snow did not become a freeman, other indentured servants, like Edward Doty, did, and the possibility that servants could eventually become free citizens of Plymouth Colony meant that the task of governing and controlling servants was a complex one. It was very much in the interests of the colony to see that when they completed their indentures they would become part of a God-fearing, law-abiding community, contributing to its general welfare.

Periods of servitude began with a written contract between master and servant or apprentice, recorded before the court. Contracts, or indentures as they were more often known, ranged from a simple pact to detailed agreements that specified the length of service (on average between four to ten years), the master's responsibilities during the period of indenture, including the trade to be learned in the case of apprentices, and payment to be received at the end of the contract. This last varied widely, and there appears to have been no legal requirement. If a land grant, usually five to twenty acres, was part of the contract, in the early years of the colony it had to come from land owned by the servant's master, but the system had problems, and by the early 1660s the court had undertaken to allocate land to a number of former servants who had not yet received what was due to them. The law required masters to be fully responsible for their servants during the time of their indenture even if they were to be "diseased, lame or impotent by the way or after they come here." In other words, they were not to be a charge upon the town where they lived. Despite this, there were some instances where the court had to require a town to take care of sick or mistreated servants. Plymouth Colony had from its inception provided for the poor, and this provision for welfare needs appears a logical and compassionate expansion.

What of laws relating explicitly to servants? Between 1636 and 1650 four laws were passed that regulated the social and moral behavior of servants. In 1636 children and servants were not permitted to drink at taverns or to spend time at them. It was the innkeeper, however, who would be punished, not the offenders. In 1638 a law was passed that in effect required servants to obtain permission from their master before they could marry, but the law also recognized the right of servants to bring before the magistrates any master who "for any sinister end or covetous desire" would not consent to their wish to get married. In 1645 a new law required servants who purloined, stole, or embezzled their master's goods to make double restitution by payment or increased servitude, for the first default. In the case of a second default, double restitution was again required, plus sureties for good behavior, failing which they would be whipped. Some ten years later children and servants were forbidden to play cards or dice, as this had been a growing problem in various towns in the colony. For a first offense, correction was to be administered by parents or masters; the second would bring a pubic whipping. What emerges from these laws governing behavior is that masters were responsible for the behavior of their servants, and that in some ways servants and children were regarded as less responsible for their actions than their masters or parents. No laws were passed that specifically protected servants from physical abuse, but cases that came before the courts show that servants had the right to bring charges against their master for mistreatment or failure to honor the provisions of their contract.

Apart from the laws mentioned above, it appears that indentured servants were subject to the same laws and punishments as all Plymouth colonists, as were apprentices. Apprentices differed from servants in that they were bound to a master to learn a specific trade over a period of years. They may well have had a higher status than servants, and the six recorded apprentice agreements in the court records show that at the end of their indenture they received more material goods than did servants. Trades to be learned by apprentices included carpentry, tailoring, joinery, nailing, and milling. There were no specific laws governing the behavior of apprentices, but clearly some of those indentured as apprentices were quite young and would have come under the laws governing the behavior of both children and servants.

There are occasional references to Indian servants in the colony, one such being after King Philip's War ended in 1676, when a law was passed that "no Indians that are servants to the English shall be permitted to use guns for fowling or other exercises," but the majority of cases that came before the courts relate to the English. After the war, only Indians who were servants prior to it could remain in the colony, subject to good behavior. It may well be, though, that the apparent decrease in the number of people entering indenture agreements from the 1640s was due to an increasing number of Indian servants, as the colony still required cheap labor. From 1676 Indians were sold into slavery outside the colony, but there are indications that this also occurred within its boundaries. There is an instance in the court records of an Indian being sold into slavery and remaining in the colony, the case of Thomas Wappatucke. In 1685 he was found guilty of burglary and "sold for a perpetual servant." There is also reference to Margaret, "an Indian slave" in the 1678 estate settlement of Steven Paine Jr. of Rehoboth. Margaret is given to his widow together with half of all the movables in the house, as well as cattle, leather, and other items "to have and to hold to her and her assigns for ever." There were no laws in Plymouth Colony that dealt specifically with black slaves or servants. There were some black servants, though. As early as 1653, John Barnes had a "neager maid servant," who came before the court to accuse John Smith Sr. of Plymouth "for receiving tobacco and other things of her which were her master's, at sundry times, in a purloining way." Captain Thomas Willett's negro servant, Jethro, was taken prisoner by the Indians, released by the colonial army in 1676 and ordered to return to his former service, and remain a servant to the now late Captain Willett's successor for a further two years, after which he was to be:

> Freed and set at liberty from his said service, provided, also that during the said term of two years, they do find him meat, drink, and apparel fitting for one in his degree and calling, and at the end of his said service, that he go forth competently provided for in reference to apparel.

Also in 1676, in a will drawn up by Walter Briggs, he requested his executors to "allow my said wife a gentle horse or mare to ride to meeting or any other occasion she may have, & that Jemy, the neger, catch it for her. Also, I will my said wife, Mariah, the little neger girl, to be with her so long as my wife lives. . . ." Whether Jemy and Mariah were servants or slaves we do not know, but it may be that there is a fine semantic line to be drawn here, and some "servants" were in fact slaves, particularly when a value is assigned to them. In 1674, the inventory of the estate of John Dicksey of Swansea included "1 Negro Maid servant named Malle" valued at £24, and John Gorum's 1675 inventory included "1 Negro man," no value assigned. Were these servants to be freed after their indentures, or were they slaves? Captain Willett's 1674 estate included eight Negroes valued at £200, and in his will, drawn up three years earlier, he listed Negroes along with "Goods horses Cattle Clothes household stuff money plate. . . ." John Demos comments that the pattern with white servants was to bequeath the remainder of their contract, a period of time in service, not the person with a value set upon them. There was no legislation relating to slavery as there was governing servants, and so it seems likely that there were very few slaves in the colony, and that black servants were set free after serving their indentures.

The position of female servants was not as good as that of their male counterparts. Only three of the indenture agreements in the court records relate to women, and at least two involve young girls. Alice Grinder was bound to Isaac Allerton for five years in 1633, during which time he was to provide her with "food & raiment competent for a servant, & at the end thereof . . . to give her two suits of apparel." It appears that Grinder arrived on the *James* in 1633. Mary Moorecock was bound to Richard and Pandora Sparrow as an "apprentice" in 1639, but the indenture makes no mention of any trade to be learned, or any payment once she had completed her time of service, although she would be provided with the usual basic necessities during it. If she wished to be married before the end of the nine years, two impartial arbitrators would decide what Sparrow was owed for any outstanding time. The terms "servant" and "apprentice" seem to have been used interchangeably, blurring the images that emerge. In 1642, Francis and Christian Billington "put Elizabeth, their daughter, apprentice to John Barnes and Mary, his wife, to dwell with them and to do their services until she shall accomplish the age of twenty-three years (she being now seven years of age . . .)." Like Mary Moorecock, Elizabeth was to have all her basic needs provided, but nothing after her contract ended.

The Billington case raises the whole strange, unresolved question of the "putting out" of children to live in families other than their own that seems to have occurred on a fairly regular basis in Plymouth Colony. In part it may have had its roots in an English custom that dated back at least to the sixteenth century, when children were brought up in families other then their own due to the belief that they would learn better manners than at home. Demos believes that children formed the greatest proportion of persons in servitude in the colony, and given the economic hardship of many families, and the shortage of cheap labor, these factors must also have played into the system. Children were apprenticed to learn a trade, obtain a general education in the household, including literacy, or were simply servants. Insofar as hard evidence is concerned, there is not a great deal, as the motivation of parents in putting their children to grow to adulthood away from their own family circle is only glimpsed. Sometimes it was both economic and a concern for the welfare of the child, as in the case of seven-year-old Zachariah Eddy, put to live with John Browne of Rehoboth as a servant until he was twenty-one, as his parents had "many children, & by reason of many wants lying upon them . . . they are not able to bring them up as they desire." Zachary's brother, John, had also been put out to live with Francis and Katherine Goulder as their servant the previous year, 1645, at the age of seven. Elizabeth Billington's brother, Joseph, was already living with John Cooke the younger, and in trouble with the court because he kept running away and going home. In this case the court took a firm line with the Billington parents, and ordered that for every time he did this, Francis and Christian would be put in the stocks. In addition, fourteen-year-old Benjamin Eaton, living with the Billingtons (Christian's son by her previous marriage to Francis Eaton), if he should "counsel, entice or inveigle" Joseph from his master, would also be put in the stocks. Francis Billington was a *Mayflower* passenger, whose father, John, was executed in 1630 for murder. Francis had married Christian Penn, widow of Francis Eaton, four years later, and had to take responsibility for the four Eaton children, one of whom Bradford described as "an idiot," but he lived until 1651, presumably at home. With their own children as well, there seems to have been good reason as to why the two children were put out to service. It could have been a form of welfare relief that existed in the colony, but the system is not fully explained by that conclusion. Demos concludes that there is no clear pattern that emerges. Some were from reasonably well-off, educated families, some were orphans, others from homes where economic hardship necessitated it. The historian Edmund Morgan, in his study of the Puritan family, has suggested another explanation, that parents in this culture "did not trust themselves with their own children, that they were afraid of spoiling them by too great affection," and so sent them to be brought up by others. There is so much that took place in seventeenth-century society stemming from a worldview far removed from our own that it is not possible to arrive at any final conclusion, and it may be that Morgan's hypothesis indeed lies at the core of this particular custom in Plymouth Colony, as well as in the Puritan colonies to the north.

When girls in servitude became women, what was their position? Or that of women entering servitude as adults? They were not granted land, as were their male counterparts, so the only permanent prospect they had of a home of their own was through marriage. As we have seen, they had to have permission from their master before they could marry, and at least were protected by law in that he could not arbitrarily forbid it. But their position was very precarious, because if they did not marry they would have to hope that they could continue in service for the rest of their lives. In the early years of the colony marriage would have been very likely, but as the demographics changed and there were more women than men, a lifetime in service was a very real prospect.

Servants of both sexes did have the opportunity to change their indentures from one master to another at their own instigation, with the agreement of their master, but often an additional year or two would be added to the indenture, although sometimes the servant was better off in the end, obtaining a more beneficial agreement than the original contract. On the other hand, a master could sell an indenture agreement to someone else, but all parties had to agree to the change, including the servant concerned. Servants of both sexes had the legal right to take their masters to court for abuse or for not fulfilling an indenture agreement. However, most of the twenty cases brought against masters by servants were brought by men, and this does raise the question as to the extent to which abused women servants were threatened into submission. In 1637 Edith Pitts accused her master, John Emerson, of abuse, and the court ordered him to appear before the court. There is no record that he ever did, and so the outcome of the charge is unknown, but the scarcity of such cases and the vulnerability of women in positions of servitude does underline the difficulties of their situation. Out of twenty-four cases brought before the court involving the misdeeds of servants, only one involved a female servant, Jane Powell, who was brought up on a charge of fornication but acquitted when she testified that she was seduced by another servant. The court case gives some sense of the problems faced by women in servitude. In October 1655,

> Jane Powell, servant to William Swift, of Sandwich, appeared, having been presented for fornication, who, being examined, said that it was committed with one David Ogilvie, an Irishman, servant to Edward Sturgis; she said she was allured thereunto by him going for water one evening, hoping to have married him, being she was in a sad and miserable condition by hard service, wanting clothes and living discontentedly; and expressing great sorrow for her evil, she was cleared of the presentment and ordered to go home again.

What sort of reception Powell had on her return to the household in which she served we will never know, and this is her only brief appearance in the court records. The reason that there is no other mention of female servants charged with misdemeanors may well indicate that women felt they could not risk their indenture and chances of marriage by committing crimes, or by running away to escape servitude as did a number of men.

Children in servitude were also very vulnerable, and they were possibly even more intimidated by abusive masters and mistresses than women. There are two important cases of abuse of children that occurred in the mid-1650s, one of which is the only case that involved the death of a servant, directly attributed to physical abuse at the hands of his master and mistress. Fourteen-year-old John Walker was the servant of Robert and Susanna Latham of Marshfield. A detailed coroner's inquest found that Walker had died from physical abuse, exposure, forced labor, and lack of sufficient food, clothing, and accommodation, particularly in the depth of winter, and so "in respect of cruelty and hard usage he died." The details given are gruesome, and what is most troubling is that neighbors testified to having seen both Latham and his wife abusing the child, yet there is no record of their complaints on his behalf, or of John Walker having approached the court as he was in law entitled to do, which is not surprising considering the level of his abuse. The report of the coroner's jury, submitted on March 6, 1655, reads as follows:

> We, upon due search and examination, do find that the body of John Walker was blackish and blue, and the skin broken in divers places from the middle to the hair of his head, viz., all his back with stripes given him by his master Robert Latham, as Robert himself did testify; and also we found a bruise on his left arm, and one on his left hip, and one great bruise on his breast; and there was the knuckles of one hand and one of his fingers frozen, and also both his heels frozen, and one of the heels the flesh was much broken, and also one of his little toes frozen, and very much perished, and one of his great toes frozen, and also the side of his foot frozen; and also, upon reviewing the body, we found three gaules [open sores] like holes in the hams [buttocks], which we formerly, the body being frozen, thought they had been holes; and also we find that the said John was forced to carry a log which was beyond his strength, which he endeavoring to do, the log fell upon him, and he, being down, had a stripe or two, as Joseph Beedle doth testify; and we

> find that it was some days before his death; and we find, by the testimony of John Howland and John Adams, that heard Robert Latham say that he gave John Walker some stripes that morning before his death; and also we find the flesh much broken on the knees of John Walker, and that he did want sufficient food and clothing and lodging, and that the said John did constantly wet his bed and his clothes, lying in them, and so suffered by it, his clothing being frozen about him; and that the said John was put forth in the extremity of cold, though thus unable by lameness and soreness to perform what was required; and therefore in respect of cruelty and hard usage he died; and also, upon the second review, the dead corpse did bleed at the nose.

Three months later the indictment was expanded to include Susanna, the wife of Robert Latham, who was presented to the general court "for being in a great measure guilty, with her said husband, in exercising cruelty towards their late servant, John Walker, in not affording him convenient food, raiment, and lodging; especially in her husband's absence, in forcing him to bring a log beyond his strength."

Robert Latham was indicted for "felonious cruelty" and found guilty of manslaughter. The case has some unusual aspects. It is one of the rare instances in New England of a citizen claiming the ancient English custom of "benefit of clergy." In England, after a guilty verdict was returned by a jury, particularly in a trial carrying the death penalty, the prisoner was asked if he had any reason why judgment should not be given, and if he could read, he could ask for a Bible and read out Psalm 51, verse 1, "Have mercy upon me, O God, after thy great goodness, according to the multitude of thy mercies do away mine offences." This was known as the "neck verse." After this the court would proceed to judgment, but being able to read the verse would save the prisoner from being hanged for murder. Bradley Chapin, in *Criminal Justice in Colonial America, 1606–1660*, comments that benefit of clergy was "a highly irrational and erratic means of mitigating the consequences of an exceedingly severe penal law. It saved thousands from the gallows." In this case, Latham

> . . . desired the benefit of the law, viz. A psalm of mercy, which was granted him; and sentence was further pronounced against him, which was, that the said Robert Latham should be burned in the hand, and his having no

> lands, that all his goods are confiscate unto his highness the Lord Protector; and that the said sentence should be forthwith executed. . . .

Chapin is emphatic in stating that in the Puritan colonies "it would have been unthinkable" for benefit of clergy to be used within their jurisdictions. "They held it as a basic truth that literacy should lead to rectitude, certainly not be an escape from the consequences of sin and crime." The penalty for murder was death, and in the seventeenth century manslaughter was homicide. It would have been equally against the principles of the Separatist leaders in Plymouth to mitigate a sentence that carried the death penalty on the basis of such an immoral and discriminatory practice. Why was this single instance permitted? Another strange aspect of the case against the Lathams is that the indictment against Susanna for cruelty and neglect of John Walker that came before the court in June 1655 appears to have been dropped. However, there must have been ongoing gossip and talk concerning her role, for three years later she was again presented to the court "for sundry years of cruelty toward John Walker." Whoever wished to come before the court to prosecute her was given "full and free liberty" to do so at the October court, failing which the presentment against her would be struck from the records. No one came forward, despite a proclamation being read three times in the court, and so the record was erased—but fortunately only crossed out in the original, so it remains to fill out the complete account.

How can we account for the turn of events described above? The key, we suggest, lies in the identity of Susanna. She was the daughter of the wealthy Boston merchant John Winslow and his wife, Mary Chilton, one of the *Mayflower* passengers. Susanna was therefore allied to a rich and powerful family, for her uncle, Edward Winslow, had already served three terms of office as governor of Plymouth Colony. We will never know exactly what pressure was brought to bear on whom, but the kinship connection certainly suggests that no one was prepared to take the witness stand, even though the court repeatedly promised "full and free liberty to prosecute" Susanna Winslow Latham. And the court's willingness to give Robert Latham the highly dubious benefit of clergy suddenly makes sense when it becomes clear that he was John Winslow's son-in-law and married to the niece of one of Plymouth Colony's governors.

Overall, insofar as those serving under indentures were concerned, it seems that the colony's laws provided guidelines for the basic rights and regulation of their way of life. The specific laws that were framed were made to control their behavior, but in the first instance, the servants or apprentices were not the ones prosecuted for breaking these laws. It was their master, or the innkeeper, or whoever led them astray. When it came to crimes such as murder, theft, and sexual misconduct, however, servants were treated as any other colonist according to the criminal justice system, without regard to their status. The colony also enforced the law requiring masters to take responsibility for their servants and care for them, even when disabled or when they failed to give satisfaction. However, in the case of Joseph Gray, misused by his mistress (the second case of child abuse), the town was required to take responsibility for his needs when she failed to do so, an extension of the policy of providing for the poor that had been in place since the inception of the colony. Despite the number of cases of servant abuse brought before the court, the case of John Walker, even given its abnormality, and that of Edith Pitts, indicate that abuse was probably a reality for a far higher percentage of indentured women and children than the records reflect. Nevertheless, the sense that emerges is that the laws were framed with the recognition that servants and apprentices who completed their indentures would become part of society, that a servant today could one day become a neighbor, even a freeman and someone of note in the community. In a colonial society, where status was relatively fluid and upwardly mobile, a great deal was possible that would have been unthinkable in Old England.

To "Drink Drunk"—The Regulation of Alcohol

Alcohol was consumed in considerable quantities in Plymouth Colony, as it was everywhere in the seventeenth century. Wine, beer, and spirits were all easily available, most households brewing their own beer as a matter of course. Alcohol was generally consumed in the home, either that of the owner or of a friend or neighbor. There were ordinaries, or taverns, in each town, and these grew from small establishments where people were licensed to sell liquor from their homes to an extensive retail business, with the government making large profits from all aspects of the import and sale of liquor.

The merry toper with his runlet of beer.

There were a number of laws passed regulating the consumption and distribution of alcohol and the licensing of ordinaries, but it is noticeable that as the government's profits from excise taxes increased, so its treatment of repeat offenders of its laws against alcoholism grew more lax. Ordinaries were originally established for strangers and travelers, not for the use of colonists in their own towns. Servants and children were prohibited from drinking in ordinaries. Alcohol was to quench one's thirst or to treat the sick. If someone was drinking to the point of being drunk, they were breaking the law, since being found drunk in Plymouth Colony was considered criminal misconduct. A variety of punishments were given, ranging from fines between five and forty shillings to, less frequently, being set in the stocks or whipped, particularly for repeat offenses. The first case documented for alcohol abuse, or being "drink drunk," in the court records is that brought against John Holmes, who later became the messenger of the court for the government, on April 1, 1633. He was "censured for drunkenness to sit in the stocks, & amerced in twenty shillings fine." As this is the only case mentioned in the court records until 1636, it is probable that the government only took action when someone was an obvious repeat offender, public life was disrupted, and there were witnesses prepared to testify against them. As Holmes, now the court messenger, was presented again in 1639 "for sitting up in the night, or all the night, drinking inordinately, when he was sent about public business," it appears that by modern standards he had a drinking problem, but perhaps not by those of the seventeenth century, as he had been appointed in 1638 to a responsible government position that he held for several years. A graphic description of the type of behavior considered to be objectionable and against the law is that of William Reynolds, presented to the court in January 1638:

> For being drunk at Mr. Hopkins his house, that he lay under the table, vomiting in a beastly manner, and was taken up between two. . . . Mr. Hopkins is presented for suffering excessive drinking in his house . . .

Stephen Hopkins was acquitted, but William Reynolds was fined 6s 8d and discharged at the June court. Stephen Hopkins clearly had a license to sell beer at the time, as he was fined for selling wine at double its official price. Selling wine at such excessive rates was considered by the court to contribute "to the oppressing & impoverishing of the colony," an early indication of the seriousness with which the colonial government regarded its revenue from excise duty. From 1633 to 1645, the level of drunkenness that was offensive enough to warrant court action had been determined on an individual basis, when someone was found to be "drink drunk," but a significant change took place in the laws relating to alcohol in 1646. The new definition was given as follows:

> And by drunkenness is understood a person that either lisps or falters in his speech by reason of overmuch drink, or that staggers in his going or that vomits by reason of excessive drinking, or cannot follow his calling.

Drunkenness was now assessed not by the quantity that someone drank, but by evidence of drunkenness in their appearance. It appears that the court was now willing to overlook lesser behavior from excessive drinking and punish only extreme behavior patterns.

At the same time the laws governing the retailing of wine and keeping of ordinaries were tightened, and it is noticeable that here too a shift had taken place. Whereas previously no townsperson was allowed to drink at a local ordinary, the law now implicitly permitted it, allowing for fining of both the ordinary keeper and the townsperson if they spent longer than an hour drinking at the ordinary. While only three licenses to keep ordinaries were issued between 1633 and 1643, from 1644 to 1654 the number escalated to thirteen. The colony was expanding and as new towns were formed, so the demand for licensed liquor retailers increased. But what is interesting is that as the number of licenses increased, so the number of cases for drunkenness brought before the court dropped by almost half, with fifteen cases being presented for the first ten years, from 1633 to 1643, and only seven from 1644 to 1654. It is probable that with the new standards for assessing excessive drinking, as well as easing up on allowing townspeople to patronize the ordinaries, people became more tolerant of reporting any excesses, and the more liquor that was sold, the more the government profited in excise dues, and so it was in its interests to place more responsibility for keeping patrons in line with the licensed ordinary keeper.

It is evident that women also drank at the ordinaries, as there is mention as early as 1638 of the widow Palmer being present at Stephen Hopkins's house, and being called as a witness to the excessive drinking he was permitting there. Some years later, Ann Hoskins accused Hester Rickard of being "as drunk as a bitch, and found in private company in an ordinary with John Ellis, of Sandwich." And in 1669 James Clark, Philip Dotterich, Mary Ryder, and Hester Wormall were each fined ten shillings (a second offence amount) "for their staying and drinking at the house of James Cole of Plymouth, at Plymouth, on the Lord's Day, in the time of public worship of God," and it was Mary Cole, wife of James Cole, who was considered to have been responsible for having permitted it, for which she was fined £3—James Cole owned the largest ordinary in Plymouth. John Demos suggests that the fact that women could drink at taverns meant that they "had considerable freedom to move on roughly the same terms with men even into some of the darker byways of Old Colony life."

Although ordinary keepers were generally men, they were often helped by their wives, but there are instances of women being licensed to sell alcohol on their own. In some cases, it is clear that the woman concerned was widowed, in others their marital status is not referred to, possibly because there was no need to as it was well known, or

Men and women in company at an ordinary or tavern.

because it was not material to the issue. The context in which women were issued licenses appears in general to be the retailing of liquor, not the management of taverns, but in 1678 Mary Williamson of Marshfield was granted a license to keep an ordinary in the town, and in 1684 Mistress Mary Combe was licensed to keep a tavern at Middleborough that included lodging, food, and fodder for horses. Mary Combe was the widow of one of the leading men in Middleborough, Francis Combe, who had died two years previously. He had been a selectman for the town, but had also been licensed in 1678 to keep an ordinary at his house, and his widow had presumably done an excellent job of helping to run it. An indication of the importance of the role of women in the running of a tavern can be seen in the following instance, where the general court, "understanding that James Leonard, of Taunton, having buried his wife, and in that respect not being so capable of keeping a public house, there being also another ordinary in the town, do call in the said Leonard his licence."

Cases of drunkenness brought before the court overwhelmingly involved men rather than women. One of the few cases brought against a woman is that of Ann Savory, the wife of Thomas Savory, who was found with Thomas Lucas (who was brought up on more charges of drunkenness than anybody else), "drunk at the same time under an hedge, in an uncivil and beastly manner," on the Lord's day, when she should have been in church. For being with Lucas on that day, at that particular time, she was sentenced to sit in the stocks, and was fined five shillings, the first offense amount, for being drunk. It is also strongly implied that Hester Rickard was drunk at the time that she was being "uncivil and beastly" in attitude and speech to her husband, John Rickard, for which she was sentenced to be publicly whipped, because at the same time the court prohibited her to brew beer to sell "as formerly she had done, because it appeared to the Court that it was a snare to her to occasion evil in the aforesaid respects."

It does seem that the court was on the whole lax about enforcing liquor laws, as there are occasions when there had evidently been a great deal of infringement before further attempts were made to enforce the law. In March 1664, just before the court revoked James Leonard's license to keep an ordinary, it called in the license of Edward Sturgis Sr. due to "much abuse of liquors in the town of Yarmouth." Sturgis was evidently considered responsible, and the court presumably hoped to deal with the problem by removing his liquor license. At the same court, Thomas Lucas was required to give surety for his good behavior and appearance at the next session of the court to answer "for his abusing of his wife to her danger and hazard, also for his railing and reviling others, to the disturbance of the King's peace." Although it is not stated, all Lucas's court appearances were for excessive drinking and the behavior that accompanied it.

The selling of liquor to Indians was prohibited, and colonists who did so were regularly fined, but the first time that the law explicitly forbade this was in 1646. Some years later, in 1654, when legal procedures were reiterated at the June general court, it was stated that:

> Whereas there has been great abuse by trading wine and other strong liquors with the Indians, whereby they drink themselves drunk, and in their drunkenness commit much horrid wickedness, as murdering their nearest relations, &c. . . . it is therefore ordered, that no person or persons whatsoever, from this time, trade any strong liquors, directly or indirectly, to the Indians within this jurisdiction . . .

Heavy fines and the revoking of licenses were the penalties to be enforced if the law was broken. Despite this, retailing liquor to Indians may well be the most consistently committed crime recorded in the court records. This was probably because of the high prices Indians were prepared to pay for liquor, in effect creating a black market, and because the court consistently avoided bringing up Indians on charges of drunkenness, preferring instead to try to control their behavior less directly. In 1662 the court ruled that if any Indian or Indians be found drunk in any township in the colony, they were to be set in the stocks, and would have to pay the costs involved in doing so, which amounted to 2s 6d a time. They were not fined in the same way as the English colonists, as they did not have the same financial resources. Despite this law, there are no records of Indians being prosecuted for drunkenness, although it was evidently taking place. It seems clear that at least until the early 1670s, after the deaths of William Bradford and Massasoit, the governing authorities in Plymouth were very cautious about disturbing the delicate balance of relations that existed with the Indians. The only instances in which they were charged with alcohol-related offenses was when the case was extremely serious, such as stealing or embezzling.

The government in Plymouth continued to restate its laws against alcohol abuse, and in 1677 reiterated the need for "prevention of the growing intolerable abuse by wine, strong liquors, &c., both amongst the Indians and English." Despite their attempts to keep townspeople out of ordinaries, or at least to restrict their access, they were also steadily issuing more and more licenses to open additional outlets, as taverns and inns continued to be central to the social life of the community. By 1691, when Plymouth was absorbed into Massachusetts Bay Colony, it appears that the court had been allowing taverns to expand on purpose, as the larger the ordinaries were, the more excise taxes went to the government. Throughout the colony's history it seems as though there was a real tension between the recognition of alcohol's problems and attempts to control its abuse, and the need of the government for the lucrative revenue that the trade in liquor brought.

Society in seventeenth-century Plymouth was on the whole regulated by a framework of law and order that gave stability to the community, but as in all societies, there were those whose activities disrupted and threatened that order to a lesser or greater degree. Against the regular changing seasons that shaped the lives of everyone who lived in the colony, there were the disruptions caused by heated exchanges with neighbors over boundaries, straying livestock, infringement of mowing rights, as well as the more colorful misdemeanors and crimes, the eruption of violence as husband and wife or neighbors resorted to physical instead of verbal abuse of each other, and the sudden unexplained deaths. From its inception the colony, through its laws and town regulations, had been trying to exercise control and yet give greater liberty within certain bounds than the settlers had known in Old England. One area over which there was far greater control than is experienced in the present century was that of sex-related crimes and acts, and accusations and prosecutions for illicit sexual acts were a regular part of the business of the court, as were inquests following deaths in suspicious circumstances.

Chapter 4

In an Uncivil Manner

Sex-Related Crimes, Violence, and Death

As in all small agrarian communities, people in Plymouth Colony lived largely uneventful lives, dictated by the change of seasons, year after year. Thomas Tusser's sixteenth-century admonitions reflect the rhythm of life:

> In February, rest not for taking thine ease
> get into the ground with thy beans and thy peas.
> In March, and in April, from morning to night
> in sewing and setting, good housewives delight . . .
> In June get thy weed hook, thy knife and thy glove
> and weed out such weeds as the corn does not love.
> In July—
> Thy houses and barns, would be looked upon
> and all things amended, or harvest come on.
> Things thus set in order, at quiet and rest
> thy harvest goes forward and prospereth best.

This was the normal pattern of days, and nights were for the most part times of much-needed rest from a day beginning with first light and ending at sundown. The majority of the population were engaged in working the land, but also, as the towns spread through the colony, became caught up in the commitments that come with growing communities. The men developed the structures of small-town life, training for the militia, serving as constables, highway supervisors, selectmen, and deputies, selling land, marketing cattle and produce, trading with the Bay Colony for goods not easily available locally, building harbors and highways, establishing churches and schools. The women, apart from childbearing and -raising, worked at spinning, weaving, cooking, planting and harvesting garden crops, tending livestock, acting as midwives, supervising servants. Generally law-abiding, what percentage of this population was in fact involved in crimes and misdemeanors? Population figures are difficult to obtain with accuracy, but by 1643 there were probably between 1,500 and 2,000 people living in the colony, of which some 600 were men between the ages of sixteen and sixty, and from 1660 it has been calculated that the population increased to around 3,000 and remained fairly static for the next three decades. It has been estimated that the incidence of crime in Plymouth between 1633 and 1669 was in fact very low. Bradley Chapin, in *Criminal Justice in Colonial America, 1606–1660*, suggests that in the seven colonies, each year on average only one in 750 appeared in court accused of a crime or misconduct, a remarkably low crime rate compared to England at this time. Plymouth law, however, required that there be at least two witnesses before there could be a conviction, so the number of people involved in any one case may have included several members of the population as well as the accused and plaintiff. In addition, crime and its punishment were very public matters. Punishments were designed for maximum publicity, for the humiliation of an individual before a close-knit population. It was a form of social control, so court days were often a mix of business and carnival atmosphere with more drinking than normal and the spectacle of public branding, whipping, and the ridicule and abuse that came with being placed in the stocks. There were many land transactions that also came before the courts even if there was no litigation, and there was a continuous stream of men and women whose lives were influenced by court decisions concerning inheritance, land boundaries, mowing rights, management of ordinaries or taverns, road construction and maintenance, to name only a few, as well as all the town regulations that affected the daily lives of everyone in the colony. It is probable that even though the incidence of crime was low, the names of the majority of male colonists and at least a fair number of the females appear in the court and town records, providing some insight into their lives.

The court records are the largest body of data available, and are well supplemented by wills and probate inventories. The laws of the colony, published as part of the court records, provide an image of the ideal that those in authority strove to achieve for society while the court cases show something of the realities with which the government was dealing. A sufficient number of men and women were caught up in crimes to afford us a glimpse of this aspect of life in Plymouth and how it was handled. Sex-related crimes are considered first as they were the most frequently committed.

The seventeenth century was a time of robust interactions between the sexes. It was commonly supposed that when a man and woman were alone together they would be involved in sexual activity of some sort or another, and the laws that governed sexual behavior are to be seen against this background. The 1636 codification of laws of Plymouth Colony incorporated laws already in existence in the colony from its inception, but which had been enforced by the governor and his assistants on a discretionary basis up until that date. When the first code of laws was published in 1636, the list of "capital offences liable to death" were, in order, treason, willful murder, solemn compaction or conversing with the devil by way of witchcraft, willful burning of ships or houses, sodomy, rape, buggery, and "adultery to be punished."

Though it now strikes us as a private matter rather than a crime, sexual misconduct was viewed by government in the seventeenth century in criminal terms. Neighbors actively kept a watch on the activities of those who lived close to them, and a woman's reputation among her neighbors was often the most important possession she had in the close-knit, small world in which she lived. It was a society in which privacy as we know it was unknown. Men, women, and children slept in the same room, crowded onto the same benches at mealtimes, and worked closely together in the house and fields. Conversation was frequently spiced with sexual allusions and bawdy comments and jokes. Children were used to hearing sounds and seeing dim shapes at night as their parents or married siblings enjoyed sexual relations. The close proximity in which everyone lived made it all too easy for a woman to "beguile" a man, just as it was for a man to demand from a woman sexual favors that she might well not wish to give. A woman's reputation could be easily destroyed if she was thought to be morally loose; and if she was involved in a compromising situation, or claimed to have been sexually attacked, it would be assumed that she was a willing partner unless she had cried out for help. The requirement of a cry for help was based on the Mosaic law, and was critical in distinguishing cases of rape as opposed to willing fornication (sex with an unmarried woman) or adultery. Illicit sex in all its facets was regarded as a crime not only in Plymouth Colony but in all the American colonial jurisdictions. In addition to what were capital crimes—sodomy, rape, bestiality, adultery, and incest—sex-related crimes and misconduct included fornication, attempts and propositions, and various provocative types of behavior viewed as lascivious or suspicious conduct.

Sex-Related Capital Crimes

Buggery, or Bestiality

The winter's day was shrewdly cold. It was January 1643, and John Holmes, Messenger of the Court for the Colony was waiting to present his account to the Court for the last three years. They were in arrears again—how he was expected to support his wife and son God alone knew, and now he had begotten her with child again, and she was soon to be delivered. Little John was but six and a rare handful. He scanned his account again. Only £1 he was charging for ten weeks of diet for Granger while he had him in prison, and £2 10s for executing him and eight beasts. A nasty business that. He had drunk so much strong water before the execution that it was a wonder of wonders that he had been able to go through with it all, even though he had help, and of course they knew that he had been drink drunk before. What was the odd bottle when your job kept you busy at the whipping post, repairing the prison, and traveling, like he had to Scituate to fetch young Granger's parents? But that business of having to say which of the sheep he buggered, show he had actually known them so they could be killed along with him, the gossip had not stopped for weeks, even though some folks wouldn't talk, said they were too disgusted even to mention any particulars. A terrible time he had bringing the sheep into town from Duxbury. Love Brewster had helped, only a few years older than himself he was. But young Thomas he knew those sheep. "No, no, not her," he'd said, and before the whole court and all, "Oh there's that auld bitch, yes, I know her." And "Yes, 'Tis auld Blackie; I'd know her anywhere," and so on until all the sheep had been brought into the court. Only eight beasts not twelve as they said in the indictment, had to be killed, but that was eight too many. A nasty job all round, but worst was having to hang young Granger. He'd talked with him many a time while he was in prison all those weeks. And just when you'd think the crowd would be yelling and shouting, there was just this silence, as he sweated there appalled by what he had to do, by what young Granger had done. And then he had to go back home, tell young John to shut his mouth, he'd know soon enough what he'd had to do that day. And he'd got drunk, drunker than he ever had been, but no one said nothing to him. They knew.

Although there were five sex-related crimes for which the sentence was capital punishment, the sole execution for a sexual offense in Plymouth Colony was in 1642 when seventeen-year-old Thomas Granger, servant to Love Brewster, was executed for buggery. Buggery (sexual relations with animals) was regarded with the greatest abhorrence and hostility. In New England's judicial practice, particularly in the Puritan colonies, pressure was brought to bear on suspects to confess their crime in open court or even on the gallows, in the hope of obtaining mercy from God, even if not from man. Although Thomas Granger denied the charges against him at first, he later confessed and provided the court with a remarkable number of details.

The account of Granger's trial in the court records is very brief, but William Bradford, in his history of Plymouth, describes the situation at length:

> And after the time of writing of these things befell a very sad accident of the like foul nature in this government, this very year [1642], which I shall now relate. There was a youth whose name was Thomas Granger. He was servant to an honest man of Duxbury, being about 16 or 17 years

> of age. (His father and mother lived at the same time at Scituate.) He was this year detected of buggery, and indicted for the same, with a mare, a cow, two goats, five sheep, two calves and a turkey. Horrible it is to mention, but the truth of the history requires it. He was first discovered by one that accidentally saw his lewd practice towards the mare. (I forbear particulars.) Being upon it examined and committed, in the end he not only confessed the fact with that beast at that time, but sundry times before and at several times with all the rest of the forenamed in his indictment. And this his free confession was not only in private to the magistrates (though at first he strived to deny it) but to sundry, both ministers and others; and afterwards, upon his indictment, to the whole Court and jury; and confirmed it at his execution. And whereas some of the sheep could not so well be known by his description of them, others with them were brought before him and he declared which were they and which were not. And accordingly he was cast by the jury and condemned, and after executed about the 8th of September 1642. A very sad spectacle it was. For first the mare and then the cow and the rest of the lesser cattle were killed before his face, according to the law, Leviticus xx.15; and then he himself was executed. The cattle were all cast into a great and large pit that was digged of purpose for them, and no use made of any part of them.

Bradford's account and the few details from the court records are graphic enough to convey a sense of the deep sorrow that he, and without doubt others in positions of leadership and in the colony, felt about the crime. It is important to note that it is described very factually, there are no bizarre details or circumstances as there were in one case in particular in the Bay Colony. And in Bradford's record of the Granger case he is at pains to show that the punishment is directly according to the Mosaic law. Leviticus 20:15 from the King James Bible reads: "and if a man lie with a beast, he shall surely be put to death; and ye shall slay the beast." While the Mosaic law dictated that a buggerer must be put to death, and the unfortunate animal must be killed, no mention is made of restitution to the owner of the beasts. So in Love Brewster's case, Granger's master, he was almost certainly out of pocket as well as without a servant. Bradford follows his account of the Granger case by raising the question of how it came about that the knowledge and practice of "such wickedness" could have taken place

"Hanged by the neck until his body is dead."

in the young colony. Granger was evidently questioned as to where he learned the habit, and it appeared that he was taught by a fellow servant who had heard of such things in England.

The governor and his assistants who were initially responsible for the maintenance of law and order in the colony were very careful to implement a mixture of English common law, the Mosaic law, and local law where they saw need to revise the common law or establish new laws due to circumstances peculiar to their situation in America. The source of their legitimacy paralleled that of the monarchy in England, which ultimately invoked the service of God as the origin of its authority and power. In cases of capital offenses, the Plymouth court was extremely careful to make it quite plain as to the final source of its own authority, hence Bradford's reference to biblical precedent.

Less than a year later, in June 1643, there is the odd case of John Walker, son-in-law of Arthur Howland Sr. of Marshfield, who was presented to the court "to answer to all such matters as shall be objected against him . . . concerning lying with a bitch. . . ." It is not clear whether the bitch was a canine or a lewd woman. The case was apparently dismissed as it did not come up at a subsequent court. The next instance of buggery to come before the court was in March 1656 when William Honeywell's case was considered. He had been jailed on suspicion of buggery with a beast, but was released as there was lack of sufficient evidence to convict him. He had "stiffly" denied the charge against him.

The final case of bestiality to appear before the court was that of Thomas Saddler. In October 1681 he was arraigned for buggery with a mare, and indicted as follows:

> Thomas Saddler, you are indicted by the name of Thomas Saddler, of Portsmouth, on Rhode Island, in the jurisdiction of the Providence Plantations, in New England, in America, laborer, for that you, having not the fear of God before, nor carrying with you the dignity of human nature, but being seduced by the instigation of the devil, on the third of September in this present year, 1681, by force and arms, at Mount Hope, in the jurisdiction of New Plymouth, a certain mare of blackish color then and there being in a certain obscure and woody place, on Mount Hope aforesaid, near the ferry, then and there you did tie her head to a bush, and then and there, wickedly and most abominably, against your human nature, with the same mare then and there being feloniously and carnally did attempt, and the detestable sin of buggery then and there feloniously you did commit and do, to the great dishonor and contempt of Almighty God and of all mankind, and against the peace of our sovereign lord the King, his crown and dignity, and against the laws of God, his majesty, and this jurisdiction.

Saddler denied the charge and claimed his right to be tried by a jury of his equals. The case went to trial and the jury of twelve men found him guilty "of vile, abominable, and presumptuous attempts to buggery with a mare." They did not find him guilty of the final act, only of attempted buggery, and so the verdict was not death, but that he be severely whipped at the post, and as the strongest reminder to him and to the public in general, he was to sit on the gallows with a rope about his neck and also be branded on the forehead (the most conspicuous place possible) with a Roman P "to signify his abominable pollution," then banished from the colony. In a letter to Richard

Bellingham, governor of Massachusetts Bay Colony, written on May 17, 1642, William Bradford makes it plain that he and his fellow magistrates distinguished between "high attempts and near approaches" to "gross and foul sins," just as they did cases of attempted willful murder when the victim did not die even though "a man did smite or wound another with a full purpose or desire to kill him (which is murder in a high degree before God), yet if he did not die, the magistrate was not to take away the other's life." In the same way, the magistrates did not feel that sodomy or bestiality should be punished by death if there was no penetration. Bradford also made it clear that not all the magistrates agreed that adultery should be punishable by death, which would explain why in Plymouth Colony no one was ever sentenced to death for this crime.

Sodomy, or Homosexuality

There are few cases of sodomy in the court records. Edward Michell, who propositioned Lydia Hatch, "attempting to abuse her body by uncleanness," was accused at the same court in March 1642 "for his lewd & sodomitical practices tending to sodomy with Edward Preston, and other lewd carriages with Lydia Hatch, is censured to be presently whipped at Plymouth, at the public place, and once more at Barnstable. . . ." Edward Preston, in his turn, "for his lewd practices tending to sodomy with Edward Michell, and pressing John Keene thereunto (if he would have yielded) . . ." was given the same sentence. John Keene was required to watch the sentences being carried out, as it was through him that their actions had been discovered, and he was suspected of not being completely without fault himself. The death penalty was not given in this instance, as it appears that the actions of the men only "tended towards sodomy."

Throughout its history the governing officials in Plymouth were slow to implement the death penalty, not only for the reasons described above by Bradford, but due to the fact that they did not hold a royal charter and so had to be very sure as to their authority to exact it. Perhaps the greatest single reform in early colonial America was the decision to abolish the death penalty for crimes against property, introducing more humane (by seventeenth-century standards) punishments such as branding, whipping, being set in the stocks, multiple restitution, and fines.

Where changes took place in the colonial legal systems the move was generally in a more rational, lenient direction. It appears that life was valued more highly in the colonies than in England. In New England this was largely because of the belief that people were created in the image of God and had worth in themselves. For Puritan and Separatist alike, the death penalty could only be implemented when sanctioned by God. In the southern colonies, Virginia in particular, the reason was largely economic. Labor was scarce and valuable. Hang someone for theft and they would no longer be part of a productive labor force.

A second case is an instance of behavior that was clearly homosexual but was not directly referred to as sodomy. It took place in 1637, five years earlier than the Michell, Preston, and Keene case, but after the law had been codified, clearly stating that sodomy was a capital crime. It involved John Alexander and Thomas Roberts, and the text of the record reads as follows:

> John Alexander & Thomas Roberts were both examined and found guilty of lewd behavior and unclean carriage one with another, by often spending their seed one upon another, which was proved both by witnesses & their own confession; the said Alexander found to have been formerly notoriously guilty that way, and seeking to allure others thereunto. The said John Alexander was therefore censured by the Court to be severely whipped, and burnt in the shoulder with a hot iron, and to be perpetually banished [from] the government of New Plymouth, and if he be at any time found within the same, to be whipped out again by the appointment of the next justice, &c, and so as oft as he shall be found within this government. Which penalty was accordingly inflicted.
>
> Thomas Roberts was censured to be severely whipped, and to return to his master, Mr. Atwood, and serve out his time with him, but to be disabled hereby to enjoy any lands within this government.

John Alexander's punishment for sodomy was severe, but stopped short of death, despite the fact that he was a known homosexual. He probably owed his life to Plymouth Colony's lack of a royal charter. Had the same incident taken place in Puritan Massachusetts Bay Colony to the north, it is probable that the death penalty would have been implemented against Alexander, for only four years later, in 1641, a servant between eighteen and twenty years of age was indicted for sodomy and hanged.

Rape

There are only two rape cases in the Plymouth Colony court records. The first is that of Ambrose Fish, indicted in 1677 for the fact that he "by force carnally did know and ravish Lydia Fish, the daughter of Mr. Nathaniel Fish, of Sandwich . . . against her will, she being then in the peace of God and of the King." Lydia Fish may have been the sister of Ambrose Fish, as suggested by Eugene Stratton, or his niece. Plymouth law required that there be at least two witnesses or "concurring" circumstances before there could be a conviction. As there were not the two witnesses required by law, and it was just Lydia's word against his, Fish was only whipped at the post. A strange twist to the story is that despite this incident and his conviction some years afterward in 1685 for trespass, abuse, and vile language toward Thomas Tupper, Ambrose Fish was appointed constable of Sandwich only months later, reflecting a pattern, somewhat tenuous but real, of a shift in the character of town constables in Plymouth Colony from upstanding citizens to less promising ones who showed an increasing disregard for the law in their personal lives. In a survey of constables in Duxbury between 1642 and 1665, for example, five of the six constables toward the end of the survey had a criminal record of some kind.

The second rape case raises some interesting issues. It involved the alleged rape of an English girl (her age is not given, but as she is described as a girl, she must've been young and single), Sarah Freeman, by an Indian known as Sam. The trial took place in October 1682, and due to the severity of the crime and involvement of an Indian, the jury of twelve Englishmen was expanded to include four Indian men—Keencomsett, Lawrance, Captain Daniel, and Concoquitt. Sam was found guilty of rape by his own confession, and found guilty by the jury of

> . . . wickedly abusing the body of Sarah Freeman by laying her down upon her back and entering her body with his, although in an ordinary consideration he deserved death, yet considering he was but an Indian, and therefore of an incapacity to know the horribleness of the wickedness of this abominable act, with other circumstances

considered, he was sentenced by the Court to be severely whipped at the post and sent out of the country.

What strikes one immediately is that this case took place after King Philip's War, which ended in 1676. Following that conflict, English attitudes to Indians had hardened. The Wampanoags, the Narragansetts, and Nipmucks had been almost totally subjugated (but due to disease and starvation more than through fighting), and hundreds of Indians had been enslaved, although fighting continued between the Abenakis in Maine and the settlers there until well into 1677. In 1682 Philip's severed head was still on display in Plymouth as it had been since that Thanksgiving day in August 1676 when Benjamin Church rode into town with it, remaining as a grim reminder of the tensions and hostilities between the native peoples and the English. So why was special consideration given to Sam the Indian? Eugene Stratton sees this case as evidence of differential treatment of Native Americans and Europeans by the colonists. Although this did appear to be a pattern earlier in the life of the colony, the later court records show that Native Americans were often treated as equals with the English before the law. Had Sam been English, would the punishment have been the same? To what extent was Sarah involved in the whole incident? What were the "other circumstances" considered by the jury? Colonial historians agree that rape was not a cultural practice of Native Americans. Laurel Thatcher Ulrich comments that Puritan writers were amazed at the sexual restraint of Indian men, who never raped their captives, although their assaults on nursing or pregnant women were cruel in the extreme. How was it that Sam "confessed" to a crime that should have been unknown to him? There appear to have been no witnesses other than Sarah to speak to the truth of the occurrence, and witnesses would have weighed strongly with the jury. Is it not possible that Sam was not guilty so much of rape as fornication?

Adultery

Adultery was defined by the married status of the woman. A married man who had sex with an unmarried woman risked whipping or a fine for fornication, but a man, be he married or unmarried, who had sex with a married woman risked death, as did the woman. No one was ever executed for adultery in Plymouth Colony, but it did happen in three instances in the Massachusetts Bay Colony. Historian Keith Thomas has shown that this double standard that discriminated against women had a medieval origin in which "the absolute property of the woman's chastity was vested not in the woman herself, but in her parents or her husband," and so adultery was violation of the property rights of the husband, just as fornication would be violation of the rights of the parents of an unmarried woman who was a valuable marriage commodity.

In the definition of adultery stated above, it appears that the core of the crime was the infidelity of a woman toward her husband, as adultery was sex with a married woman, whether the man involved was married or single. Sex between a man and a single woman, even if the man was married, was not considered to be adultery but fornication. So the double standard that was part of the cultural baggage brought to the New World continued in Plymouth, and although there are fewer than a dozen court cases in the records that refer to adultery, they probably cannot be regarded as a true reflection of the degree of marital infidelity that was taking place in the colony—it was likely to have been considerably more frequent than the small number of cases implies.

The sanctions against adultery were severe. In the first codification of the law in 1636, adultery fell under capital crimes liable to the death penalty, although the wording appears to be deliberately open to wide interpretation, stating simply, "Adultery to be punished." By 1658 the law had been amended to decree that

> . . . whosoever shall commit Adultery shall be severely punished by Whipping two several times; viz: once whiles the Court is in being at which they are convicted of the fact and the second time as the Court shall order, and likewise to wear two Capital letters viz: AD cut out in cloth and sewed on their uppermost Garments on their arm or back; and if at any time they shall be taken without the said letters whiles they are in the Government so worn to be forth with taken and publicly whipped.

Americans are familiar with Nathaniel Hawthorne's *The Scarlet Letter*, the enduring classic account of Hester Prynne's adulterous affair with the local minister, set in seventeenth-century Salem. To those living in America in the twenty-first century, the concept of having to wear "the scarlet letter" prominently every day of one's life in public and private is completely contrary to cultural norms, but adultery was seen by Separatists and Puritans alike as striking at the heart of the nuclear family and breaking down the whole structure of society, and it therefore required the severest sanctions.

Divorce was not a normal option, nor was living together condoned as in today's society. Few divorces were granted in Plymouth Colony, as there were not many reasons considered adequate to end a marriage. Adultery was one of these, and there were three others: bigamy, willful desertion, and impotence. Adultery was a difficult crime to prove, and as it always involved a married woman, and contraceptives were not available, if a woman became pregnant it might not have been easy to prove that the child was illegitimate.

The first case of adultery to come before the court was in September 1639. It involved Mary Mendame, wife of Robert Mendame of Duxbury, and Tinsin, an Indian. Mary was given the very severe sentence of being "whipped at a cart's tail" through the streets of the town, and had to wear an "AD" (abbreviation for adultery) badge on her sleeve for as long as she lived in the colony, and if she was found without it she would be branded in the face. Tinsin was given a lighter sentence, he was "well whipped with a halter about his neck at the post," the reason being that his crime came about "through the allurement and enticement" of the woman.

There are three instances of husbands appearing before the court to petition for a divorce on grounds of adultery. In 1670, Samuel Halloway of Taunton came before the court on several occasions to ask for a divorce from his wife, Jane. She had been sent to Plymouth by the Taunton authorities and jailed there awaiting trial in December the previous year. Her behavior to her husband had been "so turbulent and wild, both in words and actions, as he could not live with her but in danger of his life and limbs, and also her behavior before the Court was so audacious as was intolerable." She was committed to jail, but after one night in "close durance" was so repentant that she was permitted to return to Taunton with her husband "in hopes of better practices for the future." Three months later she was in court again, for "other horrible and abusive speeches and actions" toward her husband and others, and this time she accused Jonathan Briggs of having committed adultery with her. The grand jury investigated her

"Whipped at the cart's tail."

accusation but could find no evidence against Briggs. Nevertheless, Samuel requested a divorce on the grounds of both her abuse of him and her confessed adultery. The case came up twice more, but there is no record of a divorce being granted in the end.

On July 4, 1673, John Williams of Barnstable was granted a divorce from his wife, Sarah, as she had "violated her marriage bond by committing actual adultery with another man, and hath a child by him." In June 1686, John Glover, also of Barnstable, petitioned the court for a divorce from his wife, Mary,

> By reason of her false and treacherous dealing in her violating the marriage covenant by entertaining some other man or men into bed fellowship with her, and did by her filthiness and baseness infect him, her husband, with that filthy and noisome disease called the pox [usually syphilis], to his great sorrow and pain, ruin of his estate, and hazard of his life.

The court granted the petition, but in both cases the wife was not punished for her actions, the divorce itself presumably being considered sufficient, as there was no acceptable place in society for a divorced woman.

There is one case where a woman appeared before the court requesting a divorce due to the adulterous behavior of her husband. In 1661 Elizabeth Burgis petitioned for and was granted a divorce after the court had sentenced her husband, Thomas Burgis Jr. of Sandwich, to two severe whippings for his "act of uncleanness" with Lydia Gaunt but suspended his having to wear "AD" dependent on his future behavior. There is the implication that there could have been additional reasons for the divorce as Elizabeth's position was taken very seriously and there was no hesitation, as in other cases, about granting the divorce immediately.

Incest

There are two definite cases of incest in the court records for Plymouth Colony. The colony did not pass a specific law against incest, but when the case of Christopher Winter and his daughter Martha Hewitt came before the court in March 1669, the court record declared incest to be "a very heinous and capital crime." As neither case brought a full guilty verdict, it is impossible to say how this crime would have been punished. There was a third case of probable incest and rape, referred to above under the discussion on rape, when in 1677 Ambrose Fish was indicted by the court because "in his own house in Sandwich, in this colony of New Plymouth, [he] by force carnally [did] know and ravish Lydia Fish, daughter of Mr. Nathaniel Fish . . . against her will." The court appointed a jury of life and death to deal with the case, but as there were no witnesses other than Lydia herself, Ambrose Fish did not receive the death sentence but was publicly whipped at the post. The first case of incest involving fornication to come before the court was in 1660, when Thomas Atkins, who lived at the Plymouth trading post on the Kennebec River, was accused by his daughter, Mary, of committing incest with her several times. Atkins was held in prison in Plymouth for two months before the case came to trial before the grand jury. It appeared upon examination that he had not actually committed the act of fornication with his daughter, but had, while drunk on several occasions, tried to do so. He was acquitted but whipped for his behavior. Nine years later, in 1669, Christopher Winter and his daughter, Martha, came before the court to answer charges of incest raised when Martha refused to give the identity of her child's father and it became obvious that her husband was not involved. The examination by the jury of the parties concerned in the case was extremely thorough. Winter had been responsible for the hasty marriage, but despite the suspicions of Martha's husband, John Hewitt, and although there was circumstantial evidence—Winter was very lenient toward Martha after she conceived, even with no husband, whereas an innocent father would make his daughter confess who the father was and force marriage; Winter also had the reputation of being austere with his children, and so the fact that he was being indulgent to Martha was suspicious—Winter was not found guilty. Martha, on the other hand, as she had had a bastard child and refused to name the father, was convicted of whoredom (illicit sexual relations in general) and sentenced to be whipped at the post, which was done. Her husband petitioned the court for a divorce, but the court dismissed his application for want of clear evidence. There appears to have been a pattern of sexual laxity in the family, for thirty years earlier Winter and his wife, Jane Cooper, were found guilty of fornication before marriage. He was "censured to be whipped at the post at the Governor's discretion; and the said Jane, his wife, to be whipped at a cart's tail with . . . Mary Mendame," who had been convicted of adultery with the Indian Tinsin. Whipping at a cart's tail as it was driven through the town of Plymouth was a much more severe punishment than simply being whipped at the post, harsh though that sentence was, and the court evidently considered Jane Winter to be more at fault than her husband, although no reasons are given in the record.

Sex-Related Non-Capital Crimes

Fornication

Fornication was a criminal offense, and a distinction was made in criminal law in the colony between "fornication and other unclean carriages to be punished at the discretion of the Magistrates according to the nature thereof," and "fornication before contract or marriage." At first glance it appears that listing fornication twice is redundant, but a closer reading of the wording shows that in the colony, fornication *before* a contract or engagement or before marriage as well as *after* entering an engagement was considered a crime. As stated above, fornication was sex before marriage insofar as a woman was concerned. The man involved could be married or single, but adultery was viewed as the breaking of the marriage bond by the fact that the woman was married. The husband was not bound by the same constraints.

There were at least two reasons for the severity with which fornication was viewed. Separatists and Puritans alike regarded it as breaking a specific commandment, although considered a crime of a lesser degree than adultery that broke the Mosaic law, as set out in Leviticus 20:10. A second argument for having and enforcing strict laws against fornication was the need to limit the number of illegitimate children who could become a charge on society. Often a pregnant girl would refuse to reveal who the father of her child was, and considerable pressure would be put on her to do so, particularly by midwives when she went into labor, as the court wished not only to punish the offender but to hold him financially responsible for the upbringing of the child.

Punishment for fornication, or "carnal copulation," was set out in the 1646 code of laws, although clearly in operation before that date, and was different for those who committed the crime before or without a lawful marriage contract than for those who had already entered into one or were in fact already married—the length of time a pregnancy lasted after marriage was closely monitored. In the former case, punishment was by whipping or a fine to be paid by each person involved of £10 plus up to three days in prison. Time in prison in early years of the colony meant being locked into a cage in which the prisoner was exposed to public view and abuse. Where there was a contract to marry, or a marriage had already taken place, whipping was generally omitted, but the fine (reduced to £5 each) and imprisonment stood. If, however, the fine could not be paid, or if the couple refused to pay it, they would be publicly whipped. This law remained unchanged in subsequent codifications, but it does appear that in practice in later years imprisonment fell away.

Fornication was by far the most common sexual offense to come before the Plymouth courts. Between 1633 and 1691, sixty-nine cases of fornication were presented. These included "carnal copulation," "unclean" behavior, and the birth of illegitimate children where fornication was involved. What is remarkable is that although the law distinguished between sexual acts committed before and after a marriage contract had been made, only 6 percent of cases were during contract. The highest percentage was 48 percent, those who never married, closely followed by 46 percent, those who had sex before contract but eventually married. There was, of course, in all probability, a much higher incidence of fornication, as there would have been of adultery, that never came to the attention of the court.

In 1646 the general court ordered each town to maintain registers of marriages, births, and deaths. Assuming that an act of fornication would be discovered within a year of the marriage, an examination of the registers published in the Plymouth Colony Records for the years 1645 though 1686 shows that in this twenty-one year period 11 percent of marriages involved premarital sex. The registers are not complete, and so this figure is approximate, but it is sufficient to confirm that despite the fact that fornication was regarded as a criminal offense, it was not a sufficient deterrent to curb the sexual pleasures of many of the young women and men in the colony.

In a number of cases of fornication it is of interest to see that while the law held both man and woman equal insofar as punishment was concerned, prior to 1646 women did not always receive a whipping, but had to stand by or sit in the stocks while the man did. After that date sentences appear to have been more equal. The court did, however, take individual circumstances into account, as in the case of Jane Powell, mentioned above, who admitted to fornication with another servant but was not punished, as she pleaded that she had allowed herself to be taken advantage of due to her desperation to get married and so improve her circumstances.

Typical entries in the court records for cases of fornication read:

> *March, 6, 1648.* We present Peregrine White, and Sara his wife, both of Marshfield, for fornication before marriage or contract. Cleared by paying the fine.
> *March 2, 1652.* We further present Katheren Winter of Scituate, for committing the sin of fornication with her father in law [step-father?], James Turner.
> *March 2, 1652.* We further present Thomas Launders of the town of Sandwich, for having a child born within thirty weeks after marriage. Not appearing, fined according to order.

> *March 2, 1652.* We further present Nicolas Davis, of the town of Barnstable, for having a child five weeks and four days before the ordinary time of women after marriage.
> *June 3, 1652.* David Linnet and Hannah Shelley, for unclean practices with each other, are sentenced by the Court to be both publicly whipped at Barnstable, where they live.
> *March 3, 1663.* Nathaniel Church and Elizabeth Soule for committing fornication with each other, were fined, according to the law, each of them, £5.
> *June 1, 1663.* Nathaniel Fitzrandal, for committing fornication, fined ten pounds; he has liberty until the next October Court to pay the fine, or suffer corporal punishment.
> *October 5, 1663.* William Norkett, for committing fornication with his now wife, fined five pounds.

When a bastard child was involved and paternity could be proved, the court would order the father to provide regular support for the upkeep of the child, usually until it was seven years old, "if it live so long." There are occasional instances of cross-cultural unions. In one case, in 1685, Hannah Bonny was convicted of fornication with John Michell and also "with Nimrod, [a] negro, and having a bastard child by the said Nimrod, is sentenced to be well whipped." In his turn, Nimrod was also sentenced to be "severely whipped," and required to pay eighteen pence per week to Bonny toward the maintenance of the child for a year, "if it live so long," and if he, "or his master on his behalf neglect to pay the same," was to be found work by the deputy governor so that he could raise the money, a demonstration that the court was prepared to go to considerable trouble to see that illegitimate children did not become a charge on society.

Attempts and Propositions

There are at least fifteen cases of attempts and propositions in connection with fornication and adultery in the Plymouth court records. In all instances only one person was prosecuted, as had the attempts been successful, they would have resulted in cases of fornication, adultery, or rape. The only written law governing such attempted acts is that of 1636 where the magistrates were given liberty to determine punishment for "unclean carriages" according to the nature of the

"Presented for lascivious carriages . . ."

offense. Punishment was usually a fine or whipping. A typical case to come before the court was that of John Peck of Rehoboth, who was presented "for lascivious carriages and unchaste in attempting the chastity of his father's maid servant, to satisfy his fleshly, beastly lust, and that many times for some years space, without any intent to marry her, but was always resisted by the maid, he confesses." He was fined fifty shillings, and one wonders if that was really a deterrent after years of harassment. A slightly more unusual case was that of Nathaniel Hall of Yarmouth, who was convicted of "giving writings to . . . Elizabeth Berry to entice her, although he had a wife of his own." He was sentenced to pay a fine of £5 or to be publicly whipped. There is no indication in the court records as to whether Elizabeth Berry was married or single, but seven of fifteen cases of attempted acts and propositions were definite attempts at adultery, as in each case a male colonist attempted the chastity of another man's wife. When conviction occurred, the usual sentence of fine or whipping was given. In October 1668, Samuel Worden laid a charge against Edward Crowell and James Maker "for going in his absence into his house in the dead time of the night, and for threatening to break up the door and come in at the window, if not let in, and going to his bed and attempting the chastity of his wife and sister, by many lascivious carriages, and affrighting of his children." The men had to give sureties for their good behavior (£40 each), were each required to pay a £10 fine and the costs of the case, or be publicly whipped. On their "humble petition," the court reduced their fines to £6.

The law put the onus on a woman to cry out for help if she was being propositioned against her will. In March 1656 John Gorum was fined forty shillings "for unseemly carriage towards Blanche Hull at an unseasonable time, being in the night." Blanche Hull was fined *fifty* shillings "for not crying out when she was assaulted by John Gorum in unseemly carriage towards her." Both Gorum and Hull were married at this time, Gorum to Desire Howland, eldest daughter of John Howland and his wife, Elizabeth Tilley Howland, and Blanche to Tristram Hull. The court evidently considered Hull to have been a willing participant rather than a victim.

Attempts and propositions most commonly concerned men approaching women, but there is one instance where it appears that the woman was the instigator in what appears to be a case of incest. In 1642 Lydia Hatch appeared before the court not only "for suffering Edward Michell to attempt to abuse her body by uncleanness" and not letting it be known, but for "lying in the same bed with her brother Jonathan." Her brother was not directly accused of incest, although he was in court on other charges. Lydia was publicly whipped for both offenses, no option of a fine being given, and Jonathan was also whipped for vagrancy and "for his misdemeanors."

Lascivious and Suspicious Conduct

In a small community it was probably very difficult to conceal suspicious conduct between men and women, and several cases of lascivious and suspicious conduct emerge in the court records. The actual act of fornication or adultery could not be proved in these instances, although it was suspected, the sexual intentions between a man and woman being the immediate issue. The law did not define lascivious "carriages," but it appeared to cover everything from improper flirtations to a flagrant display of wanton sexual behavior. In June 1655, Hugh and Mary Cole of Plymouth were fined twenty shillings for "keeping company with each other in an indecent manner, at an unseasonable time and place, before marriage." The odd thing about this case is that the couple were married on January 8, 1654, eighteen months earlier, and their first child appears to have been born in November

1655. An early miscarriage could have taken place, but it does draw attention to the lack of privacy, the gossip and rumor that must have swirled and eddied around Plymouth for months before they were summoned by the court to account for their behavior. This was a society where everyone's conduct was the business of their neighbors.

A more unusual feature of the responsibilities of members of the grand jury that runs counter to modern concepts of liberty of conscience and the freedom of the individual, was the requirement that while in office they had a duty of surveillance of their neighbors. They had to monitor the behavior of everyone in their respective towns, married or living on their own, to prevent "idleness and other evils occasioned thereby . . . and to require an account of them how they live." In the event of someone not being able to give good reason for their behavior, a grand juror could make charges against them and take them before the local constable, who would then arrest the accused and present the case for adjudication to the governor and assistants at the next session of the court. The court was limited in what it could do to prevent immoral conduct from actually taking place, other than impose fines or require the parties concerned to avoid the company of each other, as they did in the case of Marcy Tubbs and Joseph Rogers, as well as Jonathan Hatch and Frances Crippen.

In March 1666 two cases involving Frances Crippen came before the court. First John Robinson was convicted of lascivious speeches and actions toward her and was fined, and then Jonathan Hatch was convicted of "unnecessary frequenting of the house of Thomas Crippen, and thereby giving occasion of suspicion of dishonest behavior towards Frances, the wife of the said Crippen." The Court admonished him and warned him in the future not to give such occasion of suspicion, "by his so frequently resorting to the said house or by coming in the company of the said woman, as he will answer it at his peril." Thomas Crippen may well have been the instigator of the whole affair, as he was convicted of lascivious speeches "tending to the upholding of and being as a pander [go-between in sexual affairs] of his wife in lightness and lasciviousness," and was required to give bonds for his good behavior to the value of £40. Until the mid–nineteenth century men did not recognize that a woman had rights to her body—she was the property first of her parents and then of her husband—and in seventeenth-century England there were cases of men "selling" their wives' sexual favors, and even selling the wife herself. The case against Thomas Crippen is evidence that the husband in the role of pander was certainly not unknown in Plymouth Colony. No mention of any admonishment or punishment of Frances Crippen is given, which is silent testimony to the reality of her situation. The court did not recognize that she had any rights that had been violated by Crippen's actions. Despite the greater freedoms that women were obtaining in the colony, the husband's right of property in his wife remained.

Alongside the common accusation of lascivious or unclean carriage, there is a case that appears to border on lesbianism. On October 2, 1650, the wife of Hugh Norman was presented "for misdemeanor and lewd behavior with Mary Hammon upon a bed, with divers lascivious speeches. . . ." She was sentenced for her "wild behavior," forced to make a public acknowledgment of her unchaste actions, and was warned "to take heed of such carriages for the future, lest her former carriage come in remembrance against her to make her punishment the greater." Mary Hammon was cleared and discharged with an admonition by the court.

Domestic Violence

While there certainly were a number of families where relationships, however strained they may have been at times, did not include violence as part of their way of solving problems, the court records show that domestic violence did occur more frequently than the general images of family life in the small colony convey today. Violence is any physical assault upon a person or property, and domestic violence as discussed here refers to any violent action occurring between persons within a household. There were different types of domestic violence, not very different from those found in modern America—a child cruelly treated, parents attacked by their son, a father attempting the rape of his own daughter, a husband using his wife harshly, abusing her, a turbulent wife attacking her husband physically, a servant forced to carry logs beyond his strength and beaten because he could not do so.

In most of the cases discussed, the type of violence is essentially authoritarian, defined by Laurel Thatcher Ulrich as that used by a superior to enforce obedience by an inferior. It lay at the heart of the system of justice brought from England and was modified in the American colonies. The state had the power to kill, whip, and brand,

A husband cuckolded (as shown by his horns) and beaten by his wife with a skimming ladle.

all violent physical punishments, but parents and employers also had the right and power to administer physical correction. The position of husbands in regard to any physical correction of their wives was less clearly defined, and wife-beating was not legal, but the court was very reluctant to grant divorce in an abusive situation. In general it appears that in New England, if there was any evidence that there was undue provocation by the wife, an abusive husband was treated leniently. When a case involving violence came before the court, the question would not be concerned with the right of a superior to use force, but rather: What degree of provocation was there?

The law in Plymouth Colony concerning the violent crimes of willful murder and rape, both capital crimes, although the latter was not ever punished in this extreme way, has already been mentioned. The other laws covering the punishment of violence are set out in three short passages in the 1636 and 1658 codes of law. Those in the 1636 code read as follows:

> That concerning misdemeanors as any shall be convicted in Court of any particular to be Censured by the bench according to the nature of the offence as God shall direct them.

> That all such misdemeanors of any person or persons as tend to the hurt and detriment of society Civility peace and neighborhood to be enquired into by the grand Inquest [Jury] and the persons presented to the Court that so the disturbers thereof may be punished and the peace and welfare of the subject comfortably preserved.

The first law gave the court considerable power to exercise control over behavior in the colony, and the second gave the grand jury the right to investigate any case in which harm was done to any person, so that justice could be served and the peace preserved. The 1658 code appears to have revised the law, establishing that in *all* matters of misdemeanor the magistrates had full power to act without a presentment by the grand jury, and to determine the amount of the fine or other punishment as the nature of the offense may have required "as if presented." As the colony had not as yet built a house of correction, the court further established that any two magistrates, having examined the case of any such as "live Idly and unprofitably but are otherwise vicious and wicked in their carriage towards their parents or otherwise," should have the power to punish the offender by "stocking [placing in the stocks] or whipping according to the nature of the offence."

Abuse of Servants

The worst example of the abuse of servants in Plymouth Colony has already been discussed in the case of John Walker, whose death was brought about by the treatment he received from Robert Latham and his wife, Susanna Winslow Latham. There are six cases of servant abuse that did not end in death in the court records, and the pattern that emerges is that the master was not punished by the court; the abused servant was instead transferred to another master, who bought out the remaining time. A good example of this can be seen in a case brought before the general court in March 1640, where Roger Glass had been abused by his master, John Crocker:

> Forasmuch as John Crocker, of Scituate, is proved to have corrected his servant boy, Roger Glass, in a most extreme & barbarous manner, the Court upon due consideration has taken the said Roger Glass from the said John Crocker, and placed him with John Whetcombe of Scituate, to serve out his time with the said John Whetcombe, which is six years

> from the fourteenth of June next; the said John Whetcombe paying the said John Crocker three pounds, deducting five shillings for his charges, & the said Crocker to deliver up his clothes to the said Whetcombe.

Crocker, despite the "extreme & barbarous manner" in which he corrected his servant, was not punished by the court, and he was, in effect, reimbursed for the loss of his servant. In a similar case, when Robert Ransom appealed to the court, accusing his master, Thomas Dexter Jr. of Sandwich, of using him hardly and unreasonably, even though the charge was not proved, Ransom was removed from Dexter's service and his remaining time was bought out by Thomas Clark of Plymouth. Likewise, when John Hall of Yarmouth complained that his son, Samuel, had been abused by his master, Francis Baker, "by kicking him and otherwise unreasonably striking him," the court's response was to have Hall pay Baker out for the remainder of his son's time. There is no record of Baker having been fined or warned in any way not to repeat the offense. It is quite possible that in all these cases the court considered that there had been provocation from the servant as well as an excessive response from the employer, and that the solution was to remove the servant but to uphold the prerogative of masters to correct their servants by not punishing the former.

The case of Joseph Gray, servant to Mr. and Mrs. Thomas Gilbert of Taunton, is of particular interest, as its solution involved the town officials. Thomas Gilbert left the colony on a visit to England, and in 1653 his wife appeared before the court to ask that "her servant, who has received some hurt, and is now in Mr. Street's family" remain with the Streets until her husband returned from England. The court allowed her servant, Joseph Gray, to remain with the Streets as long as they could keep him. If Mr. Street refused to continue to do so, the constable of Taunton was obligated to find Gray another place to live until his mistress was able to accommodate him. Gray did return in time to the Gilbert home, because in June 1657 he complained to the court that he was ill-used and lacked adequate clothing. The court ordered the selectmen of Taunton to take note of the boy's condition and to provide him with shoes and stockings before the winter. The court made this a matter of urgency as the boy's foot was "in danger of perishing." In addition, the selectmen were required to remind Gray to give his mistress "due respect and obedience." The case was not over, however, as the following March saw Mrs. Gilbert brought before the court on charges of neglect. She had not yet provided Gray with the proper clothing and accommodation. The court records state that Gray was "sometimes since frozen on his feet, and still is lame thereof." The town of Taunton was required to see that he was cured, even if it meant taking him to the Bay Colony, as Thomas Gilbert was still in England and Mrs. Gilbert was not in a financial position to pay for the costs of a cure. This case demonstrates that not only was Mrs. Gilbert not admonished for her neglect of her servant, preserving the balance of master-servant relations, but that the town was expected to assume corporate responsibility for the welfare of an abused servant. From its establishment in 1620 the colony had actively taken on its obligations to the poor, and clearly by mid-century the towns were being held responsible for those in need, even though employers of servants were legally responsible for the welfare of those indentured to them.

Child Abuse

There are not many reports of child abuse in the Plymouth Colony court records, but that does not necessarily mean that there were only a few times when this occurred. It is quite possible that a distinction was not made between harsh discipline and abuse, and so much that would today be considered abuse was not reported. When it did come to the attention of the authorities, it tended to go unpunished, for much the same reasons as servant abuse—the need to hold in proper balance the authority of parents to correct their children. The concept of family in Plymouth was that of a strong nuclear unit under the authority of the husband as head of the household. It was his duty to exercise that authority, punishing when necessary his children and servants, as well as correcting his wife. Discipline really had to have gone well beyond what would be considered today as very harsh, but what in the seventeenth century was a norm, before the matter was brought to the attention of the authorities. A case in point came before the court in March 1669:

> At this Court, Mary, the wife of Jonathan Morey, and her son, Benjamin Foster, appeared, being summoned to answer a complaint against the said Mary, for that she, by her cruel, unnatural, and extreme passionate carriages so exasperated her said son as that he

> oftentimes carried himself very much unbeseeming him and unworthily towards his said mother, both by words and otherwise; yea, so was her turbulent carriages towards him, as that several of the neighbors feared murder would be the issue of it; she, the said Mary, being examined respecting the premises, and owned her fault, and seemed to be very sorry for it, and promised reformation; the youth, her son, likewise owned with tears his evil behavior towards his mother, which gave the Court such satisfaction as they passed his fault by with admonition; and in reference to the said Mary Morey, the Court, upon her engagement of better walking, are willing to take further trial of her, and therefore condescended to let her son remain with her until the next June Court, and then further to do in the case as occasion shall require.

There is no further mention of Mary Morey being brought before the court, so presumably her neighbors were satisfied that she really had altered her behavior toward her son, and vice versa. In this instance it appears that Morey's husband did not involve himself in any meaningful way with the manner in which she treated the son of a former marriage. The case also provides an excellent example of the discretion that the court exercised in deciding whether or not admonition was adequate, or whether the removal of the child was a necessity to prevent murder.

Murder *was* the issue, however, in the case of Alice Bishop. On July 22, 1648, twenty-three-year-old Rachel Ramsden visited the house of Richard and Alice Bishop on an errand. While there, she saw Alice's four-year-old daughter, Martha Clark, lying asleep on a bed. After she returned with some buttermilk she had fetched for Alice Bishop at Goodwife Winslow's, she found Alice, who had earlier seemed quite as usual, "sad and dumpish." Rachel saw blood at the foot of the ladder that led to an upper chamber, and asked what blood it was she saw. Alice pointed to the chamber and told her to take a look, but Rachel, seeing the empty bed, realizing Alice had killed her child, was afraid, and ran for help. The murder was gruesome. According to the jury's report, the child had had her throat cut, "with divers gashes cross ways, the wind pipe cut and stuck into the throat downwards, and a bloody knife lying by the side of the child, with which knife . . . the said Alice has confessed to five of us at one time, that she murdered the child."

Alice Bishop was found guilty of willful murder and was hanged. As in the case of Mary Morey, the child, Martha Clark, was Alice Bishop's daughter from a previous marriage, and although two cases do not make a pattern, it does make one realize that while children in second-marriage situations were generally integrated into a stable new family unit, as was certainly the case among remarriages in the early years of the colony, there were also instances where there were considerable stresses between spouses and stepchildren that ended in violence.

Abuse of Parents

An incident that reinforces the suggestion that there were deep tensions caused by second marriages is the abuse of George Barlow, who was marshall of Sandwich, Barnstable, and Yarmouth, by his stepdaughters Anna, Dorcas, and Mary Bessey. Barlow is referred to as their "father-in-law," but this term was at times used differently in the sixteenth and seventeenth centuries to mean "stepfather." It is clear from the context that Barlow was the stepfather of the three women. They were required to appear before the court to answer for their "unnatural and cruel carriages" toward George Barlow, and the case was heard on March 4, 1662:

> Anna Bessey, for her cruel and unnatural practice towards her father in law, George Barlow, in chopping of him in the back, notwithstanding the odiousness of her fact, the Court considering of some circumstances, viz. her ingenious confession, together with her present condition, being with child, and some other particulars, have sentenced her to pay a fine of ten pounds, or to be publicly whipped at some other convenient time when her condition shall admit thereof.

Dorcas and Mary Bessey, "for carriages of like nature towards their said father in law, though not in so high a degree," were sentenced to sit in the stocks. George Barlow and his wife were both severely reproved for their "most ungodly living in contention one with the other, and admonished to live otherwise." What appear to be strange, violent actions if Barlow was indeed their father-in-law make much more sense when one realizes that the women were caught up in tensions between their mother and stepfather. It is more than possible that while their actions could be construed as those of women taking control of their own circumstances in what is often perceived as a male-dominated society, what we are actually seeing is a glimpse of what Ulrich describes as "a violent underside in the village culture of New England." In the seventeenth century the level of physical and verbal violence was high, and a visceral, violent response to problems is the most likely explanation for this case and the two previous ones.

There is a second instance of abuse of parents in the court records that gives an insight into how seriously this offense was regarded. The 1679 record is succinct:

> Edward Bumpas, for striking and abusing his parents, was whipped at the post; his punishment was alleviated in regard he was crazy brained, otherwise he had been put to death or otherwise sharply punished.

It may be that the Bessey sisters' lighter treatment was due to the fact that Barlow was not their natural father.

Spousal Abuse

In comparison to other types of domestic violence in Plymouth Colony, spousal abuse was a problem that occurred more frequently, and was probably also underreported. Although women in Plymouth Colony were given greater opportunities and freedoms than many of their peers in Old England, and some women were moving out from their traditional place to involve themselves in religious, social, and even political spheres (witness the religious dissenter Anne Hutchinson in the Massachusetts Bay Colony, and other women persecuted there for their beliefs), society was still under male control, and a certain degree of spousal "correction" was probably accepted as the norm. Out of ten instances found in the court records of violence toward a spouse that did not end in murder, eight charges were against men and two were against women. There could have been more instances of wife abuse of husbands, but generally an abused husband would be unlikely to wish to lose respect by this being known.

It is the nature of the offenses and the insight they can give us into relations between spouses that are of particular interest in the following cases of wife abuse. Punishment was at the discretion of the magistrates, and varied from fines to corporal punishment, or a combination of the two. One of the strangest cases in the court records is that of Ralph Earles, who was fined twenty shillings for "drawing his wife in an uncivil manner on the snow." Was he dragging her physically through the snow? Or drawing lewd cartoons of her? We will never know. In August 1665, John Dunham was brought to court "for his abusive carriage towards his wife in continual tyrannizing over her, and in particular for his late abusive and uncivil carriage in endeavoring to beat her in a debased manner, and for frightening her by drawing a sword and pretending therewith to offer violence to his life. . . ." Dunham was sentenced to be severely whipped, but "through the importunity of his wife" the execution of the sentence was suspended, depending on his future behavior, and he had to provide bail of £20 against his continued good behavior and appearance at the next court. There is no further mention of continued wife abuse by Dunham, so presumably this one public court appearance was sufficient to keep him in line. Some years later, Richard Marshall was sentenced to sit in the stocks "for abusing his wife by kicking her off from a stool into the fire."

It has been said that "a man's honor depended on the reliability of his spoken word; a woman's honor on her reputation for chastity." So publicly calling a woman a whore, particularly when she was your wife, was probably worse than physically abusing her. John Williams Jr. was guilty of "disorderly living with his wife . . . [of] abusive and harsh carriages towards her both in words and actions, in special his sequestration of himself from the marriage bed, and his accusation of her to be a whore, and that especially in reference unto a child lately born of his said wife by him denied to be legitimate." The case of Williams vs. Williams dragged on from June 1665 to July 1668, ultimately ending with Williams being required to provide Elizabeth, his wife, with an income of £10 a year, clothing, a bed, bedding and all such necessities, and also to secure one-third of his estate for her livelihood and comfort. He himself was sentenced to "stand in the street or market place by the [whipping] post with an inscription over him that may declare to the world his unworthy carriages towards his wife," although "at the earnest request of his wife, this part of the sentence was remitted and not executed." Elizabeth was a generous woman. Williams was also fined £20 for his "wicked carriages" and the expense to which he had put the government through the length of the trial. What does emerge from the Williams case is the lengths to which the court went to try to secure a reconcilement between husband and wife, and only when that failed due to Williams's recalcitrance did they then agree to a separation, not a divorce, and Elizabeth Williams's name was publicly cleared in the court. As we have seen, divorce was never lightly granted in Plymouth Colony, however abusive a spouse might've been, although early in the case Elizabeth was encouraged to move in with relatives and so be out of danger until the situation was resolved.

There are two cases of wives physically abusing their husbands in the court records. Both are revealing in their reflection of attitudes toward women who responded violently toward the spouse whom they perceived as the cause of their marriage problems. In March 1655, Joan Miller was brought to court for "beating and reviling her husband, and egging her children to help her, bidding them knock him in the head, and wishing his victuals might choke him." A marginal note states briefly, "Punished at home." In a society where the husband was considered to be the head of the wife, it is evident that the court regarded this as a case for the husband, Obadiah Miller of Taunton, to deal with himself, but under this type of punishment great abuse was possible. The prospect of the abused person being given the right to punish his abuser is troubling to the modern mind, but in a seventeenth-century marriage context the court was simply upholding the husband's rights. Given the court's ruling in other instances, it seems likely that, had it really considered that the woman's life was threatened, other measures would have been taken.

The case of Jane and Samuel Halloway that first came before the court in October 1669 is another instance of the court refusing to grant a divorce due to its high view of marriage, even though Samuel Halloway could not live with his wife "but in danger of his life or limbs," and she had "most horribly abused him," and, moreover, she admitted to having committed adultery with several people. The court instead advised Samuel to "take mature advice and deliberation about it, as behooves such weighty a matter." At a later court it was agreed that after Jane had given birth to the child she was carrying, if she still admitted to adulterous acts, Samuel could be divorced from her. The couple did not, however, return to court to pursue the matter and in fact continued to have children (three sons and a daughter) at least until 1674, according to the Taunton register of births contained in the court records.

Then there is the case in 1685 in which Betty, an Indian woman, killed her husband, Great Harry, by throwing a stone at him. Was this simply a case of spousal abuse gone wrong, was it willful murder, or simply an accident? Betty at first denied that she had killed Great Harry, then later owned to it but said that it had been an accident. She had not meant to kill him, but by throwing a stone at a bottle of liquor she missed the bottle and hit her husband on the side of the head, as a result of which he died. Betty was found guilty of homicide by misadventure and was set at liberty.

The world in seventeenth-century New England was in many respects very different from that of America in the twenty-first century, and the exact nature of life in Plymouth Colony will never be known, but there are certain commonalities. Violence is one, whether it occurs at home or abroad, as is death, whatever the circumstances.

Death

Between 1633 and 1685, the magistrates in Plymouth ordered the investigation of at least sixty-four deaths that took place in suspicious circumstances, excluding cases that were clearly homicides, but the colony was not alone in the number of deaths that it required to be investigated. All the colonial court records reflect a similar pattern. In Plymouth it was comparatively rare for an accidental death to be a homicide, but nevertheless sudden death required an inquest and the determination of the jury as to whether there was willful, premeditated intent on the part of someone to cause death, or whether it had been homicide by misadventure, manslaughter, accidental death, or suicide. Of the sixty-four inquests surveyed for Plymouth Colony, only 3 percent proved to be homicides, although there was the unresolved case of Anne Batson's child, which appeared to be a case of death through sexual abuse notwithstanding the fact that the jury returned no verdict, simply stating they saw nothing that might be the cause of its death, presumably apart from the suggestive bruising. Various accidents were responsible for 22 percent of deaths investigated by juries, death by exposure for 13 percent, suicide accounted for a further 9 percent, and natural causes 8 percent, but almost half, 47 percent, were by drowning if two suicides are included with the figures.

"Time doth stay for none. Death cuts down every one."

Considering the majority of deaths were by drowning and exposure, and that many of the settlers came from inland villages in England and were not used to water, it is not surprising that being out of doors was a hazardous experience, for Plymouth Colony had rivers, streams, and ponds in abundance, not to mention the ocean. All the deaths by exposure took place in winter, as one would expect, from December through February, and the severity of the New England winters is explanation enough, although in one instance the circumstances were much more complex. In February 1651 a nine-year-old boy, John Slocum, who lived in Taunton, was gathering cranberries at Fowling Pond about two miles out of town with twenty others, but on his return, about a mile from home, strayed into the woods and was never seen alive again. His father made an immediate search for him as soon as it was realized that he was missing, raised the town, and a massive hunt was undertaken, "with drum, guns, and loud voices," and it continued for a further three days with no success. The recorded dates are confusing at this point, as the inquest was held in June, the child went missing on "the 25th February last," and on "5th January" one John Lincoln found John Slocum's skull, "having the brain not wholly consumed." On January 9, he found "some other parts of the corpse, with parts of his clothes scattered in small pieces about the place by a certain pond at the head of the Mill River, 3 miles from the town, and two miles from the said Fowling Pond." The jury concluded that John had died of exposure, and his dead body was "devoured and torn, and the parts of his carcase dispersed with ravenous creatures." There seems no doubt that the child's body was found fairly soon after his death, as it was identifiable, although a number of weeks could have passed between the death and discovery of the pieces of the body. The question does arise as to what animals had devoured the body. It does not seem likely that it was wolves, on which there was a regular bounty paid by the colony, as it has been argued that wolves do not drag their prey away from a kill site and consume it in a different place. It is possible that wolves did devour his body initially, but that other scavengers were responsible for the dispersal, as the verdict suggests. There is, of course, the possibility that it could have been a grisly murder. How easy was it to make a murder appear to be an accident? How many of the coroner's inquests could really be unsolved murders? The case of John Sassamon, discussed above in the context of trial by ordeal, whose death was first assumed to be accidental, is still witness to the possibility that an apparent death by drowning could indeed have been murder.

Accidental deaths other than by drowning varied from a child choking to death on a piece of pumpkin shell, to a servant being run over by a runaway cart, to death by eating a poisonous root, to death by riding at speed into the low branch of a tree, to being gored by a bull. This last is worth recounting, as it gives a vivid sense of immediacy:

> We, whose names are underwritten, being summoned together by order from the Gou. to view the corpse of Mr. John Barnes, and to give in a verdict how we judge he came by his death, do judge, that being before his barn door in the street, standing stroking or feeling of his bull, the said bull suddenly turned about upon him and gave him a great wound with his horn on his right thigh, near eight inches long, in which his flesh was torn both broad and deep, as we judge; of which wound, together with his wrench of his neck or pain thereof, (of which he complained) he immediately languished; after about 32 hours after he died.

"Gored while stroking or feeling of his bull . . ."

Looking at the spread of inquests over the years, no particular patterns emerge, which is not surprising as the deaths were mostly accidental and unpredictable. The majority of deaths that took place in the colony were in the natural course of events, apart from the abnormal number that occurred during King Philip's War. Writing in July 1676 of the losses sustained in the whole of New England during that conflict, Nathaniel Saltonstall commented:

> And, as to Persons, it is generally thought, that of the English there hath been lost, in all, Men Women and Children, above Eight Hundred, since the War began: Of whom many have been destroyed with exquisite Torments, and most inhumane Barbarities. . . .

Eugene Stratton notes that it has been estimated that the colony lost more than a hundred colonists in the conflict, men, women, and children. One incomplete list that has been compiled of documented casualties from Plymouth Colony contains some ninety names. In addition there were almost certainly some who survived the war but died of wounds received at a later date. The number of Indian casualties has never been precisely calculated for all the areas involved in the war, let alone Plymouth Colony, and they too would have to be supplemented by those who died in captivity after the war, while being shipped as slaves to the West Indies, or while working on the plantations.

Turning from death to the final acts that surround it, the rich tradition of gravestone carving that was to emerge in New England in the closing years of the seventeenth century and through the eighteenth, shows some distinctive regional characteristics in Plymouth County that Peter Benes, in his detailed study of folk gravestone carving in Plymouth County, Massachusetts, from 1689 to 1805, *The Masks of Orthodoxy*, sees as evidence of the emergence of a regional cultural identity in the colony during the years of its existence from 1620 through 1691. The grave markers used in the burying grounds in Plymouth Colony were not of stone but of wood, and none have survived to give evidence of whether or not they were painted with images of death's heads, the earliest form of gravestone carvings, but it is quite probable that they were. What there is is some evidence as to the cost of funerals—the coffin, winding sheet, and grave—that were minimally required to bury the dead. The details are found in a few probate inventories, supplemented on occasion by instructions contained in a will, such as Christopher Wadsworth's request in 1677 that "my body have a Comely and decent burial; and that my funeral expenses together with all my Just and due debts be paid out of my personal estate." The desire for "a decent burial" was standard language in wills, part of an overall often-repeated phrase, as expressed in the will of Elizabeth Hoare of Yarmouth: "I bequeath my soul to God that gave it me & my body to the earth from whence it was in decent burial. . . ." Entries in probate inventories indicating the method of marking graves are rare. One such is an entry in the inventory of Samuel Chandler of Duxbury, taken on November 17, 1683: "It[em] Posts & Rails—15s." Funeral expenses are not often mentioned as separate items in the Plymouth Colony inventories, probably being subsumed under "Debts due from the estate." Where they are mentioned they range from £1 due to be paid from the estate of *Mayflower* passenger George Soule, who died in Duxbury around January 1680, to his son, John, to £6 4s 6d in the 1684 estate of Daniel Combe of Scituate, due to his executor, Timothy White, for "funeral Charges & tendance and doctors." William Wetherell, a pastor in Scituate, had funeral charges amounting to £5 3s 5d, and Peter Collamer, also of Scituate, had funeral charges of £3 2s 4d. His estate

Shrouded for burial—minimal requirements for burying the dead.

amounted to £507 11s 8d. In each case the funeral charges did not form any significant percentage of the value of the estate, which is consonant with the findings of Peter Benes, that in Plymouth County in the late seventeenth century the average cost of a funeral was 1.3 percent of the estate value.

In the light of the scarcity of details concerning funerals and funerary practices in Plymouth Colony, the court records following a 1656 inquest are particularly valuable. The inquest was held to determine the cause of death of Titus Waymouth, and the verdict reads:

> Having viewed the dead body of the said Titus Waymouth, and finding neither wound, or bruise, or other thing which might cause his death, only that, as is testified, he having been a man often troubled with stoppings [constipation], together with drinking of cider, we conceive might be the cause of his sudden death.

As Titus Waymouth was not a resident of Plymouth, the court records include not only a brief inventory of the few possessions he had with him, but the charges for his burial:

	£	s	d
It[em], for a winding sheet, 5 yard of lockram [a linen fabric] & thread	00	08	05
It, for a coffin	00	08	00
It, for digging the grave	00	03	00
It, to the clerk of the Court	00	02	06
It, to the charges of the ordinary	00	12	00
It, paid in money to him of his wages	00	12	00
	02	03	11

What is significant here that is not found in the probate inventory accounts for burial costs is the tavern charge for beer or wine consumed by those attending the funeral. Benes shows that after 1700 "generous quantities of rum were consumed by the participants and celebrants" at funerals, and although it was customary in rural England as much as in New England to have plenty of liquor at funerals, Benes does suggest that later in the seventeenth century proportionately "more money seems to have been spent [in Plymouth County] on burial drinking than elsewhere." Gloves are occasionally listed in the probate inventories of the Plymouth County as gifts for mourners, and as the seventeenth century came to a close, and moved into the early eighteenth, it appears that increasingly mourning clothes and rings, often carved with death's heads, were part of the expense account of an estate. It is probable, though, that for the majority of colonists, important though the final rites of death were, funerals were simple and even frugal events, wooden markers being placed upright in the ground at each end of the coffin, grave rails defining the length of it, probably with a painted epitaph as well as the name and date of death, and possibly a grim death's head. A common seventeenth-century epitaph that captures the macabre humor that pervaded images of death might also stand for all the people who lived and died in Plymouth Colony:

Remember me as you pass by
As now you are, so once was I.
As now I am, soon you will be.
Prepare for death, and follow me.

Chapter 5

A Few Things Needful

Houses and Furnishings

For most people, history is a narrative recounting of major events in the past, and their causes and effects: the Norman conquest, the voyage of the *Niña*, *Pinta*, and *Santa Maria*, the founding of Jamestown, the establishment of Plymouth Colony, Bacon's Rebellion and King Philip's War, the American Revolution, the Declaration of Independence, the Civil War, the Depression, the bombing of Hiroshima, or the counterculture of the 1960s and 1970s, to name only a few. Important names are linked with these events, such as Christopher Columbus, John Smith, William Bradford, Thomas Jefferson, Ulysses S. Grant and Robert E. Lee, Franklin D. Roosevelt, Harry Truman, and the Beatles. But between these moments of rich drama and far-reaching effects, most Americans went about their daily routines in the usual way, and the further back in time one goes, documents become increasingly scarce and uninformative regarding everyday life. For this reason, it is necessary to rely on other sources of information, and material culture is a vital contributor to our knowledge of the ordinary people of the past. Material culture is that part of the physical world that we modify according to a set of cultural plans and includes all that is wrought by human activity, whether something as simple as a thimble or as complex as a space station. Even when the records are unusually rich, as is the case with seventeenth-century Plymouth, a complete picture of life at the time cannot be constructed from them alone. In his *Folk Housing in Middle Virginia*, Henry Glassie makes the point with his usual eloquence:

> A knowledge of Thomas Jefferson might be based on his writings and only supplemented by a study of Monticello, but for most people, such as the folks who were chopping farms out of the woods a few miles to the east while Jefferson was writing at his desk, the procedure must be reversed. Their own statements, though made in wood or mud rather than ink, must take precedence over someone else's possibly prejudiced, probably wrong and certainly superficial comments about them.

In the case of seventeenth-century Plymouth, and for that matter anywhere in the world at any time, we must place the people in their physical setting, and pay close attention to all of the objects used by them to cope with the world in which they lived.

Almost all secondary accounts of Plymouth are people-centered, and with two exceptions, Darrett Rutman's *Husbandmen of Plymouth* and John Demos's *A Little Commonwealth*, only make passing reference to the material setting in which the events recounted took place. The four most useful and significant bodies of material that provide us with access to the material world of Plymouth are the few early descriptions available, standing houses, probate inventories, and those findings obtained through excavating the homesteads of the early settlers of Plymouth.

Vernacular Houses

We will begin our construction of the physical world of early Plymouth by considering the houses in which the people lived, and then move on to probate inventories, all of which tells us what was in those houses, and some of which gives us an indication of the type of house. It is important at the outset to distinguish between vernacular and academic architecture. Vernacular architecture is not made with plans, but is rather the idea of what a house should be like that is carried in the minds of the people and passed down through the generations by word of mouth and by example. This is not to imply that house carpenters in the past were not highly skilled; they most certainly were, but the mode of transmitting the knowledge needed to construct a well-built house resulted in the slow changing of house forms over time. To paraphrase Henry Glassie, houses don't change, but ideas do. Such slow change in all aspects of people's lives is the hallmark of traditional cultures, often referred to as folk cultures, which are conservative, largely rural and agrarian, and exhibit regional variation, the result of differing degrees of isolation from one another. In contrast, academic architecture, referred to in the United Kingdom as "polite architecture," is the product of plans drawn by professional architects. Such a process leads to academic architectural forms showing a great degree of similarity over broad regions, but that also tend to undergo more rapid change.

Domestic buildings throughout Plymouth Colony during its entire period of existence fall into the vernacular or folk category. The archetypical house type of any structure of two or more rooms is the so-called hall and parlor house, the idea for which was carried by the settlers from England, where it had been the most common form since the late Middle Ages. As the name implies, a hall and parlor house had at least two ground-floor rooms. The hall was the more public of the two, and a variety of activities took place in it. The parlor was more private. A visitor might enter directly into the hall, but access to the parlor required passing through an interior door. The parlor was usually the sleeping room for the owner of the house and his wife, and more highly valued objects were located there. It was used for more formal events, such as entertaining important guests. With rare exceptions, hall and parlor houses in Plymouth had their chimneys placed centrally between the two rooms, with a fireplace in each. Should an addition be made to the basic two-room plan, it almost always took the form of a third room built onto the rear of the house, often called a lean-to, which over time was used increasingly as a kitchen. Such an arrangement was well suited to the harsh New England winters, since the chimney, which was rarely if ever cold, would serve as a core source of heat.

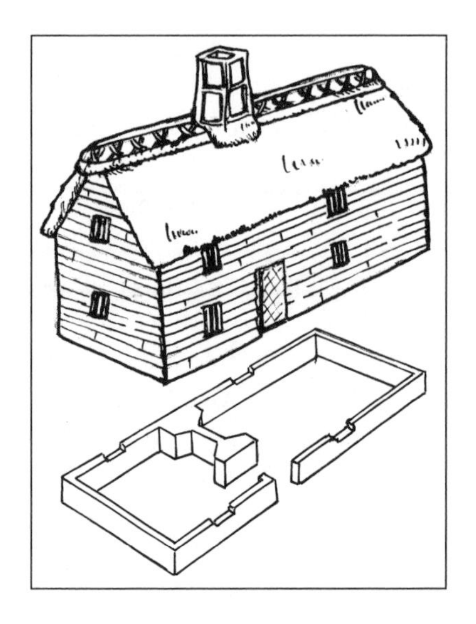

Seventeenth-century hall and parlor house.

Second-story rooms reflected the plan of the first, usually with two or three rooms, each named after the room that was located beneath it: hall chamber, parlor chamber, and, when there was a kitchen beneath, kitchen chamber. However, the earliest hall and parlor house in Plymouth for which evidence exists is figured in the probate inventory of Will Wright (1633), in which rooms are named along with the objects located in them. What type of dwellings made up the first fortified village is more difficult to determine, but there are enough scattered bits of evidence in the early accounts to allow us to make an informed guess as to what they may have been like. One thing can be stated with absolute certainty; they were *not* cabins made with round logs, notched for joining at the ends. Prior to 1939, it was generally believed that the first houses built in Plymouth were such buildings, but this is yet another small part of the Pilgrim Myth, and as is the case of the other parts of this story, it reflected widely held ideas about the American past, in which log cabins symbolized the spirit of the pioneers, and how humble beginnings can sometimes lead to greater things. William Henry Harrison made the log cabin a part of his presidential campaign, having grown up in one. This type of house is also associated with Abraham Lincoln. In the 1940s a popular toy was "Lincoln Logs," which came in a cylindrical container and could be used to construct log cabins. At about the same time, Log Cabin syrup was packaged in a tin shaped and printed like such a house, with the chimney serving as a pouring spout. So it was entirely logical at the time to perceive the log cabin as a metaphor for great things arising from humble beginnings, analogous to the "Pilgrims" themselves. Earlier renderings of the first Plymouth village show neat rows of log cabins lining the street of the village.

It was not until 1939 that this perception was challenged by Harold Shurtleff in a book entitled *The Log Cabin Myth*. Shurtleff explores the symbolism of the log cabin in the American collective consciousness, and proceeds to demonstrate in a most convincing way that houses of this type were introduced in North America by the Swedes who were the first permanent settlers of the Delaware Colony in 1638. Log construction of the type here in question is unknown in the English vernacular architectural tradition, so it would be highly unlikely for the Plymouth settlers suddenly to begin to employ a construction technology of which they had no prior knowledge. Later in the seventeenth and early eighteenth centuries, buildings were occasionally constructed in New England using *squared* timbers, joined by corner notching, but these buildings, known as garrison houses, were so built for defensive purposes, although they could serve as dwelling houses as well. So, however modest the first dwellings erected at Plymouth may have been, they were constructed in the timber framing tradition that characterized English domestic buildings.

Timber framing of the period involved setting large vertical posts into sills that in turn rested on some type of level footing, either stone or brick. These posts were tied together by equally sturdy horizontal beams that marked the division between floors in houses of two stories, or the floor of the space under the rafters, which was termed the garret or loft, or what we would call today the attic. The rafters rested on the uppermost pair of horizontal beams, known as plates. Each pair of major vertical posts, usually spaced a distance of the order of sixteen feet, though that dimension could vary somewhat, defined what is termed a bay. A hall and parlor house then would consist of two bays, one defining each room, with a third located in the center, known as the chimney bay, since the chimney would be built within it; this bay would be roughly half the width of the other two, on the order of eight feet. Between the large vertical posts were smaller vertical members known then, as today, as studs. The studs would be slotted or notched in such a way that flexible branches could be set into them, and these in turn would have other branches woven at right angles. These sticks were known as wattles, and would be given a thick coating of clay, known as daub, giving rise to the term "wattle and daub construction." In more elaborate houses, brick could be used in place of the wattle and daub, but all evidence points to the exclusive use of wattle and daub in earlier seventeenth-century dwelling houses, and it is almost a certainty that this type of wall construction characterized the houses that lined the first street in Plymouth. The exterior of such houses would be sheathed with clapboards, narrow strips of wood produced by a process known as riving, in which a tool known as a frow, its blade at right angles to the handle and driven with a mallet known as a beetle, is pounded into the end of a log to produce rough boards that would then be finished and smoothed with a spokeshave, or draw knife. The rough clapboard would be held firm in a shaving horse, a long wooden frame slanted in such a way that a worker could sit at one end and smooth the board by pulling the knife toward him with the sharp edge facing him.

Plymouth's First Houses—Early Written Descriptions

It is probable that most if not all of the first houses constructed at Plymouth were single bay, one-room structures, in which case the chimney would be located at one end. These early chimneys were probably framed of wood, and filled with wattle and daub, and possibly, as a protection against the weather, sheathed also in clapboards. With all of this in mind, we can turn to the various mentions regarding houses in the early records, which when combined with what is known of contemporary English building practices, will permit at least a generalized description of the first houses built in Plymouth. The first structure erected was the common house, twenty feet square, slightly larger than a one-room dwelling house would have been. But then, it was built for a special purpose, and as such needed not conform to a "typical" dwelling house in dimensions. If most if not all houses were single-bay structures, the door would've been on the same end as the chimney, so that upon entering, one would be confronted by the side of the chimney. Chimneys were a recent innovation in England, part of the changes brought about during the housing revolution of the late sixteenth century. Before that time, the usual vent for smoke from a fire on the hearth was a simple hole cut in the roof. It is possible that some

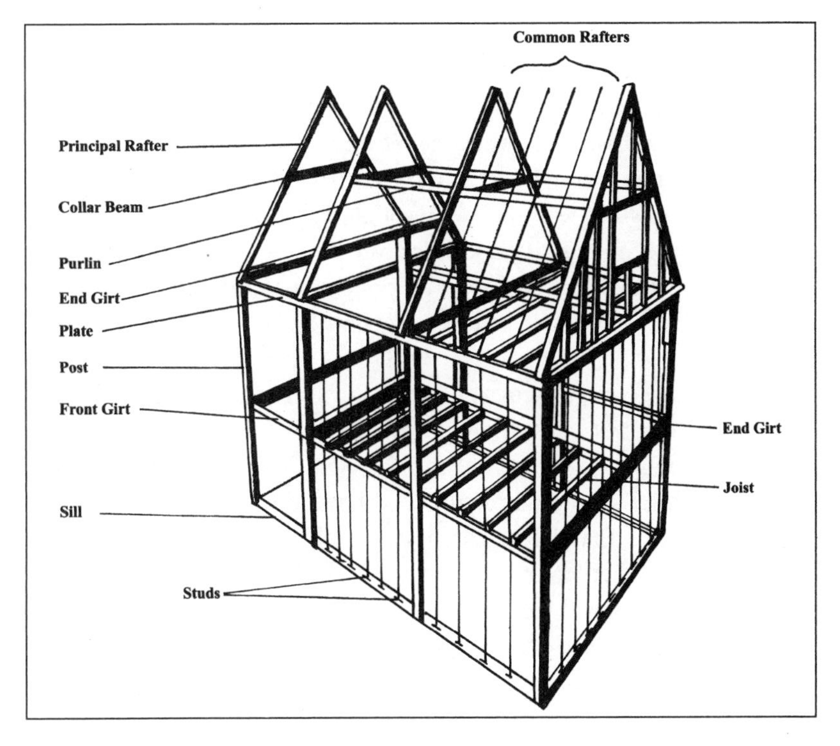

Framing diagram showing main structural pieces.

of the first dwelling houses, especially those built during that first terrible winter, followed this older pattern. Windows would have been few and small, if a set of instructions for building a house in Ipswich, Massachusetts, in 1638 is typical, as it probably is: "For windows, let them be not over large in any room, and as few as conveniently may be. . . . Leaded diamond-shaped glass panes mounted in iron casements, typical of seventeenth-century houses at a later time, almost certainly had not yet made their appearance, for Edward Winslow in his letter of December 11, 1621, gives a lengthy description of "a few things needful" to be brought by newcomers to the colony. Among such "needful things," he advises prospective settlers to "bring paper and linseed oil for your windows." By paper he is referring to some type of parchment, which when oiled would become translucent, admitting more light. It is virtually certain that roofs were thatched, as was the first building, the common house. Bradford describes a fire that broke out in one of the houses, which clearly had a thatched roof:

> This fire was occasioned by some of the seamen that were roistering in a house where it first began, making a great fire in very cold weather, which broke out of the chimney into the thatch and burned down three or four houses . . . when the vehemency of the fire was over, smoke was seen to arise within a shed that was joined to the end of the storehouse, which was wattled up with boughs, in the withered leaves whereof the fire was kindled; which some running to quench, found a long firebrand of an ell long . . . which could not possibly come there by casualty, but must be laid there by some hand. . . .

This passage is particularly informative. Aside from its being an early case of arson (another "Pilgrim first"?), it provides several important details concerning the houses. The fire occurred on November 5, 1623, after Emmanuel Altham's letter to his brother, in which he wrote that he counted twenty houses. By that time, all houses probably had chimneys, if some lacked them in 1621. Furthermore, it is clear that the chimney was framed in wood, and certainly filled with wattle and daub, and of course, the roof is explicitly referred to as thatched. The date of this fire could well have special significance. "Seamen that were roistering" can be taken to mean partying seamen, and November 5 was and still is Guy Fawkes Day, declared an official day of thanksgiving by the English Parliament in 1606 to commemorate the unsuccessful attempt to blow up Parliament and the royal family on November 5, 1605, by planting twenty odd barrels of gunpowder in the cellar beneath the palace at Westminster. Guy Fawkes Day was celebrated by the English with fireworks and bonfires, among other things, and by 1623, "making a great fire," cold weather or not, could well have been a good excuse for an uproarious and somewhat malicious reveling as much in New England as England.

What other details regarding the form of the earliest dwellings in the village can be gleaned from the slim number of references from the period? Winslow's mention of the storm of February 4, 1621, causing "much daubing of our houses to fall down," would suggest that wattle and daub were used on the few houses that had been constructed by that date, since he mentions houses in the plural. The daubing that Winslow refers to must have been on the exterior of the houses, and that in turn would indicate that it was not covered with some type of sheathing, either vertical boards or clapboards. Whether such covering had not been used because of constraints of weather, time, and labor, or because it was not thought necessary by people accustomed to the more gentle climate of England and Holland, is impossible to determine.

Emmanuel Altham's letter of September 1623, written before the November 5 fire that destroyed three or four houses, mentions "about twenty houses four or five of which are very fair and pleasant, and the rest (as time will serve) shall be made better." Most agree that the four or five "fair and pleasant" houses were fully framed in the English tradition, with footings and sills. But what of the other fifteen-odd buildings? Three possibilities suggest themselves, and each has merit.

The first is that they may have been semi-subterranean dwellings as described by Cornelius van Tienhoven, secretary of the province of New Netherland, in 1650:

> Those in New Netherland and especially in New England, who have no means to build farm-houses at first according to their wishes, dig a square pit in the ground, cellar fashion, six or seven feet deep, as long and as broad as they think proper, case the earth inside all round the wall with timber, which they line with the bark of trees or something else to prevent the caving in of the earth; floor this cellar with plank and wainscot it overhead for a ceiling, raise a roof of spars clear up and cover the spars with bark or green sods, so that they can live dry and warm in these houses with their entire families for two, three and four years, it being understood that partitions are run through those cellars which are adapted to the size of the family.

Although architectural historian Abott Lowell Cummings correctly states that no parallel for such dwellings can be found in the records of Massachusetts Bay, there are at least two examples of just this type of construction known from Virginia in the 1620s from archaeological excavation. At Martin's Hundred, some seven miles southeast of Williamsburg, the archaeologist Ivor Noël Hume excavated what he terms a "cave house," which fits van Tienhoven's description (which he cites) very closely, and which permitted him to create a conjectural reconstruction that, when combined with the archaeological evidence, is probably quite accurate.

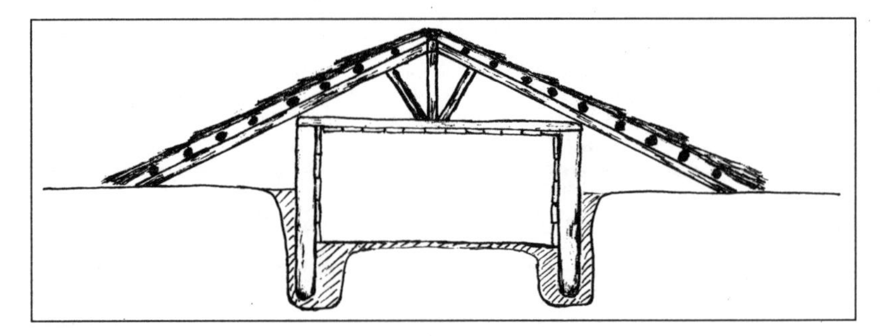

Semi-subterranean "cave" or cellar house.

A second such pit was excavated at Flowerdew Hundred, on the south side of the James River thirty miles below Richmond. There were some differences between it and its counterpart at Martin's Hundred; there was no clear evidence of post holes, but in size and form the two are very similar. Its rectangular shape clearly indicated that it was not the result of simply digging clay for daubing, for pits of this type are roughly circular. In the fill of the pit were a number of tenterhooks, used to stretch fabric over a wooden frame, suggesting that the roof may have been made of canvas or some similar sturdy fabric. It predates a very substantial house, with stone footings brought from England and constructed in 1625. So in combining van Tienhoven's description with the archaeological evidence from Virginia, the possibility that such houses were used at Plymouth in the early years becomes a very attractive one.

A second and equally plausible explanation of what form the less than "fair and pleasant" houses took is that they were constructed using a technique termed "earthfast" or "post-in-ground" construction. Building a house in this fashion involved setting large vertical posts directly into the ground, eliminating the need for both footings and sills. Such houses are relatively impermanent, since the posts decay in the ground after ten or so years, but they are easier to build and require less expense and less time than does a fully framed house resting on sills with footings beneath. Although this method of house construction was known in England, but not used to any great extent, by 1600 it became the commonest construction technique in the Chesapeake region, where it has been convincingly argued that it was standard practice employed by planters who had limited capital, most of which was invested in the lucrative but highly labor-intensive production of tobacco; it was simply a matter of where one might best invest one's limited resources. In 1972, archaeologists excavated the house site of Isaac Allerton, who had moved to Duxbury shortly after 1627. The remains of a small earthfast dwelling were uncovered. It measured twenty by twenty-two feet, and consisted of a single bay with a wooden chimney on one end. While in the case of the use of earthfast building in the Chesapeake the motivation was economic, it also seems reasonable to suggest that it could have also been used as a matter of expediency, and this certainly would be the case in the early years of Plymouth.

The third type of dwelling may have been what is referred to as a puncheon or palisado house. This type of building is known archaeologically from the Chesapeake, but in New England only from documentary sources. Houses of this type are constructed by setting rounded vertical posts into the ground, very closely spaced, sheathing them on the exterior with clapboards, and covering the interior walls with daub. While there is no evidence for this kind of house having been among those that "as time will serve, shall be made better," a

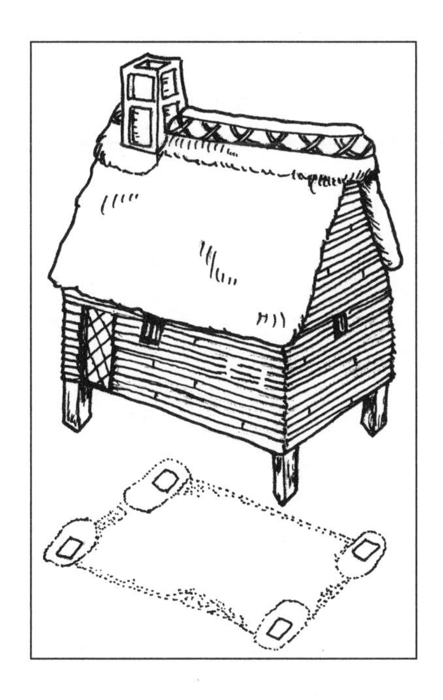

Earthfast or post-in-ground construction.

description of the town of Scituate some twenty-odd miles north of Plymouth, and founded by a separate group of settlers from Kent in 1628, makes explicit mention of houses of this type. In 1637, the Reverend John Lothrop, writing in the Scituate church records, listed fifty-seven Scituate houses and occupants from 1634 through 1637. He first listed nine houses that had already been built when he arrived at the end of September 1634, bracketing them together with the annotation, "all which [were] small plain pallizadoe Houses." Perhaps the most reasonable suggestion regarding the first houses in Plymouth would be that in addition to the four or five "fair and pleasant" fully framed houses, some combination of pit or cave houses, earthfast structures, and palisado buildings were erected, these being those that Emmanuel Altham said would in time be improved.

Regardless of the exact types of houses in the first settlement, it is clear that at least some substantial dwellings were constructed, which was not in accord with the wishes of the London adventurers, who wanted the colonists to devote as much time as possible to acquiring those commodities which would turn a profit for the company, the Merchant Adventurers of London. What they perceived as excess labor in building substantial houses would divert labor from such activities. Remembering that at the time of departure from England this issue had not been satisfactorily resolved, it was never fully adhered to in the first years of the colony's existence. People were allocated lots of land and allowed to build their own houses. Apparently, some invested more effort in such construction than others, and this was not viewed favorably by the merchant adventurers. Robert Cushman, a a member of the Leiden congregation and a close friend of William Bradford, was in the unenviable position of acting as intermediary between the colonists and the merchants, and when Bradford expressed in a letter that such terms (equal sharing and ownership of both property and work) "will hinder the building of good and fair houses, contrary to the advice of politics," Cushman responded:

> So we would have it; our purpose is to build for the present such houses as, if need be, we may with little grief set afire and run away by the light. Our riches shall not be in pomp but in strength; if God sends us riches we will employ them to provide more men, ships, munition, etc. You may see it amongst the best politics that a commonweal is readier to ebb than to flow, when once fine houses and gay clothes come up.

Clearly the planters and adventurers were at cross purposes on this issue, and the problem was doubtless further complicated by the arrival of those who came on their own "particular," since they were not bound by any agreements between the London company and the settlers.

One final insight regarding the appearance of the earliest houses in Plymouth is to be found in a law enacted on January 6, 1627:

> It was agreed upon by the whole Court . . . that from hence forward no dwelling house was to be covered with any kind of thatch, as straw, reed, &c. But with either board or pale & the like; to wit: of all that were to be new built in the town.

This law has almost universally been interpreted as outlawing thatched roofs, but architectural historian Dell Upton has made a very convincing argument to the contrary. The two operative words in the passage are "covered" and "pale." "Covered" could be taken to mean a kind of sheathing, and a pale, by definition, is inserted in the ground, and therefore could in no way be used as a roof. Although apparently not used in seventeenth-century England, covering a house with bundles of reed would be a very efficient method of sheathing, and would afford a significant degree of insulation. Two types of reed were abundant in the coastal regions of North America, Norfolk reed (*Phragmites sp.*) and cattail (*Typha sp.*), growing in coastal marshes and along streams. Plymouth was not the only colony where such a method of house covering was used. In a 1586 letter to London, Ralph Lane, governor of the ill-fated Roanoke Colony in the Carolina Sounds, wrote that his own house was covered with reeds, and the others may well have been also. Reed covering would be particularly suitable for a palisado house, since its vertical posts were set close enough together to chink the spaces with daub and cover the interior walls with daub as well. The date of the enactment of the law regarding house covering is not without significance. It was passed in 1627, the same year as the cattle division, the last remnant of the "common course," which was never strictly adhered to since the laying out of individually owned house lots in 1621, and was a disputed point not successfully resolved between the settlers and the adventurers when the *Mayflower* set sail in 1620. The law seems to have been an effort to improve the quality of all the houses in the settlement, since the twelve undertakers had assumed the colony's debt, freeing everyone else to tend to their affairs as they saw fit. But there is an irony in this; while the court was attempting to improve both the appearance of the village and the quality of life of its residents, by 1630, as Bradford writes, the town had become "almost desolate," as the population had begun to disperse itself across the countryside.

Surviving Seventeenth-Century Structures in Plymouth

We have no knowledge of the details of house construction in the colony between 1620 and at least 1667 that is based on surviving structures, of which there are only a scant five. Archaeology has provided some interesting data on the houses of this interim period, and, as we shall see, is at odds with conventionally held beliefs about dwellings during these forty or more years. A small percentage of probate inventories that were recorded on a room-by-room basis name the rooms and the objects that were in them, but they fall short of providing any knowledge of how the houses were built. The five surviving seventeenth-century domestic buildings in the Old Colony area, with their traditionally accepted dates of construction, are as follows:

William Harlow house	(1667)
Jabez Howland house	(1669)
Joseph Churchill house	(1672–1695)
John Bradford house	(1674)
Richard Sparrow house	(1679)

Three of these houses (Harlow, Howland, and Sparrow) are within a half-mile radius of the Plymouth town center; the Churchill house is located a mile to the south, and the John Bradford house some six miles to the north. To this list may be added a sixth, the Joseph Allyn (Robert Hicks) house, which once stood near the town center, but was torn down sometime in 1822. However, it is represented by an excellent drawing showing details that suggest a construction method quite different from the other five.

These houses pose certain problems if one is attempting to make any kind of statement regarding construction methods during the entire period of the colony's existence. First, of course, is the size of the sample. To extrapolate construction techniques to those used in raising literally hundreds of houses from such a small sample is a risky proposition at best. Second, at least two of them, Howland and Sparrow, were heavily "restored" in the 1930s. This restoration appears to have been aimed primarily at changing the roof lines, replacing sash-type windows with casements with small diamond panes set in lead, and altering the facades to give them a more "seventeenth century" look. The interiors exhibit all of the characteristics of timber-framed construction, which in New England continued to be used until at least 1725, the date Cummings assigns as the end of what he calls the first period of house framing in Massachusetts Bay Colony. Finally, one should not accept uncritically the dates traditionally assigned to the structures. Excavation of the Harlow house cellar in the 1960s produced no artifact that dated earlier than the turn of the century. Excavations done during the same period in the rear lot of the Jabez Howland house produced similar results. Perhaps the most suspect date is that given to the Sparrow house, for it derives from the transfer of the property and not necessarily the construction of the building. The house which is probably most accurately dated is that of Major John Bradford, where only minor restoration has been carried out, and excavations in the side yard produced artifacts dating to the last quarter of the seventeenth century. It is fully understandable for people who live in historic towns, such as Plymouth or Yorktown, Virginia, to take pride in having seventeenth-century houses in their community. But when tree-ring dating was sufficiently refined to be used on framed houses in the eastern colonies, it was shown that a large number of houses in Virginia that had been dated to the seventeenth century were in fact earlier eighteenth-century structures. To date, no tree-ring dating has been attempted on the Plymouth houses, but its accuracy was dramatically demonstrated when applied to the Jonathan Fairbanks house in Dedham, Massachusetts, producing a date of 1637, which was in perfect agreement with documentary sources, and which established this house as the oldest timber-framed structure in English America.

In 1969 Richard Candee published a lengthy, three-part article on Plymouth Colony architecture in *Old-Time New England*. This was a full ten years before the publication of Abbott Cummings's *The Framed Houses of Massachusetts Bay, 1625–1725*. Candee proposed that Plymouth Colony employed a construction method that was unique to and characteristic of the colony, setting it apart from its neighboring colonies to the north and west. This way of building a house is referred to as either vertical plank siding or plank framing. It entails covering the house with wide boards, set vertically and extending from the sill to the plate without benefit of studs. Lath would then be attached to the interior and covered with daub or plaster. Candee sees this method of building as having been learned in Holland during the Scrooby party's sojourn there. It *was* a method employed by the Dutch, but only on the gable ends, and never on the front or rear of the buildings. He goes on to state that since plank framing was unknown in Massachusetts Bay, the only significant difference between the backgrounds of the people of both colonies was that the Massachusetts Bay settlers had not undergone the same Dutch experience, coming directly from England, where he claims the method was not employed. There are a number of difficulties with this thesis, but before discussing them, it is important to state that at the time of writing the article, Candee was a graduate student at the University of Pennsylvania, and our knowledge of seventeenth-century vernacular building has increased considerably in the years since. Nonetheless, except for the discussion of construction methods, the article still is a valuable one, with solid data on house layouts based on probate inventories and a good accounting of building in the first seven years.

However, we now know several things that raise serious questions regarding plank framing being a unique method of house construction in Plymouth. First, the number of former Leiden residents was small from the outset, and by the time the surviving houses were built, even if we accept the somewhat suspect early dates, they would have been built by second-generation people who had never lived in England, much less Holland. Second, the technique was used in England, although sparingly, as is shown in Richard Brunskill's *Illustrated Handbook of Vernacular Architecture*. But perhaps the most telling bit of evidence is the fact that plank framing was employed in Massachusetts Bay where no fewer than twenty-five such buildings have survived that date to 1725 or earlier. Abbott Cummings attributes the appearance of this method to the rapid development of sawmills in southern Maine and New Hampshire. For if planks can be produced economically and in large numbers, construction becomes a simple and efficient operation:

> The plank frame house presented few problems in terms of erection. The carcass will have been reared according to any one of the methods we have described and the planks then raised manually into a vertical position and secured to the bracing frame.

Cummings sums it up in a succinct manner:

> It can thus be argued that the plank frame was a known structural variant by the late seventeenth century, though obviously the majority of carpenters in all parts of the area clung to the traditional English method of framing walls with stud and nogging.

Which brings us back to the Allyn house. A "fair" stud-framed house, originally known as the Hicks house, it is now thought to have been built possibly as early as the 1630s, despite its traditional 1660s date. Its most important feature is its close resemblance to those built in Massachusetts Bay, with an overhanging second floor and a gable or dormer in front of the chimney. Plank framing of such a house is precluded by the overhang, and Candee himself states,

> Although it is not impossible in theory to construct such a framed overhang using vertical planks between the horizontal beams, no such building is known. The many overhang buildings in New England are all built on the principle of studs with nogging between the interstices.

What then can be said about house construction in seventeenth-century Plymouth? Based on what we know, which is rather limited, combined with some informed guesses, chances are high that Plymouth differed only sightly, if at all, from its northern neighbor in the manner in which its buildings were raised, framed and clad. Cummings is quite correct in stating that the majority of carpenters maintained traditional English methods of framing walls with studs and nogging (the material placed between the studs, which in Plymouth was almost certainly wattle and daub). This framing method would be used for both earthfast, post-in-ground houses and those

Allyn House, Plymouth.

which had their sills on footings or brick or stone. Plank framing appears to have made its appearance late in the century in both colonies, and was never a common way of framing a house.

Probate Inventories

What we now have is an empty shell—the house—which was, of course, furnished with all of the goods its occupants would need to manage their lives and live in at least some degree of comfort. We are on much more solid ground on this subject, for we have available copies of over four hundred probate inventories transcribed from the records of the colony, up to the year 1685, when separate county courts were established, after which records were housed in the different counties and are not as readily available. Probate inventories are a very detailed listing of the contents of the house of a recently deceased person, and their value, taken both for tax purposes, as is still done today, and to provide an accounting of the estate for the survivors. A small percentage of these inventories were recorded on a room-by-room basis, which allows us to see in which room various objects were located, as well as giving the names of the rooms in the terms used by the appraisers, names which are quite consistent and inform us as to the terminology of the time, which is very different from that of today. Entire books could be and have been written on colonial inventories, but for our purposes we present only three, with detailed comment on some objects listed. Two of the three will be room-by-room inventories, and the third a more typical straight listing without mention of location. Inventories of the latter type become quite repetitious after reading them for only a short time; they are like the Oompa-Loompas in the film *Willy Wonka and the Chocolate Factory*; they all look very much alike, with only minor variations. This does not mean they are of less value; quite to the contrary, they provide data for a wide range of statistical analyses that have been carried out by a number of scholars.

It is not easy to determine how the surviving sample of inventories reflects the total number of deaths in the colony during the period of fifty-four years between 1631 (the year of the first inventory, that of widow Mary Ring) and 1685. Population estimates vary greatly for the colony at any given time, but even so, 461 deaths for that period does not seem to be too much at variance with the number of inventories to which we have access. Almost all of them are those of adult males, only 8 percent of the total being those of women, the majority of whom were widows who had not remarried; of course, children are not included. It is probably safe to say that the majority of male decedents are represented, and possibly a large majority. Economic status seems not to have been a major factor of whose estate was probated and whose was not, for the Plymouth inventories range in value from one as high as that of Captain Thomas Willett, whose estate was valued at above £2,000 in 1674, to that of the town's best known pauper, Webb Audey, who left an estate valued at a mere £7 7s 10d. Plymouth was a poor colony, and most estates were assessed at below £100–£200 in value. A factor that must also be taken into account, though, is the widow's dower, one-third of the movable estate, which was not included when the inventory was taken by the appraisers, and so omitted from the valuation.

House Types from the Room-by-Room Probate Inventories

Before we turn to the furnishings of houses in Plymouth, it is important to note that room-by-room probate inventories are the one source of information, apart from archaeology, which can give us a sense of the *type* of houses that were found in Plymouth. There are sixty inventories for the period 1633 to 1685 in which the appraisers recorded the names of the different rooms in the house that can provide a key to home type. Room names range from passing mention to detailed room-by-room coverage, but the overwhelming majority of the sample (90 percent) that contain sufficient data for the analysis of house forms reflect the traditional hall and parlor house of rural England. The remaining 10 percent appear to have been one-room houses that may have formed the majority built given that room inventories form only 12 percent of the probates. The room-by-room inventory evidence therefore suggests that the dominant house form of two or more rooms in Plymouth Colony from the 1630s through the 1680s was the hall and parlor house.

When the hall and parlor house form was carried across the Atlantic in the minds of many emigrants to New England, not only was it built with a central chimney to heat the whole house during the bitter northern winters, which became a distinctive feature, but it was repatterned to include service rooms to the rear, ultimately creating a house type that has become known as the "saltbox" house, with its characteristic steep down-swept roof covering a lean-to addition of service rooms (kitchen, dairy, buttery) that ran the full length of the house. Most hall and parlor houses in rural seventeenth-century England had service rooms to the side of the house, across the passage from the hall, but there were a good number in East Anglia in which these unheated rooms extended out from the rear of the house, to form that characteristic saltbox shape. Historian David Hackett Fischer emphasizes the number of saltbox houses in the eastern counties of England in the late sixteenth and early seventeenth centuries, and suggests that New Englanders simply built houses with which they were familiar.

Folklorist Henry Glassie, however, has drawn attention to an important, little known fact concerning the ultimate choice of this house type.

New England saltbox house.

> Some seem to argue that East Anglian people came and simply built what they were used to. Maybe at first. But in time, they chose, and they chose a very rare type for special reasons. Given what would happen, they chose the most modern house, though it was rare in the repertory.

Glassie is referring to the saltbox house, many of which he has observed still standing in England. This insight opens a whole new perspective on our understanding of the development of one of the most characteristic New England vernacular house forms and also challenges our assumptions about the choices made by the early settlers. When room-by-room inventories for Plymouth Colony are examined to see how many referred to lean-to service rooms, however, only 22 percent include lean-tos, most of which appear to have been used as supplementary storage areas. Only one lean-to was in use as a kitchen, that of Captain Nathaniel Thomas (1675). The probate inventory room-by-room evidence, therefore, would suggest that in Plymouth Colony the development of the salt-box configuration, at least during the greater part of its history through 1685, may not have been as prevalent as it was in Massachusetts Bay Colony to the north.

House Furnishings from the Probate Inventories

How, then, were the houses that we have been discussing furnished? Plymouth inventories are extremely detailed, as in the case of Will Wright's inventory, which lists such small things as two papers of hooks and eyes, and one pair of wadmore (wadmal, a coarse wool) stockings. Yet, in spite of their attention to the many small items in a household, the inventories almost always include a reference, as in Wright's inventory, to "some other smale things" in a lot with a total value of six pounds for the contents of the entire room. Similar references are commonly encountered in the listings, such as "things too trivial to mention," or "small things forgotten." The spelling of "small" as "smale" in the Wright inventory is so frequently encountered in the probate inventories that one wonders if it is an indication of the way the word was pronounced. While the transcription of Will Wright's inventory given below has been changed to conform to modern orthography, "smale" is used throughout as in the original. These catchall categories for items not thought worth listing individually may account for an otherwise puzzling matter, in that while the remains of white clay smoking pipes are among the most common artifacts found on archaeological sites, they are scarcely visible in the inventories. However, in one instance, that of the estate of John Dicksey, taken in 1674, a case containing five gross of tobacco pipes, valued at 17s 6d, is listed, and so it is possible to determine the value of a single pipe at only 0.29 pence. Items of such low value would most likely be included in the "things too trivial to mention."

It is now time to accompany the appraisers Manasseh Kempton and John Faunce as they enter Will Wright's house, and see what is inside and where it was located. Their listing is given below, the values to the right being given in pounds, shillings and pence.

Probate Inventory of Will Wright

November 6, 1633

New Plymouth 1633
An Inventory of the goods & Chattels of Will Wright late of Plymouth deceased as it was taken by Manasseh Kempton & John ffans [Faunce] the sixth of Novbr 1633 & presented in Court the 2^d^ of January 1633.

In the first Room

Impr. [First] one chest wth one sad colored suit &
cloak, one other suit the breeches being w^th^out lining,
one red bay waistcoat & one white cotton waistcoat
one old black stuff doublet, 2 hats, a black one
& a white one, 1 piece of loom-work. 4 knots

	£	s	d
of white tape. 2 pr [pair] of boothose & 2 papers of hooks & eyes 2 lb of colored thread, 2 doz. of laces, & 2 pr of old knit stockings wth some other smale [small] things at 6 £.	06	00	00
It [Item] one small Table wth a carpet, one Cupboard & a chair wth a sifting trough	01	10	00
It six kettles 3 iron pots & a dripping pan	02	03	00
It 7 pewter platters 3 great ones & 4 little ones. 1 small brass mortar & pestle 2 pint pots & one pewter candlestick. 1 pewter flagon 2 pewter cups, 1 wine & one other beer bowl. 1 beaker & 1 Caudle cup. 1 dram cup & a little bottle 2 salt cellars 3 porringers. ½ doz. old spoons, 3 pr of pot hooks. 1 old pr of tongs & an old fire shovel. 1 pr of pot hangers. 2 small old iron hooks. 1 pr of andirons. 2 old iron Candlesticks, & a pressing iron. 2 basins 1 small one & another great one all at 2 £.	02	00	00
It one fowling piece	02	10	00
It 2 pr of boot breeches an old pr of Cotton drawers an old blue coat. 2 pr of old Irish stockings. 2 pr of cloth stockings. 1 pr of wadmore stockings. 1 old red waistcoat and old black Coat.	01	10	00
It one little old flock bed & an old feather bolster, wth a pr of worn sheets, an old green Rug.	00	15	00
In the Buttery			
Two old barrels one full of salt, the other half full, 1 bucking tub, 1 washing tub & 2 empty runlets wth small trifling things.	01	00	00
In the loft over the first room			
One old half headed bedstead. 1 old bag of feathers. 1 old white Rug 2 hogsheads & a barrel.	00	16	00

	£	s	d
In the bedchamber			
One bedstead one warming pan. 1 feather bed & bolster. 2 pillows, wth 2 Rugs 1 green one & one white one. 1 trunk & a little chair table, with a small carpet & a curtain & valence for the bed. 1 small cushion five pr of sheets 4 pr of pillowbeers [pillowcases] 2 table cloths & 15 napkins. 4 towels & 7 shirts. 3 pr of linen drawers & 2 wrought silk caps & one white holland cap & one dimity waistcoat 3 bands & 4 pr of linen stockings.	13	08	00
It one great Bible & a little bible. 1 Greenham's works. 1 psalm book wth 17 other small books	01	03	00
In the loft over the bedchamber			
One broad axe & 2 felling axes & 2 hand saw. 1 thwart saw wth a wrest to it. 3 augurs 2 chisels 1 gouge. 1 drawing knife. 1 [?] 1 gimlet. 2 hammers. 1 pr of old hinges. 2 chest locks. 1 padlock. 1 splitting knife 1 old spade. 2 old hoes. 2 fishing lines 1 old hogshead. 1 small runlet half full of powder. 1 garden rake. 1 pitch fork. 1 tiller of a whipsaw. 3 iron wedges with some small implements & other lumber of small value	02	07	00
It the house & garden	10	00	00
It the Cattle being one Cow & a steer calf	20	00	00
It 2 Ewe goats & 1 ewe lamb	07	10	00
It one old Sow. 1 hog. 1 young sow of 1 year old. 1 shoat. 1 boar. 1 Canoe & a churn.	07	00	00
Debts due unto him as appear per book	20	00	00
Suma	99	12	00

Priscilla Wright allowed the Executrix and Administratrix of her deceased husband.

Mr. Will Bradford bound wth her in an hundred & ninety pounds for discharge of the Court.

Wright's inventory was the ninth taken since recording commenced in 1633, and is the first of the small number of room-by-room inventories available to us. While the names given the rooms are different, it is clear that he and his wife occupied a traditional hall and parlor, story-and-a-half house. The first room would be the hall, and the bedchamber the parlor, with a loft of half a story over each. The buttery was probably a small enclosed space within the first room, as was usually the case, although when rear kitchens became more common it was more often located there. The term "buttery" is misleading; it has nothing to do with butter, but rather refers to a space where butts, large barrels of two hogshead capacity (105 gallons), were stored, and other items such as tubs, and in this case runlets, small coopered containers for water that served much the same purpose as a canteen would today. Later we find a second similar space, the dairy, where butter, cheese, and milk would be stored.

First consider Wright's clothing, which includes a sad (deep) colored suit and cloak, four waistcoats, one red bay (reddish brown), one of white cotton, an old red one, and one made of dimity, a strong cotton that had a pattern of raised stripes and fancy figures. He also owned two coats, blue and black, as well as black and white hats. Clothing in the bedchamber included shirts and drawers, but also two wrought (worked) silk caps as well as one white holland cap. The silk caps were normally those worn by women, but in this instance they appear to have belonged to Wright. "Wrought" means embroidered, oftentimes with silk threads in multicolored floral patterns, very finely done, or with black thread, a kind of embroidery now commonly referred to as Elizabethan blackwork. The same kind of embellishment was to be found on women's bodices. The bands referred to are white detachable collars, which could have been either standing bands, supported by a wire frame, familiar to most of us in portraits of Queen Elizabeth I, or falling bands, which simply draped around the neck. It becomes clear from this listing of clothing that gives colors, a feature repeatedly found in other inventories, that, far from the somber blacks and grays traditionally thought to have characterized "Pilgrim" attire, the people of Plymouth had quite colorful wardrobes.

In addition to the two Bibles, Will Wright had a modest library of nineteen books, so he clearly was able to read, and probably to write as well, for he signed his will. The number of books in the colony was impressive. William Brewster's library of 382 volumes, sixty-four in

Latin and mostly religious works, was larger than the number given by John Harvard in the establishment of the college that bears his name. William Bradford's library contained over eighty-seven volumes, similar in subject matter to those of Brewster. Myles Standish had a smaller but more varied collection of volumes, some of them dealing with military matters.

Turning now to furnishings, Will Wright and his wife, Priscilla, had a relatively comfortably outfitted home, especially for such an early date. Particularly noteworthy is the contents of the bedchamber. Not only did he and his wife enjoy a feather bed, which were quite costly, but also a bedstead that was equipped with valance and curtains, almost certain evidence that the bedstead was of the canopied four-poster type. We must make an important distinction here; in seventeenth century terms, a bed is what we would call today a mattress, and would be placed on the bedstead, which today would be called a bed. The bolster and pillows are evidence of the standard sleeping position of the time, which was a somewhat higher angle from the waist up, similar to one of the more upright positions that we see today on television advertisements for adjustable beds. This manner of lying in bed can be seen in numerous graphics of the time. The two rugs were bed coverings, rather than used on the floor, another semantic difference from present usage. The same applies to the carpet listed in the first room along with a small table. Carpets, or Turkey carpets, were used as table coverings, not for floors. Contemporary prints and paintings suggest that they were Oriental carpets of one kind or another, hence the name "Turkey carpet." The contents of the bedchamber are valued only slightly lower than those of the first room, or hall, and both, taken separately or together, are more valuable than the house and garden. Robert Anderson makes a convincing argument for Wright having been a joiner, a furniture maker, based on the tools listed in the bedchamber loft. They are diverse and specialized, far more than those found in other inventories. So it may well be that he made much of the furniture for his house.

This brings us to the question of the source of all the furniture listed in the probate inventories. The consensus is that it was almost all locally produced, for furniture does not appear in provision lists and bills of lading for ships arriving from England. Other smaller items, from brass spoons to iron kettles, and a variety of metal tools were certainly imported, at least in the early years of the colony, for their production would have been beyond the capabilities of local artisans. However, following the establishment of New England's first blast furnace at Saugus, in the Massachusetts Bay Colony in 1640, at least some of the iron implements, and possibly items of other metals, would have been traded south to Plymouth. From the beginning, there were at least two, probably three members of the community who were skilled in woodworking. John Alden, a cooper, and Francis Eaton, a carpenter, were *Mayflower* passengers, and Will Wright arrived a year later on the *Fortune*. If we accept Anderson's identification of Wright as a joiner, as seems very likely, then three men would have been available to supply the tiny community's furniture needs, which seems quite adequate. As the colony's population grew, so did the number of craftsmen able to produce quality furniture, from cupboards to chairs, boxes, and tables. In *The Wrought Covenant*, Robert St. George provides a list of men capable of making furniture numbering 428 in Plymouth Colony alone from 1620 through 1700, counting a few from communities that were immediately bordering on the colony, such as Hingham in Massachusetts Bay. The furniture produced in the colony is very much in the English style, and includes some truly spectacular pieces, particularly elaborately turned and assembled cupboards. As early as 1924, Wallace Nutting drew attention to what he called the "serrated chests" of Plymouth, a characteristic in fact shared by other pieces such as cupboards and chairs. This feature, small notches along edges, among a number of others that we need not enumerate here, sets the Plymouth furniture-making tradition apart from that of the neighboring colonies. Within the colony, St. George further defines three subregions, which show subtle differences in the design and decoration of the furniture produced there, but in general they share in a common tradition of joinery and decoration. One of the best known Plymouth craftsmen was Kenelm Winslow, who arrived in Plymouth prior to 1633 when he was made a freeman. He moved to Marshfield just to the north in 1641, where he appears to have been living as late as 1670, although he was buried in Salem following his death two years later. He is mentioned here not only because he is probably the best known of the Plymouth joiners, but also because his name bears a striking similarity to that of the very talented tight end for the San Diego Chargers during the 1970s, Kellem Winslow.

Will Wright's inventory lists only two chairs, one of which is a chair-table. Chairs are relatively scarce in the inventories; apparently most people sat on chests, trunks, or stools, which are far more commonly encountered. Chairs were of four basic types. Joined chairs, sometimes referred to as wainscot chairs, have a solid wooden back and seat. Great chairs, referred to also as turned chairs, have backs that are made from either turned vertical pieces or horizontal slats, as well as turned supports for the arms, and usually have seats of some type of woven reed. The third type of chair had a fully upholstered back and seat, and the fourth was the chair-table, in which the back was hinged in such a way that it could be lowered onto the arms to form a small, usually round, table. In most inventories one cannot determine which type of chair is being listed, although on occasion, one of the terms cited above appears. Today, great chairs are often referred to as Brewster or Bradford chairs, since chairs of this type have survived and can be documented as having belonged to the two men. But as with much of the modern embellishment of names given to pieces of seventeenth-century furniture, this is current usage, and certainly not that of the period when the furniture was made and used. Robert St. George properly cautions: "We must discard impressionistic and modern notions of 'dower chests' and 'bible boxes' for phrases like 'chest with drawers,' 'chest and drawers,' or simply 'boxes.'" One relevant and particularly enlightening piece of information concerning the effort made in Plymouth Colony to maintain high standards of furniture making, is the request made by Josiah Winslow in his 1673 will that "the bedstead in the Parlour with the settle, Great table and form Remain in the house as standards."

Little more need be said about Wright's inventory, except to note the cluster of cooking and food-consumption items in the first room, a grouping that is always to be found in the hall at this early period, but that as time passes becomes increasingly present in the kitchen, if the house has one. Such groupings of functionally related objects are also found in those inventories that were not taken on a room-by-room basis and that form the vast majority of those available to us. But the simple fact of association at least gives a hint of the route taken by the appraisers through the house, since it is very unlikely that they went back and forth randomly between rooms. In the case of inventories that do not name rooms, another consideration should be kept in mind. At least some, if not most of them, could represent houses of only one room. In his *Reshaping of Everyday Life, 1790–1840,* Jack Larkin states that according to the 1798 Direct Tax, which described every free family's dwelling house, the American housing landscape was "striking in its small scale, its plainness and its inequality." Most families, in fact, lived in houses of one or possibly two rooms, and these have mostly disappeared. If such was the case in 1798, it is very probable that it had been that way since the establishment of the earliest colonies.

Turning now to a second inventory, that of John Rickard, taken on July 4, 1678 (no fireworks or parades; we had to wait ninety-eight more years), we find that, although the inventory was taken on a room-by-room basis, the plan of the house is not quite as clear as was the case in that of Will Wright, although the ambiguities are minor.

Probate Inventory of John Rickard

July 4, 1678

An Inventory of the estate of John Rickard senior of the Town of New Plymouth lately deceased taken and apprised by Sergeant Ephraim Tinkham Jonathan Barnes and William Crow the 4th day of July 1678 and exhibited to the Court of his Majesty held at Plymouth the fifth of July 1678 on the oath of Hester Rickard.

	£	s	d
In the Parlour			
Impr: 1 pair of old Andirons 1 spit 1 fire shovel	00	08	?
Item 1 frying pan 18 pence 1 old Iron pot 6d	00	02	00
Item 1 small Gun 7s 1 Catuce box 2s 6d shot and bullets 2s	00	11	06
Item 3 powder horns and powder 2s 6d 1 snapsak 6d	00	03	00
Item 1 old sword 2 shillings 3 Chairs 5 shillings	00	07	00
Item a Commentary on the Revelations	00	05	00
Item a book of Mr Mather's 1s a Psalm book 1 shilling	00	02	00
Item more old books 1s 1 smoothing Iron 8d	00	01	08
Item 1 pair of Stilliyards 12s 1 razor 1 stone & Comb 2s 6d	00	14	06

	£	s	d
Item a Psalter 6d a Remedy against Satans devices 1s	00	01	06
Item divers Golden Chain 1s one small silver spoon 5s	00	06	00
Item several sorts of pewter	03	05	00
Item eleven alchemy spoons 4 quart pots 1 pint pot	00	12	00
Item a small pot 4 porringers 2 Latten pans 2 [trenmells?]	00	02	10
Item Several [?] in tin and pewter	01	01	03
Item 3 Chests 1 box 1 Churn & small battens	00	14	00
Item Several small things	00	04	08
Item a silver spoon 10s Cash 1li 11s 2d	02	01	02
Item 1 Table form 2 Cushion 1 Chest 1 Cushion Cover	01	08	00
Item 1 Close bedstead 1 feather bed and bolster	04	00	00
Item 1 Coverlet and a pair of blankets	01	00	00
Item 1 old Curtain and valence and an hour Glass	00	02	06
Item a warming pan and several sorts of [linen?]	01	05	00
Item several small things	00	05	00
Item 1 brass pan a half bushel and an half peck	00	09	00
Item a piece of Red Cotton and Cotton Yarn	00	05	00
Item several sorts of linen	01	19	00
Item one Fending Iron 1 fork 1 hake 1 settle	00	17	06
Item 1 sifting trough and one Chair Table	00	09	06
Item 1 pistol and a looking Glass 1 brass kettle	02	19	00
	26	14	04
His Apparel			
Item 1 Great Coat & a pair of breeches	01	10	00
Item 1 old Coat 1 pair of [...cs] 1 serge waistcoat	00	13	00
Item 1 pair of Cotton drawers and 1 waistcoat 3 old waistcoats	00	08	00
Item 1 Cape Coat Gloves a Cap	00	06	06
Item 2 white shirts 2 blue shirts 1 pair of drawers	00	13	00
Item 1 Cap 2 Neckcloths 2 hats	00	06	06
Item 2 pair of stockings and one pair of shoes	00	09	00
	04	06	00

	£	s	d
In the Kitchen			
Item 1 spade 1 tackling for a scythe	00	05	00
Item a little old brass kettle & a skillet	00	03	06
Item 3 kettles 1 Churn 1 pail 1 Candlestick	01	01	03
Item 9 old wooden dishes & 1 Can 1 straining dish	00	05	03
Item 3 stools 2 pairs of Cards	00	02	09
Item 2 hoes 1 sifting trough and one old trough	00	06	00
	02	04	03
Item 1 feather bed 1 bolster 1 pillow			
Item 1 pair of sheets 1 pair of blankets			
Item 1[?] not appraised			
Item a small quantity of household provision not appraised [Marginal note—Kitchen total?]	05	09	03
In the Chamber			
Item 22 lbs of woollen yarn 4 old barrels	01	[10?]	06
Item 1 small feather bed and blanket	02	00	00
Item 2 yards of [?] 20 pound of Rhode Island wool	[10?]	12	06
Item a parcel of old Ropes and old Iron	00	03	00
Item 2 plains 1 auger 1 Chisel and file	00	03	00
Item 1 adze old Iron & a [line?]	00	03	[..]
Item 1 old box with grater and other small tools	00	02	[..]
Item 5 fishing leads & 2 pieces of line	00	[..	..]
Item 4 old barrels a hogshead		[..ms torn..]	
[Folio 123] Brought from the other side	38	11	10
In several places			
Item broad axe 2 old axes 3 small wedges and two little rings	00	09	00
Item 1 pair of tongs 1 sieve 1 pair of pot hooks and one pothanger	00	05	03
Item 7 pound of leaden weights 3 Gimlets 2 augers	00	05	00
Item one Glass bottle 1 auger 1 Gimlet 1 Gouge	00	01	06
Item 2 books Called the sudden appearance of it	00	01	00
Item 2 Chisels and some Fishing Craft	00	07	00

	£	s	d
Item some particulars in an outhouse	00	06	00
Item 1 Cooling tub 1 [lawn?] sieve & 1 picture	00	06	00
Item Molasses old Cask and several particulars	08	16	00
Item a pair of wooden scales 1 basket and some Cotton yarn 2 bushels of wheat four bushels malt	01	10	00
Item a little brass kettle	00	03	00
Item 1 small Chest	00	03	00
	50	04	06
Item his part of his sons Fishing voyage	06	00	00
Item his sixth half part of a small Ketch at	15	00	00
Item Cider in ten barrels and 2 barrels of beer	05	04	00
Item Tubs baskets pails butterpots & firkins	00	09	00
	77	17	07
Debts due to the estate			
Item Thomas hughes	01	10	00
Item John Andrews	01	10	00
Item Ephraim Tilson	00	03	05
Item The Town of Plymouth	00	11	06
Item Jonathan Shaw	00	04	00
Item Waterman	00	09	00
Item Gyles Rickard senr	01	15	00
Item Samuel dunham Junir:	00	08	07
debts	06	12	06
The total is	84	10	00
Debts to be payed out of the estate			
To George Watson in Money	04	00	00
To James Clarke	03	09	03
To Jabez howland	00	03	00
To James Cole	00	04	00
To Mr John Blake of Boston	00	15	03
To Elizabeth Warren of Boston	00	19	03
To Gorge Bonum	00	02	07
To William Clarke	00	12	06
To Funeral Expenses	01	03	00
	11	04	10
Debts deduced there Remains	73	05	03

Besides his house and 2 Gardens and eight acres of land lying near the Town and two acres of meadow att dolnes [?] Meadows in Plymouth, with a small parcel of meadow at a place Called hannaswapett;

Ephraim Tinkham
his E T mark
Jonathan Barnes
William Crow

Letters of Administration were Granted by the Court unto Ester the wife and Relict of the said John Rickard deceased and unto John Rickard the son of the said John Rickard Jointly to administer on the said Estate and as Concerning the disposal and settlement of the property thereof it is Referred unto them that if they Can settle it to mutual satisfaction and Give in a Clear account thereof unto the Court that so it may be entered on Record, In as much as there is none to enjoy it but they two; so far good provided it be done in some Convenient time; otherwise they are to Repair to the Court for further settlement thereof, as there may be occasion.

As we can see, Rickard's house had a parlor, a kitchen, and a chamber. In seventeenth-century usage a chamber was almost invariably an upstairs room, and so the plan was probably a standard hall and parlor house with a chamber above. It will not be necessary to comment on Rickard's inventory to the extent of that of Will Wright, but there are some items of interest. "Stilliyards" is a variant spelling of steelyard, which is a kind of balance scale. One of the more puzzling items, which required a bit of effort to define, is the "catuce box" listed on the third line of the inventory. However, in part because of the objects with which it is associated, it was possible to identify it as a cartouche box, cartouches being paper cartridges that would be used with a snaphaunce firearm, listed with the spelling "snapsak." But the most interesting grouping of items is to be found in the listing "in several places," and includes molasses, a sieve, a cooling tub, two bushels of wheat and four of malt, noted under the subheading "in an outhouse." There is no question that this assortment relates to the brewing of beer, with the possible exception of the molasses. Household production of beer was commonplace in the seventeenth century, both in England and in the English colonies of America. Seventeenth-century inventories from Essex, England, list the contents of various rooms as well as outbuildings far more frequently than those of Plymouth, and often refer specifically to a brew house, where beer was made. It is very likely that most households in Plymouth made their own beer, but the inventory evidence is far from clear. But in John Rickard's case, there is no doubt that beer was being made in the outhouse, and we might assume with confidence that it was Rickard who was brewing it. However it seems that he wasn't, for in the records of the court for March 5, 1672, we find the following:

> At this Court, Hester, the wife of John Rickard, Senir, of Plymouth, appeared, being summoned to answer for her uncivil and beastly carriages and speeches to her said husband; and the premises was fully proved against her by sufficient testimony, and she was sentenced by the Court to be publicly whipped at the post; but at the earnest entreaty of her self and others, and her promise of amendment, the said sentence was suspended from present execution, with this proviso, that if at any time for the future she be taken in the like fault, either towards her husband or in any uncivil carriages to others, she is forthwith to be publicly whipped as aforesaid.
>
> She was also at this Court prohibited to brew beer to sell, as formerly she had done, because it appeared to the Court that it was a snare to her to occasion evil in the aforesaid respects.

The fact that Hester Rickard was not only brewing but also selling beer suggests that not all families in the community were producing it, but chances are good that a high number of them were.

One final item in the Rickard inventory merits comment. In the last entry of the contents of the parlor we find listed a looking glass. By 1678, this object was probably a mirror, but in earlier inventories, those predating 1660, if English usage was used, "looking glass" was a common nickname for a chamber pot, and it is virtually impossible in most cases to determine just what was being listed. This double meaning has misled some modern editors to misinterpret certain passages in written plays. In his "Some Domestic Vessels of Southern Britain: A Social and Technical Analysis" (a classic example of British understatement; the entire monograph is about chamber pots), in a section headed "Pisspot semantics," P. Amis writes:

A modern editor was obliged to annotate the following lines from Thomas Middleton's *A Chaste Maid in Cheapside* which was printed in 1630 and perhaps first performed as early as 1613:

> Hyda [Allwit calls a servant],
> a looking glass: they have drunk so hard in plate,
> That some of them had need for other vessels.

The editor's interpretation was that Allwit was calling for a looking glass to see if one of the characters, out cold on the floor, was still breathing, whereas what was urgently needed was a chamber pot. Amis also quotes John Collop, a poet writing in the mid-1650s, who uses "looking glass" and "chamber pot" interchangeably, seeing them as analogous to a glass urinal into which a physician looks:

> Hence, looking-glasses, Chamber pots we call,
> 'Cause in your pisse we can discover all.

Physician with looking-glass urinal—
Bartolomäus Anglicus, Haarlem, 1485.

The use of glass urinals as diagnostic devices was a common practice of the time, and the term was simply generalized to any container for urine, which of course would be a chamber pot. And even in the case of the Rickard inventory, we are not on entirely safe ground, despite the late date. In 1688, Randle Holme published his monumental *Academy of Armory* in which he attempted to list, define, and illustrate thousands of devices that could be incorporated into a coat of arms, hence the name. The book is a truly remarkable record of a former material and semantic world, and an invaluable source for anyone working on seventeenth-century material culture. Holme writes:

> A chamber pot or a Bed pot . . . This is by the Jolly crew when met together over a cup of Ale; not for modesty sake, but that they may see their own beastliness, in powering in, and casting out more than suffices nature, which if it went not suddenly downwards, would force its way upwards, is called a Looking glass; But there is nothing never so useful, but it may be abused, so is this when it is called by such persons a Rogue with one ear, and a piss pot.

The last inventory to be considered is far more typical of most of those from Plymouth; it is not a room-by-room inventory, and the value of the goods is fairly standard for inventories of this type. It is of particular interest in that it represents the estate of a woman, Judith Smith of Taunton, and was taken in 1650. Since it was not the inventory of the goods of a man, the problem of the widow's dower did not present itself. In the transcription which follows, we have used seventeenth-century orthography, but included parenthetical definitions of more obscure terms.

Probate Inventory of Judith Smith, Widdow

December 15, 1650

The Inventory of mis Judith Smith widdow made and apprised the fiveteenth of the 10th Month 1650

	£	s	d
Impri: her apparrell one gowne and kertle	05	00	00
her best petticoate	03	00	00
her serge gowne	02	00	00
It one cloake one Ryding coate and hood	01	00	00
It one green apron and a payer of bodies [bodices]	00	14	00
It one wastcoate one petticoate	01	05	00
It 4 petticoates and one old wastcoate	01	08	00
It one hatt one muff	01	00	00
It her wearing linnin	01	00	00
It one skarffe	00	18	00
It shooes and stockens	00	05	00
It sheets napkins pillowbeers and other linnin	04	00	00
It pewter and Candlestick	03	00	00
It 3 kettles one copper one bakeingpan 3 skillets one brase pott	04	10	00
It one warming pan	00	06	00
It one cherne 3 cheesfatts [molds to shape cheese] one buttertubb 6 trayes one keeler [milk tub] a cheestubb 2 milk pailes 5 old barrells one beer vessell and som other old tubbs one salting trough	01	18	00
It one paier of kob Irons one fierpann 2 paier of tonges 2 hakes [pot hooks]	01	00	00

	£	s	d
It one brandlet [trivet] one Rostiron one frying pan a smothingiron with the heates one chaffing Dish	00	12	00
It 3 wedges 3 axes one longe sawe one crosecutt sawe 2 hoes	01	10	00
It 2 sithes 2 sickles 2 spits	00	12	00
It one shave 2 wimbles [gimlets]	00	04	00
It 2 morters one pestell	01	00	00
It one plow with the Irons therto belonging and 2 chaines	01	10	00
It one cart as it goeth and yokes	01	10	00
It one flock bed one Rugg one bolster and a pillow one paier of curtaines one vallence and a bedtester	01	10	00
It 2 towe combes one heckell [tool for combing flax]	00	10	00
It 2 spining wheeles	00	10	00
It 2 chist 2 trunkes 3 old boxes	00	11	00
It 3 bedsteds	00	10	[?]
scales waightes howerglasse an old hayne	00	10	00
It bookes	01	00	00
It one pillion 5 cushens and feathers	01	05	00
It 4 sacks 3 smale baggs	00	16	00
It one spade one mattacke one holborne and stub syth and a cuting spade	00	12	00
It butter and cheese	01	00	00
It Indian corn 40 bushels	06	00	00
It wheat 25 bushells	06	05	00
It Rye 9 bushell	02	00	00
It Rye upon the ground	03	00	00
It hay	10	00	00
It 2 great oxen	12	00	00
It one sow and 7 piggs	03	00	00
It one hogg	02	00	00
It eleven piggs	06	00	00
It 4 calves and pte of one [a part share in one]	08	00	00
It 2 young cattle and halfe a one [a half share]	08	00	00
It 8 bushels of hoggs corn	00	12	00

	£	s	d
It cotten woole 14	14	00	00
It in Debts	00	17	00
It Desparate Debts	01	15	00
It 2 bedsteds 2 buffet stooles [low three-legged stool] one Table and other lumber	01	00	00
summa is	120	06	00

apprised by Josepth Pecke and Thomas Cooper

Henery Smith and Judith Smith tooke theire oath to the truth of this Inventory before mee Timothy Hatherley the 22cond of September 1651

As mentioned above, while not taken on a room-by-room basis, groupings of functionally related items are found together, giving some sense of where the appraisers were in the house when valuing the goods. Note the last entry, for two bedsteads, two buffet stools, one table, and other lumber. Will Wright's inventory also lists lumber, but the term did not have the same meaning that it has today. In the seventeenth century, "lumber" referred to a hodgepodge of things, not wood or boards. Rather, it is what can be found in many modern closets or cupboards, and often garages. "Desperate Debts" were those of which there was no anticipated recovery.

There is no question that probate inventories are an extremely valuable source of detailed information concerning all of the items that were found in a house. But they do fall short at times in terms of specific detail. A set of old spoons could be of pewter or brass, a parcel of earthenware could consist of locally made utility wares, or of more refined types of pottery imported from Europe. The only source of such information comes from archaeology, which also has revealed details of housing that would otherwise not be available. Only when probate data and archaeological findings are combined can we get a more complete picture of the material world of the time. Of the two, archaeology occupies a special niche. Inventories are primary sources, as are court records and histories written by contemporaries. There is no question of their critical value in reconstructing the life of seventeenth-century Plymouth, yet they are in essence two-dimensional, and intended to be read to extract the necessary information, and while they are referred to as primary sources, there is always the slightest possibility of certain biases in them that cannot be detected. One could say that for this reason they too are secondary sources, although admittedly quite different from what is commonly referred to by that term. If this is correct, the only true primary source is archaeology, for through only it can a modern person come directly face-to-face with the past.

It would take someone with a minimal amount of imagination not to appreciate the emotional impact conveyed by the various artifacts encountered in excavating a site—someone holding a brass spoon just recovered from a seventeenth-century trash pit is the first to have handled that object in more than three centuries. Likewise, standing in a newly cleared cellar, an archaeologist is the first person to be physically present since the house under which the cellar stood was abandoned some three hundred years before. So in a sense, archaeology mainlines the seventeenth century directly into the twenty-first, with no intervening documents, however reliable they may be, standing between the archaeological data and the researcher. But there is more than just emotion involved here. As we shall see, detailed knowledge of a wide range of artifacts can be obtained only through excavated materials. The same applies to architecture, because prior to 1972, there was a void of half a century in our knowledge of specific details of house construction in the colony, with the exception of the half dozen houses still standing in Plymouth, the dates of which are to a greater or lesser degree suspect. So let us now turn to a consideration of archaeology in Plymouth Colony carried out between 1863 and 1972.

Chapter 6

Still Standing in the Ground

The Archaeology of Early Plymouth

In 1863, when James Hall first put spade to earth on the site of a then vanished seventeenth-century house in Duxbury, Massachusetts, he could not have realized that he was about to conduct the first truly systematic archaeological excavation in America, and possibly the entire world. Unlike Thomas Jefferson's excavation of a burial mound in Virginia less than a century before, Hall was not seeking answers to broader questions regarding the people, or, in this case, person, responsible for the site's creation. The house that had stood there in the seventeenth century was that of Myles Standish, and Hall was one of his descendants, so his motivations seem to have been singularly personal. Nonetheless, his recording of his findings was meticulous, even by present-day standards. He produced a scale map of the house foundation, and listed the artifacts recovered, keyed into the location in which they were found, a process known as "piece plotting," which did not come into fashion until the mid–twentieth century. In addition, he tied the plan into *two* datum points, springs that still flow on the property today. A datum point is a fixed, and one hopes permanent, feature that enables later excavators, should they have the need, to locate the site without difficulty. Considering Hall's profession as a civil engineer, this comes as no great surprise, yet it is remarkable nonetheless. This important piece of archaeological history remained unknown until 1963, when Hall's descendants, who resided in Mexico, discovered the map and artifacts among his effects and sent them to Pilgrim Hall in Plymouth. Many of the artifacts listed on the map were missing, but there were a sufficient number to permit a date to be assigned the site between 1637, the date when Standish moved to Duxbury, and about mid-century.

Hall's work marks the first example of historical archaeology, a field of inquiry that became a central component of archaeological studies in the 1960s, although as we shall see, it was carried out here and there well before that time, and Plymouth Colony sites were among the earliest investigated, and leaving aside James Hall's pioneering work in the mid-nineteenth century, had their beginnings in the 1920s. For the most part, historical archaeology involves the excavation and analysis of sites that represent the expansion of European culture worldwide, beginning in the late fourteenth century, and the impact that this had on indigenous cultures, and how that interaction produced a mix of cultural elements from both groups involved, a process known as creolization. To cite but one example, enslaved Africans in eighteenth-century Virginia manufactured handmade, burnished pottery, both aspects of West African ceramic technology, but frequently the shapes were distinctively English, including teapots, chamber pots, porringers, and punch bowls.

Admittedly the definition given above is Eurocentric, but it does describe the work of a vast majority of historical archaeologists. True, other cultures were literate, and written records exist to complement the archaeological record, ancient Sumer and dynastic China being two examples. But these have had a secure niche in other disciplines, classical archaeology and Far Eastern studies. And of course, there are cultures, such as the Zulu of South Africa, with strong oral traditions, which can serve much the same purpose as written documents. Historical archaeology, however, emerged as a part of anthropology, and in a sense had to define itself. But regardless of how we choose to define the field, it has strong and deep roots in Plymouth.

Following Hall's excavations in 1863 and 1864, no further archaeology was carried out until the 1920s, and between then and the beginning of the Second World War, four more sites were investigated. Before detailing the results of this work, some general observations should be made concerning the state of the field at this early time. Our knowledge of seventeenth-century ceramics, valuable in both dating a site and determining the origin of pottery fragments of pieces manufactured in Europe, was rudimentary at best. But in the period between 1969 and the present, our knowledge of ceramics has increased exponentially, and today we are in a position to put the analysis of ceramics in its rightful and valuable place. Such an emphasis on a single category of artifacts is not misplaced. All households used them, they are amply represented on all sites of the period, and, while relatively fragile, their fragments preserve extremely well once buried. Since they are so plentiful, they permit various quantitative comparisons among sites. They are also well represented in probate inventories, but, as we have said, usually not in the kind of detail that would permit identification of the specific type listed.

A second and equally important contribution to historical archaeologists' approach to their data was made by the late J. C. Harrington, known as "Pinky" to his colleagues and friends. In the early 1950s, he was engaged in the study of white clay English smoking pipes recovered from the excavations at Jamestown, Virginia. Fragments of these pipes, particularly their stems, are found in abundance on all English seventeenth- and eighteenth-century sites. Both men and women smoked; in fact, the incisors of two females whose burials were recovered from outside the Quaker Meeting House in Newport, Rhode Island, showed wear that was identified as resulting from holding pipes between their teeth. There is also a variety of documentary references to women pipe smokers. In earlier years, pipe bowls were quite small, and over time, as they became larger, the stems became longer. By the time Harrington conducted his study, there was a solid body of careful scholarship, conducted in England, on pipe bowls and the changes they underwent over time. The bowls could often be identified from maker's marks impressed on them, which could be traced to directories and apprentice rolls. As a result, Harrington's work with the Jamestown pipe bowls was relatively easy. But the problem was that, as is the case with any site of the period, stem fragments outnumbered bowls by a ratio in the neighborhood of one hundred to one.

It was long thought that these pipe fragments were of little significance, save to show that smoking was a prevalent practice in colonial households. Fortunately, despite this, they were retained as part of a site's collection, and properly cataloged. However, as he worked on the pipe bowls from Jamestown, Pinky noticed that the earlier bowls had larger stem bores than the later ones; there is almost always a bit of the stem still visible on the bowls, and as he pondered this, the stem fragments took on a completely new significance. We cannot do better than quote from a letter sent to James Deetz in 1992 from Pinky Harrington. As he describes it:

> You ask when it occurred to me that stem bores varied, and that the earlier ones (based on bowls) are larger. This all came gradually over several years, and not as a sudden revelation. . . . after we moved to Richmond in 1946 I finally got around to exploring this tentative observation. So I brought all the pipe bowls in the Jamestown collection to Richmond, and continued to gaze at the stem bores, finally deciding that there was something here that needed further analysis. My problem was how to quantify this; that is, how to measure the bore sizes. I couldn't just say "little," "medium," "big," etc. As we always do when faced with a problem, Virginia [his wife] and I were discussing this over a martini, or two, and she said: "Why not use those steel drills you have down on your work bench?"
>
> And thus it all began some time in 1952 or '53. I then spent evenings measuring stem bores [and] worked up my paper, but at that time there was no broad outlet for articles on historical archaeology, so I gave it to the Archaeological Society of Virginia. The rest is "history."

Harrington published his method in the *Quarterly Bulletin* of the Archaeological Society of Virginia in 1954, and since then pipe-stem dating based on bore diameter has become a most valuable technique for dating sites, and has been used in part to assign dates to various sites described in this chapter.

There are two things worthy of note in Harrington's letter. First, it was Virginia who thought of using drill bits, which are manufactured in diameters that increase incrementally from 9⁄64 inch to 4⁄64 inch, 1⁄64 inch at each step, and it was this range of sizes that applied to the stems with which Harrington was working. And while Harrington is rightfully recognized as the person who introduced this method of dating, as is so often the case a woman was instrumental in its application, but few people are aware of her role in its development. Second, while in no way condoning alcohol abuse, it may well be true that a martini (or two) sometimes connects certain synapses in the mind that otherwise would remain closed. Who knows?

In any event, once one has measured a sample of stems from any given site, and a minimum of at least fifty is believed to be a sample of adequate size, the percentage of bore diameters of each size is calculated and presented in the form of a simple histogram, as illustrated below. Such histograms serve two purposes. The first and most obvious is that they provide a date for the site in question. Less obvious, but equally important, is that they also indicate the length of occupation; the flatter the graph, the longer the site was occupied, and sites that produce a sharp peak, with little on either side, indicate

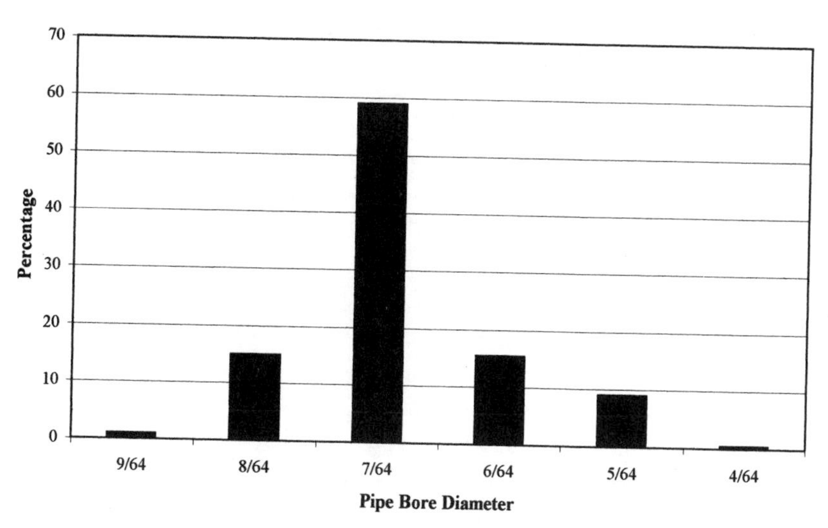

Pipe-stem bore histogram for Winslow site, Marshfield, Massachusetts.

a shorter period of occupation. Harrington assigned a date range for each bore diameter as follows:

9/64 inch	1590–1620
8/64 inch	1620–1650
7/64 inch	1650–1680
6/64 inch	1680–1710
5/64 inch	1710–1750
4/64 inch	1750–1800

With the exception of the smallest bore diameter, the dates assigned provide a date range of plus or minus fifteen years, which is a very comfortable degree of precision. It is less than clear why Harrington chose 1590 for the beginning date for bores of 9/64 inch. He may have had access to certain material from England from that early a date, but recent excavations of the original Jamestown Fort have produced very small Elizabethan pipes that have bore diameters smaller than 9/64 inch. But for the vast majority of American English colonial sites, the method seems to work very well, and certainly in the case of Plymouth, Harrington's date ranges present no problems. So the method seems to work, and we shall rely on it in helping to assign dates to those sites that have been excavated in Plymouth.

In 1926, Percival Lombard excavated what he believed to be the site of a trading post located in present day Bourne, Massachusetts, just south of the canal that today separates Cape Cod from the mainland. The post was established by the Plymouth settlers in the late 1620s, at Manomet, or Aptucxet as it was called by the Indians, some twenty miles south of Plymouth. William Bradford mentions that "they built a house there and kept some servants, who also planted corn and reared some swine, and were always ready to go out with the bark when there was occasion. All of which took good effect and turned to their profit." The building was destroyed by a hurricane in 1635, and has since been, in Samuel E. Morison's words, "faithfully restored." However, a recent reanalysis of the material recovered by Lombard has been conducted by Craig Chartier, and this study shows that far from being the Aptucxet trading post, as Lombard thought, the site was the location of the home of Ezra Perry II, occupied between the 1670s and 1720s. Given the state of knowledge of both artifacts and building techniques at the time, Lombard cannot be faulted in his misidentification.

Plymouth Colony has at least two sites supposedly representing very early homesteads, that of Robert Bartlett, a wine cooper who arrived on the *Anne* in 1623, and William Bradford, in present-day Kingston. But in both cases, excavation done in the 1960s and 1970s showed that neither site was much earlier than circa 1675. The same applies to the second site to be excavated following James Hall's early work on the Standish site, that of John Howland. The project was undertaken in 1937 by an architect, Sidney Strickland, who was also responsible for the restoration of both the Jabez Howland and Richard Sparrow houses.

Strickland had a title search carried out, and there is no question that the property was lived on by John Howland, whom we first encountered when he was swept overboard from the *Mayflower.* Howland acquired the property from John Jenney in 1638, who had already built a house there. There was an obvious depression in the soil, and a quantity of rocks protruding from the surface. Understandably, Strickland assumed that this was evidence of the location of the dwelling house on the site, and proceeded to excavate the surrounding area. He made a map of the foundation that he had uncovered, but it is simply in the form of an outline, with no detailing of the nature of the laying of the stones that served as footings. To the south of the house foundation, a curious stone-lined basin was partially exposed and identified as a source of water for cows, which could lick the dew that had condensed in it in the mornings. The "dew pond" appealed to visitors, but no such feature has ever been encountered elsewhere, and as we shall see, the identification was both speculative and incorrect.

No grid was employed in Strickland's excavation, which would have provided data on the location of the recovered artifacts, which were not cataloged, and for many years they reposed on the porch of the home of a summer resident of Plymouth, stored on paper plates. They were cataloged in a general way only in the 1970s, and by that time, there is little question that there had been considerable mixture of pieces from different locations. Strickland published a brief account of his work in 1939, but provided little detail regarding his findings. The ceramic collection is a strange assortment of pieces, many of which date to the early nineteenth century, a quantity of later seventeenth-century pieces, and a very small number of pieces that date to the earlier occupation, perhaps that by John Jenney rather than John Howland.

Three years later, in 1940 and again in 1941, two sites were excavated in a thoroughly professional manner by Henry Hornblower, who later became president of Plimoth Plantation, that museum having been his original concept. Hornblower, an undergraduate at Harvard at the time, received his initial training under the direction of one of the foremost archaeologists of the period, John Otis Brew, at Awotovi Pueblo in Arizona. Having been a summer resident of Plymouth for most of his life, Hornblower had developed a strong interest in the history of the colony, and made an innovative and far-reaching decision, that archaeology had something to contribute to our understanding of life in those early years. There is no question that he was correct in this judgment, and his work was to have important implications in the growth of historical archaeology.

Hornblower combined the rigor of James Hall with the proper questions to be asked of the data. What types of houses were typical of the period? What can we learn from the artifacts recovered concerning everyday life in the colony? At the time, Hornblower was a member of the Harvard Excavators' Club, made up of anthropology undergraduates under the sponsorship of J. O. Brew, and it was this club that undertook the two excavations under his direction. The first site to be investigated in 1940 proved to be one of the two oldest to be excavated in the Old Colony. It is known as the R. M. site, the name deriving from a set of initials scratched on the end of a spoon handle. Later research suggests that it was the site of the Clark garrison house, located on the Eel River two miles south of the town center, which was burned by the Indians during King Philip's War, but ceramic and pipe-stem dates suggest a pre-1650 construction date. It is located on property presently owned by Plimoth Plantation, and has recently been paved over to provide additional parking for visitors to the museum, so it is not available for reexamination.

The following year, in 1941, Hornblower and the Excavators' Club worked at the Josiah Winslow site in Marshfield, Massachusetts, a hundred yards or so behind the standing Winslow house, Careswell, which was built in 1699. The same careful standards of excavation were employed as at the R. M. site. In both cases, a large area surrounding the house remains was exposed, a practice that was not to become standard procedure until the 1960s, and then in the Chesapeake region. As a result, we are in a position to make some important observations concerning the construction methods employed in raising the houses.

Further archaeology in the Plymouth area did not resume until 1959, when an archaeology program was established by Plimoth Plantation under the direction of James Deetz. The first site to be investigated was that supposedly occupied by Robert Bartlett in the early seventeenth century. However, it was a disappointment, since the data recovered indicated occupation in the late seventeenth century at the earliest, and the artifact yield was very low. While the crew of only two people was working at the site, which abuts the Plymouth golf course (tees and golf balls were among the artifacts encountered), the greenskeeper for the course, Arthur Vantangali, an amateur archaeologist himself, told them that he knew of a site that was not only very rich, but from artifacts on the surface dated to the seventeenth century. The site was located on the opposite side of Howland's Lane, directly across from that excavated by Sidney Strickland in 1937. It was covered by a dense growth of greenbrier, poison ivy, and other underbrush; it was possible to gain access only by crawling through the mass of greenery. However, once cleared of all the growth, it was indeed an impressive site, at the time identified as the house of Joseph Howland, John's son, and built just after the end of King Philip's War. More crew members were recruited, and most of that summer was devoted to excavating the site.

Excavations continued during part of the summer of 1960. From that date onward, Plantation archaeologists investigated six more seventeenth-century sites. The first five of these were the William Bradford site in Kingston, the cellar of the Harlow house in Plymouth, the Jabez Howland house lot in Plymouth, a whaler's tavern in Wellfleet on Cape Cod, and two contiguous rear yards in the heart of downtown Plymouth. The sixth one, the home site of Isaac Allerton, was excavated in 1972. Archaeological investigation of seventeenth-century sites in Plymouth then came to a halt over the next quarter century, due to policy and personnel changes at Plimoth Plantation, but in the summers of 1998 and 1999, a team of archaeologists from the University of Virginia reexposed the foundation originally excavated by Sidney

Strickland at the John Howland site at Rocky Nook. As a result of these excavations it is possible to make a number of observations regarding both house construction and artifacts used in early Plymouth.

The Archaeology of Early House Construction in Plymouth Colony

Allerton Site, Kingston, Massachusetts

On a chilly April morning in 1972, Christopher Hussey, a Plymouth architect, and his client Orfeo Sgarzi, also from Plymouth, were inspecting a parcel of land on which Sgarzi planned to build a house. It was a pleasant site, on a small rise in Kingston, some six miles from Plymouth, overlooking the Jones River to the north. While examining the property, Hussey noticed a thin scattering of artifacts on the surface, concentrated in an area about thirty feet across. He placed them in a paper bag and took them to Plimoth Plantation to show to the archaeologists there. It took but a cursory examination to realize that they were indeed early, with many of the pipe stems dating before 1650, using the Harrington method. Clearly Hussey had stumbled on what would become one of two of the earliest sites investigated in the Old Colony area.

There is a significant subtext to this account. Few, if any, architects at this time would have taken note of the artifact scatter, and even if they had, most would have assumed they were relatively modern bits and pieces. Chris Hussey, however, grew up in a family deeply interested in seventeenth-century Plymouth. His maternal grandfather was Sidney Strickland, who had excavated what he thought to have been the house site of John Howland in Kingston in 1937 and was also responsible for the restoration of the Jabez Howland and Richard Sparrow houses in downtown Plymouth. Hussey's father, Rodman, conducted the title search on the Howland property, and his mother was the horticulturist at Plimoth Plantation. As a result Chris Hussey grew up surrounded by things "Pilgrim," and had an eye trained to observe evidence of earlier land occupation.

A title search confirmed that the property was acquired by *Mayflower* passenger Isaac Allerton sometime around 1628, when he built the dwelling house there. Robert Anderson states that he:

> Was one of the busiest and most complicated men in early New England. . . . Records for Allerton may be found in virtually every colony on the Atlantic seaboard and in the Caribbean, including Newfoundland, New Netherland, New Sweden, Virginia, Barbados, and Curacao.

Allerton doubtless had business dealings in all of these locations and, based on his record at Plymouth, had no hesitation in robbing Peter to pay Paul. There seems little doubt that many of his ventures in these various ports of call had their shady side, based on what we know of his involvement in affairs at Plymouth, but as far as the folks there were concerned, it may have been of little moment, or they may well not have known the true nature of his activities. However, when their lives were involved directly, it became a very different matter. Allerton was the first assistant appointed under William Bradford when he became governor after John Carver died in the spring of 1621. One of Allerton's major responsibilities was to serve as Plymouth Colony's agent representing the undertakers (the twelve men who had assumed the colony's debt) in dealings with the London merchant adventurers. He was one of the undertakers himself.

From the beginning, it was apparent that some better choice could have been made. Bradford writes in his history:

> Concerning Mr. Allerton's accounts. They were so large and intricate as they could not well understand them, much less examine and correct them without a great deal of time and help and his own presence, which was now hard to get amongst them. And it was two or three years before they could bring them to any good pass, but never make them perfect.

Such obfuscation seems almost intentional, and in the light of later developments probably was. Each time Allerton met with the merchants in London, on his return to Plymouth the Colony's debt had grown larger instead of smaller as it should have. As James Sherley, one of the London merchants wrote to Bradford, "if their business had been better managed they might have been the richest plantation of any English at that time." Allerton was not beyond treating his own family in a cavalier way, as witnessed by the following passage from Bradford, which refers to William Brewster, lay leader of the church at

Plymouth, and most senior of the Scrooby church members who emigrated to Leiden and then Plymouth in 1620. Brewster's daughter Fear had married Isaac Allerton sometime between 1623 and 1627. She was his second wife.

> Yea, he screwed up his poor old father-in-law's account to above £200 and brought it on the general [Plymouth] account, and to befriend him made most of it to arise out of those goods taken up by him at Bristol, at £50 per cent, because he knew they would never let it lie on the old man; when, alas! he, poor man, never dreamt of any such thing, not that what he had could arise near that value, but thought that many of them had been freely bestowed on him and his children by Mr. Allerton. Neither in truth did they come near that value in worth, but that sum was blown up by interest and high prices, which the company did for the most part bear (he deserving far more), being most sorry that he should have a name to have much, when he had in effect little.

The first sentence of this passage initially comes as a surprise to the modern reader for "to screw up" is not a seventeenth-century expression. However, once one has completed reading the sentence and the remainder of the quotation, clearly Bradford meant Allerton increased or raised up William Brewster's debts to an exorbitant amount under the pretense of inflated values. Part of the increased debt that the undertakers faced as a result of Allerton's mismanagement, or worse, was the result of a fishing and trading expedition to which he committed the undertakers without their knowledge. It was Bradford's suspicion that Allerton had so rigged it that if these ventures were successful, the undertakers would not obtain any profits, but should they fail, the losses would be underwritten by them. By 1627, Allerton was trading on his own initiative, and by 1630 was using the undertakers' credit to pay for his own speculative business dealings. The same year he hired a ship, charging the cost to the undertakers. This was the proverbial straw that broke the camel's back, and Bradford dismissed Allerton from his position. But the damage was done. Under Allerton's control, the debt that had been some £2,400 in 1627 had mushroomed into close to £6,000.

Allerton apparently acquired the property on which he built his house in 1628, and by 1646 had deeded it to his son-in-law Thomas

Cushman. Whether he remained in Plymouth in the intervening years is unclear. At some point he moved to New Haven, where he apparently lived until his death on February 12, 1658.

The location of almost any site found on the coastal plain of the eastern United States has been cultivated in most cases for many centuries. This applies equally in those areas that today have a forest cover, for the trees there are mostly third and fourth growth, and earlier these places were probably free of forest. For example, today along the track of the Baltimore and Ohio railroad from Washington, D.C., to Cumberland, Maryland, there is now a heavy growth of trees, yet in the 1930s, this same area was open farmland. The turning of the earth by the plow mixes organic substances such as stubble, roots, and other materials, producing a rich brown layer known as the plow zone. Beneath this layer is what is referred to as subsoil, which is often tan in color, although in parts of the south it can take on a distinctly reddish hue. Any intrusion into the subsoil can occur from one of two sources, burrows of animals such as woodchucks, gophers, and mice, or intentional cutting into the subsoil by the former inhabitants of the site. They are easily distinguished by the irregularity in shape of animal burrows and the more regular, usually circular or rectangular shape of those created by human agency. These human-excavated pits are known as features, and can include cellars, trash pits, postholes, fence lines, wells, and graves. They are filled with darker soil, the result of a mixture of topsoil and the deterioration of the contents of the pits. By definition, features are those things that cannot be collected and taken into the laboratory for analysis; they must be carefully recorded and mapped in place.

Work began at the Allerton site in late April 1972, and the first step was to establish a grid of five-foot squares, delineated by string, that was thirty feet square over the area's richest artifact scatter, producing thirty five-by-five-foot units. A soil auger, a long rod with a thick screw end and a T-shaped handle, was used to take samples at each grid square intersection. When withdrawn, it produces a core that shows the depth of the topsoil and the point at which it meets with the subsoil. With two exceptions, the entire area produced cores that indicated a topsoil depth of nine to twelve inches, but in the north-central part of the thirty-foot square, the auger encountered very hard objects, only a few inches below the surface. Being New England, they were almost certainly rocks, and subsequent excavations bore this out. The second area lay to the west of this, and showed a depth of dark organically stained soil of between two and three feet.

A word or two about measurement is in order here. More often than not, historical archaeologists measure their sites in feet and tenths of feet. This is only logical, since they are generally excavating sites that are the remains of structures using the English system of measurement. Prehistorians, on the other hand, prefer the metric system; one can only assume that it is perceived as more "scientific." Granted, it does accommodate international comparisons, but we can rest assured that early humans were not equipped with metric tapes. Given a choice between a measurement system based on the wavelength of light as opposed to the distance between the king's nose and the tip of his thumb, no true humanist would choose the former.

The initial test pit was a five-foot square unit located on the deep topsoil deposit. Within an hour, a complete spoon was discovered of a very early type. Known as a seal-top spoon, it dated between the late fifteenth and early seventeenth centuries. The knopf, or knob, at the end of the handle opposite the bowl was formed by a sphere not much larger than a pea, atop which was attached a small flat disk, which was the seal itself, and which gives the spoon its name. This was a most encouraging find, for it was in close agreement with the documented site dates. The left side of the bowl was far more worn than was the right, indicating that its user had been right-handed, which of course comes as no surprise. More important, this degree of wear shows that the spoon was used for purposes other than just food consumption, perhaps to stir various dishes in kettles of brass or iron. It was made of latten, a type of brass that, along with pewter, survives in the ground, while seventeenth-century horn spoons that also may have been buried have long since decomposed. Several latten spoons from the Old Colony area show a similar wear pattern to that of the seal-top spoon excavated at the Allerton site.

Plans were made to excavate each five-foot square unit with either a trowel or by shaving very thinly with a sharp-edged shovel, placing the artifacts recovered into individual bags marked with the square's designation. On occasion, depending on circumstances, such five-foot squares can be further subdivided into four two-and-a-half-foot squares each, providing a horizontal location of artifacts recovered that is four times more precise. Equally important is control over the vertical location of objects that, if there are visible layers (strata), are segregated one from the other. In the absence of visible strata, the best one can do is to excavate in arbitrary levels, usually six inches in depth.

This type of control is particularly critical in prehistoric archaeology, where sites can be deep and represent successive occupations by the same or different groups. But it also has its place in historical archaeology when one is dealing with deep pits such as cellars, trash pits, and wells. A good example, although not typical, was a filled abandoned icehouse located behind the remains of a late-eighteenth–early-nineteenth-century dwelling house at Flowerdew Hundred, Virginia. Fifteen feet in depth, when excavated it was found to have seven discrete visible deposition layers. Each was removed separately, its contents placed in appropriately marked bags. What is unusual about this deposit is that fragments of pottery from the lowermost level could be matched and mended to those from the topmost layer. This process is known as cross-mending, and in the case of this feature showed that it had been filled very rapidly, perhaps in only a day or so. Additional cross-mends between the icehouse fill and artifacts on the floor of the dwelling house established the contemporaneity of both features. Normally, however, deposition is a far slower process, and cross-mends are rarely, if ever, encountered. One must keep in mind that trash pits, privies, and, on occasion, wells, are filled over a long period of time by the occupants of the house with which they are associated, and thus represent objects that were in use in the household at the time. Cellars, on the other hand, are usually filled with refuse following the destruction of the house, and therefore their contents may not as reliably represent objects in use during the lifetime of the house.

Investigation of the Allerton site began with the excavation of the five-foot-square units, but, as Robert Burns said, "the best laid schemes o' mice and men gang aft a-gley." They went agley in very short order. The contractors announced that they would begin excavating the cellar for the new house in two weeks' time. As is so often the case with contractors, two weeks came and passed and nothing happened, followed by a promise that it would happen two weeks later. This postponement continued from May through August, but it was necessary to operate in two-week increments, without knowing just when the new cellar would be dug in the center of the excavation. In view of the shortness of time, two choices offered themselves: abandon the effort entirely, or employ more unconventional field methods. The second, deemed the more desirable course of action, was followed. The grid was dismantled, and one Saturday a large crew of student volunteers from Brown University removed the plow zone from the entire area and loaded it into wheelbarrows, depositing it neatly in a heap some distance from the site in the hope that at least it could be screened and whatever artifacts it contained could be recovered. Some weeks later, when a group of students from a local middle school wanted to visit the site and involve themselves in some aspect of the work, they were given the job of screening the removed plow zone, which they did in the course of a single day. The artifact yield was disappointingly sparse but did produce two types of pottery that had not yet been recovered at this site.

When the plow zone had been removed, the material struck by the auger in the north-central part of the unit indeed proved to be rock, some pieces weighing over a hundred pounds, filling to the top an eleven-foot-square cellar. The cellar was initially cross-trenched to establish the precise limits of its edges, and in the process numerous artifacts were encountered, the most significant of which were a King James I farthing and a lead bale seal with the initials "IR" stamped on it. Bale seals, before used, were dumbbell-shaped pieces of lead about half an inch in diameter that were then folded or clipped onto bundles of cloth or fur. The initials "IR" probably stood for Jacobus [James] Rex, uppercase J's and I's being used interchangeably. Following the cross-trenching, the entire top of the rock fill was exposed. Parallel to the cellar, on either side, spaced sixteen feet apart, were two dark stains with very blurred edges, indicating that they were extremely old. Such blurring is a result of the activity of a variety of burrowing insects, which makes the definition of the edges less than sharp. These stood in contrast to a number of plow scars where the plow had penetrated more deeply than usual and that had very sharp, clearly defined edges.

Sixteen feet is a common dimension for the width of a seventeenth-century house. The best guess that could be made at the time was that the stains had originally held stone footings that had been removed and possibly were among those in the fill of the cellar. It was decided to follow the northerly one to see how far it went by the simple technique of cutting short cross-trenches at right angles to it, spaced roughly three feet apart. The trench continued for at least three hundred feet to where it crossed the road and may well have reappeared in the neighbor's yard. But a three-hundred-foot-long house was unheard of in any English vernacular building tradition, so at this point it was not clear what this trench represented. Someone jokingly suggested that it might have been a seventeenth-century bowling alley, but in truth, when the intervening sections were removed, it became evident that what had been uncovered was the evidence of a so-called palisade fence in which pales set in contact with each other were buried with one end in the ground and secured by a stringer along the top. Evidence for such fences, built early in the seventeenth century, had been found at Martin's Hundred in Virginia, and they clearly show in the foreground of a 1666 engraving of the city of London where there were still farmsteads on the south side of the Thames. No more evidence that might have been used to further characterize the fence, however, would be uncovered at the site.

The crew, which by now numbered between nine and twelve people, depending on the number of volunteers present on any given day, then turned their attention to the deep fill that had produced the seal-topped spoon. It was their good fortune to have defined it at its northernmost limit: half of the five-foot square being tested showed the typical tan subsoil, demarcated by a straight line roughly in its center, perpendicular to the grid. A lengthy section was cleared and it was shown to be a V-shaped trench, the bottom foot of which, it emerged, was vertical. The faintest traces of the bases of large round posts could be seen, making it obvious that it had at one time held a substantial palisade. The same technique used to trace what turned out to be the fence line was employed to trace the extent of this trench with cross sections cut every five feet. In several locations, accumulations of stone and artifacts suggested that after the palisade had been removed, buildings contiguous to it had been razed, and the associated building materials and artifacts deposited in the now empty trench. Slightly over 300 feet south of where the trench began, it abruptly ended. Some structure had stood at this point and had burned, as evidenced by heavy carbon staining and charcoal flecks.

The expectation was that it would turn a corner at this point and proceed onward to enclose the entirety of the house site. Again due to time constraints, a bulldozer was used to scrape the area over a broad expanse on all three sides of the V-shaped trench. But no such evidence came to light. For whatever reason, someone had built a one-sided fence, probably abandoning it when it was no longer needed. Could this have coincided with the Pequot War in 1637, when upon the cessation of hostilities a fully fortified enclosure may have been deemed no longer necessary?

By this time, the construction workers had finally arrived and began excavation of the cellar for Orfeo Sgarzi's house. In what was clearly a magnanimous act, even if he was aware that he had an important archaeological treasure in what would become his backyard, Sgarzi moved the proposed location twenty feet closer to the river, thus removing the threat to the site. After the excavations had been completed and the artifacts processed, Plimoth Plantation installed a small exhibit in his basement of some of the more significant artifacts that had been discovered on the site. These were returned to the plantation as a part of his estate following his death.

By now, there was a lengthy fortification fence, fence lines, and a cellar still filled with a mass of heavy rocks. These had to be removed by hand for there were artifacts intermingled with them. When the last of the rock was removed, the cellar was shown to have been walled skillfully with dry laid stone, and had a deposit of several inches of rich dark fill on its floor.

Three people were assigned to clear the floor to subsoil, very carefully, using trowels. The three included two of the regular crew members and a volunteer, Hartman Lomawaima, a Hopi from the village of Shipaulovi on Second Mesa, a part of the Hopi Reservation in Arizona. Hartman, a student in the Harvard American Indian Program, assisted in the efforts from time to time, and in all probability was the first and only Native American from the American Southwest to assist in the excavation of a European home site in Massachusetts, an obvious reversal of the usual order of things, since Southwestern archaeology had been pioneered by A. V. Kidder of Harvard University. As such, it was indeed a historic moment. However, as the floor was scraped clear, the excavators were shocked by a discovery. Fragments of a nearly fully restorable wine bottle that dated to circa 1690 were recovered, along with smoking pipes and ceramics from the same period. Obviously something was amiss. For some curious reason artifacts of this late date had not been previously encountered on the site, but their presence directly on the floor of the cellar associated them unquestionably with both the cellar and whatever structure had stood above it. To reverse an old expression, "We wouldn't have seen it if we hadn't believed it."

The evidence, although subtle, was unmistakable and demonstrated that the site did not represent a single occupation as had been thought, but two sequential ones. The southwestern corner of the cellar had barely cut into a rectangular hearth of European construction that was clearly part of an earlier house built ca. 1630, soon after Allerton acquired the property. The stones of the hearth were heavily fire-blackened while those of the cellar wall were not, even though they were touching. The obvious explanation for this was that the cellar had been constructed after the original house had been torn down, sometime around 1650 according to artifact and pipe-stem dating. With only three days left to complete excavations, the site was examined for evidence of other remnants of the earlier building.

The effort was rewarded when the presence of three large ten-inch-diameter post molds, which had been cleared for weeks, suddenly fell into place. Along with a curious rectangular trench, on the northeast corner of the square, the four formed a rectangular pattern twenty by twenty-two feet square that had evidently formed the frame of the 1630s house. What had been excavated was, at the time, the first earthfast or post-in-ground dwelling house to be discovered anywhere in New England.

A posthole is excavated at the time of building for the purpose of seating the post, and its definition centuries later is exceptionally subtle, consisting of a slightly marbled mixture of subsoil from different layers from which the hole was dug. In 1972, recognition of postholes and molds, the mold being the actual decomposed remnants of the post, was not a developed skill and practiced largely in the Chesapeake region. In order to see a posthole at all, the surface has to be meticulously scraped and sprayed with water, otherwise it is not visible. Alternatively, the posts could have been driven directly into the ground without benefit of an accommodating hole, but unless reexcavation is carried out on the site, which seems unlikely, we will never

Allerton site, Kingston, Massachusetts. Excavation of 1630s house site, the first earthfast or post-in-ground structure to be discovered archaeologically in New England, intersected by the cellar of a post-1650 house. Palisade trench to the right. Hearth of 1630s house cut by cellar at upper left corner.

know for certain. Excavations were extended to the east for a sufficient distance to determine whether additional such post molds existed, but none came to light. The conclusion finally arrived at was that the original house was a single bay earthfast structure with a fireplace and wooden chimney at one end. It now seems apparent that post-in-ground construction was more common in New England than previously thought.

We must close on a cautionary note. Title searches inform us as to the owner of the property, but not necessarily of the occupants of the residence built on it. It is highly probable, however, that Allerton did reside in the original house for a short while, but we will never be certain.

Myles Standish Site, Duxbury, Massachusetts

One of the earliest of the seventeenth-century archaeological sites in Plymouth Colony is that of the Myles Standish house in Duxbury, excavated by James Hall, a civil engineer, in 1863–1864. We are extremely fortunate in that his map of the Standish house site was preserved amongst his effects, even though the quantity of artifacts that survived and accompanied it is scant. Standish moved to Duxbury, which is adjacent to Plymouth, in 1637, and one can only assume that he set about having a house erected on the land granted to him right away. It is not likely that he built the house himself; he probably concluded a detailed building contract with a house carpenter who in fact may have never before seen a house of the type that Standish planned to build. It was customary at the time for a landowner to retain the services of a house carpenter to do the actual construction, but building contracts were drawn up to ensure that what was being built suited the tastes and needs of the owner of the new house. Such contracts were common, some being more detailed than others. By way of example, we have chosen a 1638 contract from Ipswich in the Massachusetts Bay Colony that, in its specific detail, left little to the carpenter's imagination or initiative, although latitude was given him in matters considered by the owner to be more within the carpenter's area of expertise than his own. Samuel Symonds left the following instructions:

> Concerning the frame of the house . . . I am indifferent whether it be 30 foot or 35 foot long; 16 or 18 foot broad. I would have wood chimneys at each end, the frames of the chimneys to be stronger than ordinary, to bear good heavy load of clay for security against fire. You may let the chimneys be all the breadth of the house if you think good; the 2 lower doors to be in the middle of the house, one opposite to the other. Be sure that all the doorways in every place be so high that any man may go upright under. The stairs I think had best be placed close by the door. It makes no great matter though there be no partition upon the first floor; if there be, make one bigger than the other. For windows, let them not be over large in any room, & as few as conveniently may be; let all have current shutting draw-windows [probably casement windows], having respect both to present & future use. I think to make it a girt house will make it more chargeable [expensive] than need; however, the side bearers for the second story, being to be loaded with corn &c. must not by pinned on, but rather either let into the studs or borne up with false studs, & so tenented in at the ends; I leave it to you and the carpenters. In this story over the first, I would have a partition, whether in the middle or over the partition under, I leave it. In the garret no partition, but let there be two lucome [lucarne, meaning skylight or

> dormer] windows, if two, both on one side. I desire to have the spars reach down pretty deep at the eves to preserve the walls the better from the weather, I would have it cellared all over & so the frame of the house accordingly from the bottom. I would have the house strong in timber, though plain & well braced. I would have it covered with very good oak-heart inch board, *for the present,* to be tacked on only for the present, as you told me. Let the frame begin from the bottom of the cellar, & so in the ordinary way upright, for I can hereafter (to save the timber within ground) run up a thin brickwork without. I think it best to have the walls without to be all clapboarded besides the clay walls. It were not amiss to leave a doorway or two within the cellar, that so hereafter on may make comings in from without, & let them be both upon that side which the lucome window or windows be.

Symonds is describing an end-chimney house, rather than one with a chimney in its center. It is more typical of what we know of later seventeenth-century houses in hot, humid Virginia than of those in New England, where a central chimney installed between the hall and parlor became common, and, as we have noted earlier, the central position was chosen to provide maximum heating for the house in the cold, northern climate. Both end and central chimneys were in use in England, and the colonists evidently selected that which met their needs. It probably took at least a generation or two before any pattern emerged, however, as traditional cultures are innately conservative, and it would have taken time before such selective changes became apparent in the New World.

Samuel Symonds's request for a few small casement windows in his new house reflects a very different perception concerning lighting from that found in America today. We take air conditioning and central heating so much for granted and the presence of spacious, large windows giving maximum light is a norm. But in the seventeenth century, the framework of the house was the only buffer against the elements, and every effort was made to keep houses snug and warm in the winter and dark and cool in the summer; that meant installing only a limited number of small windows as well as having a central chimney. The placement of the windows did not seem to be of great concern to Symonds, apart from the fact that if there were to be two dormer windows, he wanted them on the same side of the house.

Irregular placement of windows was common in seventeenth-century vernacular architecture, where builders were not influenced by classical principles. It is also noteworthy that Symonds requested the house carpenter to put in "current shutting draw-windows." As in the case of the selection of the saltbox house type, it is another pointer to the fact that the colonists were well aware of modern developments in England, and some at least were selecting the latest fashions to shape the vernacular building tradition that was emerging on the New England landscape.

The request that the house be "cellared all over" would make it an archaeologist's dream. The footprint of a long vanished house is most informative when the outline of a house with a full cellar is discovered, providing definitive evidence as to the layout of the building that stood above it. The site of Standish's house was known and visible because of a depression in the ground that marked the location of an old filled-in cellar. Once James Hall had excavated the depression he drew a meticulous scale plan of the foundation that he exposed. The plan shows that Myles Standish had a house type that, in the state of our present knowledge, is unique in seventeenth-century Plymouth and adds one more house type, the byre house, to fill the gap between 1628 and the end of the century.

The Standish house takes the form of two elongated rectangles, Section A being fifteen feet wide and sixty feet long, and Section B of identical width but some five feet shorter, and abutting Section A at the southeast corner approximately one third of the distance from the southeast end, and set at a forty-five degree angle to Section A. Upon first glance, one might assume that this plan represents two sequential construction episodes. However, that such is not the case is indicated by a doorway connecting the southeastern room with the northwestern room. Clearly, they are all part of a single building constructed at one time.

Houses of this type sheltered both cattle and people under the same roof, or at least in contiguous connecting structures. The space for the cattle was known as a byre, and M. W. Barley, in his pioneering study of English farmhouses and cottages, describes houses that contained livestock byres at one end as long houses or byre houses. They date from the fourteenth and fifteenth centuries, and are particularly associated with South Wales, and Devon in the West Country.

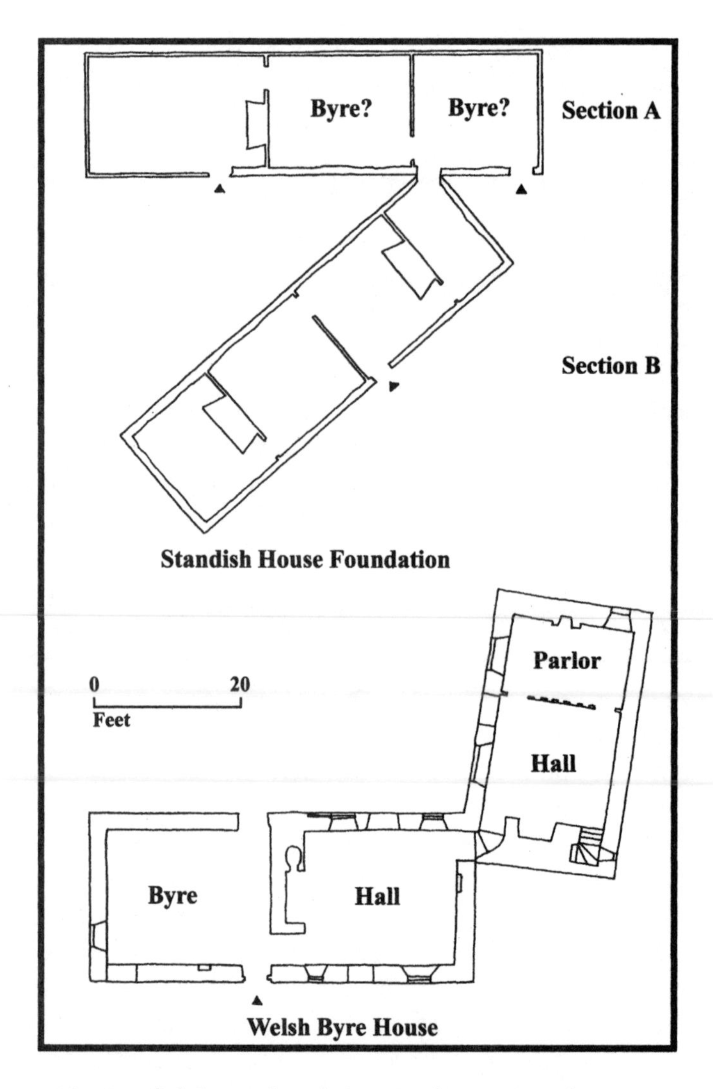

Myles Standish house foundation, Duxbury, Massachusetts (*top*), with a similar Welsh byre house antecedent (*bottom*). The Standish house foundation (after James Hall's manuscript map of excavations on site 1864) was constructed between 1637 and 1650. Section A shows the upper range of three rooms, one with hearth, two possibly used as byres. Section B shows the lower range of four rooms, two with hearths. The Welsh byre house plan from Llanfihangle Cwm Du, Breconshire, Wales.

This type of house had largely disappeared from England by the turn of the seventeenth century, replaced by houses accompanied by at least two other separate structures, a barn for the storage of cereal grains, and a cow house in which to shelter cattle. However, in the so-called "Celtic fringe" (Scotland, Wales, and Ireland), houses with byres continued to be erected into the seventeenth century.

Standish must have been familiar with this type of structure from his Lancashire home near the Welsh border. Certain details shown by Hall on the site map confirm this interpretation. The slightly shorter extension to the east (Section B) was clearly a four-bay building erected on stone footings that had two fireplaces. Approximately forty feet of Section A also rested on stone, but the remainder on the right-hand side was made simply by setting heavy beams horizontally directly into the ground into which posts and studs would have been mortised. As folklorist Robert St. George says, "this would satisfy creature comforts, but not those of humans." He identifies the two right-hand rooms in Section A as possible byres; the left-hand room, which contained a hearth, could well have served as a makeshift dwelling for people while Section B was under construction. The Welsh connection does appear to be clear and incontrovertible, as St. George has shown that a standing house in Llanfihangle Cwm Du, Breconshire, Wales, is very close in form to the Standish structure. The connecting doorway between Sections A and B would provide easy access to the livestock living inside during the winter months, and the two rooms tentatively identified as byres are also connected by a doorway. So there the matter stands, but St. George's argument is a compelling one and it seems virtually certain that Myles Standish had constructed a house unlike any known from other archaeology, probate inventories, or surviving structures.

R. M. Site, Plymouth, Massachusetts

Along with the Allerton site, the R. M. site has the distinction of being one of the two oldest sites to have been excavated in the Old Colony. Both produced spoons of the "short seal" type most typical of the sixteenth century, a complete one from the Allerton site, and a handle from the R. M. site. In addition, both yielded ceramic types typical of the pre-1650 period. Harrington pipe-stem histograms from the Allerton site match those of the R. M. site quite closely, with the latter appearing to be slightly earlier, although this might not be statistically significant. While in both cases the largest number of pipe stems date to the period 1650 to 1680, a large number date to its occupation from 1620 to 1650, 37.8 percent in the R. M. case, and 30.5 percent from the Allerton site.

The R. M. site was the first to be excavated by Harry Hornblower in 1941, but he certainly was aware of its existence well before that date. Harry's archaeological mentor, Jesse Brewer, amateur archaeologist and gardener for the Hornblower estate on the Eel River, two miles south of Plymouth, spent countless hours with Harry carefully examining every part of the estate for evidence of prior human occupation. Harry would have been in his teens at this time, and Jesse Brewer, by Harry's own admission, was the person who awakened his interest in archaeology, which he then pursued when he attended Harvard. While some prehistoric sites were identified, only one that dated to the historic period was located, and is in fact the only seventeenth-century site anywhere on the entire Hornblower estate. The R. M. site was situated overlooking the Eel River on a rise that commanded good views in all directions. The site takes its name from a set of initials scratched in the seal of a spoon that Harry Hornblower excavated. The initials do not match those of any of the occupants of neighboring houses described in detail in an order to lay out a highway from the town of Plymouth to the Eel River in July 1637. The roadway ended at the property of Thomas Clark, who held forty acres of riverfront land that probably included that on which the R. M. site was located.

Hornblower cleared the entire area surrounding the house site in 1941, and again in 1959 the area was further expanded in a search for postholes that might indicate the existence of what had been a palisade, but to no avail. Hornblower excavated the remains of a house approximately thirty-eight feet long and sixteen feet wide. Since no evidence of footings was encountered, these measurements must of necessity be somewhat approximate. The house had a hearth in its center and one at the eastern end. At the western end was a cellar four by six feet in size and remarkably deep, extending six feet into the subsoil, which would have made it closer to seven feet in depth at the time the house was occupied, allowing for a foot of topsoil. Almost all other cellars excavated on Old Colony sites are more of the order of three to four feet in depth. These more shallow cellars were no doubt used for storing various root crops such as potatoes, beets, and parsnips, as well as foodstuffs such as cabbage and apples that would last through the winter. The only other cellar of similar depth is that of the John Alden site in Duxbury, but it was not as long or wide as that at the R. M. site. A cellar of this depth appears to have been intended to store other things in addition to foodstuffs. Deposits of charcoal in the cellar and surrounding the house remains strongly suggest that the house had burned. This, then, is the full extent of factual information that we have concerning the R. M. site. We will never be certain of who occupied it, or of how it met its end. However, there is a large body of quite strong circumstantial evidence that permits us to venture some well-informed speculation concerning these matters.

When hostilities began in 1676 between the Indians and the English colonists in what became known as King Philip's War, the council of war for the jurisdiction of New Plymouth issued a number of directives to towns ordering the establishment of garrisons to protect their inhabitants. In some cases a garrison house was hastily built, with a guard of ten to twelve men appointed, and in other instances existing houses in strategic positions were fortified to a greater or lesser extent, with only a couple of men appointed to guard them. A watch had to be strictly maintained in every garrison, which also acted as a repository for arms and ammunition. William Clark, the eldest son of Thomas Clark who had arrived in Plymouth on the *Anne* in 1623, was living on the Eel River property owned by his father. William had married Sarah Wolcott in 1660, and became actively involved in civic life in Plymouth. In 1674 he was appointed one of the deputies for Plymouth, and in 1676 one of four selectmen. Clark's house on the Eel River was "slightly fortified," according to the testimony of Keweenam, an Indian, who was one of several accused of the murder of Sarah Clark and the destruction by fire of the Clark house on the morning of March 12, 1676. The primary sources are in conflict as to the details of what took place on that Sunday morning, but the salient points are as follows. William Clark was on good terms with the Indians and despite the fact that in December 1675 the council of war had issued a directive that Indians were not permitted to come nearer to Plymouth than Sandwich on pain of death or imprisonment, it appears that one at least was permitted by Clark to visit his house and get some sense of its occupants and contents, as well as pattern of life. This emerged when three Indians were brought before the council of war accused by an Indian woman of being present and active in the murder of Sarah Clark. The Indians implicated a fourth, Keweenam, who when questioned said he knew that on a Sunday the family, of which there were "but three, the most of them would be gone to meeting . . . and in case they left a man at home or so, they might dispatch him, and then they would meet with no opposition but might do as they pleased."

Contemporary accounts provide a few more details than the court record, and differ from it in referring to two families living in the house at the time, to the murder of eleven people, only a young boy surviving, and to the fact that the house was burned after the attack. Nathaniel Saltonstall, in an account published in London in 1676, mentions that the house was "plundered of Provisions and Goods to a great Value, Eight complete Arms, 30 *l.* of Powder, with an answerable Quantity of Lead for Bullets and 150 *l.* in ready Money," which would indicate that the house was indeed one of those that had been directed to be used as a garrison, even though it was "only slightly fortified" at the time of the attack. Benjamin Church and his wife, who was expecting a child, were in Plymouth at the time, and although they had been urged by his wife's parents to stop for the night of March 11 at "Mr. Clark's garrison (which they supposed to be a mighty safe place)," they instead pressed on to Taunton, narrowly escaping the attack. The court records note that there were eleven Indians who were "copartners in the outrage committed at William Clarke's house at the Eel River," a strange coincidence with the number of those killed (eleven) given by William Hubbard in his 1677 account, when the court records referred only to the murder of Sarah Clark.

And so the matter stands: two sets of solid empirical data connected by evidence that, although circumstantial, seems nonetheless very strong. Only one house site on the Eel River in the vicinity of where the Clark garrison stood was located and excavated. Pipe-stem dates and other artifactual evidence agree with the house having been attacked in 1676. Additionally, the R. M. site showed evidence of burning, which serves to reinforce this interpretation. As far as "slightly fortified" is concerned, this admits of a number of interpretations, but certainly does not refer to any kind of heavy palisaded enclosure. Considering the fact that discussions concerning fortifying outlying houses were only begun at the end of February 1676, William Clark would have had little time to erect any kind of substantial fortification. It could have been as simple as a heavy brush enclosure, perhaps secured to posts not set deeply into the subsoil, similar to those erected by the exploring parties of 1620 as described in *Mourt's Relation*. Such an enclosure would be an effective defense against arrows, but by 1676 the Indians possessed firearms that would have made it far easier to breach any perimeter fortification and attack the house. The large, deep cellar may have been expanded at this time to store firearms and food supplies.

Unlike other cellars on Plymouth Colony sites the R. M. cellar was not lined with dry laid stone, suggesting that it might have been enlarged in response to the directive of the council. Whether the R. M. site is that of the Clark garrison or not, it is still a very important one, both in view of its early date and, as we shall see, lack of evidence of footings of any kind. It is unlikely that any more can be added to this account.

John Alden Site, Duxbury, Massachusetts

John and Priscilla Alden and their young family moved to Duxbury in the early 1630s. The exact date does not appear to be documented, although Robert Charles Anderson in *The Great Migration Begins* gives 1632 as the date of Alden's move from Plymouth to neighboring Duxbury. In 1960 what was believed to be the site of Alden's first house (he built a second house nearby in 1653) was excavated by Roland Wells Robbins near Eagletree Pond, some two miles from where Myles Standish built his Duxbury house. The house plan that emerged was clear and unambiguous and needs little extensive commentary in this section on the archaeology of early house construction in Plymouth Colony. Robbins exposed the foundation outlined by stone footings of a curiously elongated house measuring thirty-eight by ten and a half feet, nearly four times longer than its width. At the western end was a six-foot-deep cellar, carefully lined with dry laid stone. While as deep as the cellar at the R. M. site, the stone lining permits us to suggest that the cellar at the R. M. site was hastily enlarged in response to the war emergency. Traces of a central chimney were encountered in the middle of the house foundation, which would make it the remains of a two-bay hall and parlor house with the cellar possibly beneath a lean-to on the western end. Pipe-stem dates support a peak occupation between 1620 and 1680, and other artifacts recovered indicate a first third of the seventeenth century construction date.

Howland Site Complex, Kingston, Massachusetts

The Howland sites are located on a twenty-one-acre parcel of land, situated on both sides of Howland's Lane in Rocky Nook, Kingston, Massachusetts. It is but a portion of a much larger property sold to John Howland by John Jenney, who arrived in Plymouth on the *Little James* in 1623. The deed is dated February 2, 1638, and records that Jenney sold "All that his house barn & outhouses at Rocky Nook" to Howland. It has not been possible to find any record of Jenney's acquisition of the property, but it had to have taken place between 1627, when the initial dispersal from the original fortified village took place, and 1638. Rodman Hussey's detailed title search in 1938 established that two parcels of the original property conveyed to John Howland are the location of not only the original, pre-1638 house, but of at least four others, an occupation which lasted until 1735, when the last parcels (O and P) were conveyed to Benjamin Lothrop. When John Howland died in 1672, his property passed to his widow, Elizabeth Tilley Howland, for her life, and then to their son, Joseph. Joseph in turn devised the property to his widow, Elizabeth, daughter of Thomas Southworth, and after her death to their son, James. Between 1721 and 1735, James Howland sold all his land, with the two parcels (O and P) being the last to be disposed of, in 1735.

Since parcels O and P contain the remains of at least four structures, it would seem a reasonably straightforward matter to identify these with each of the successive occupants of the property. However, recent work carried out by the University of Virginia Field School, and examination of the collections recovered by Sidney Strickland, present a series of apparent contradictions that must be resolved through further excavation. There are extensive collections from two areas of the site that form a readily accessible database, and the Howlands are very well represented in the primary records of the colony, making it possible to place them in the context of the community of which they were a part. The site was occupied by the same family for

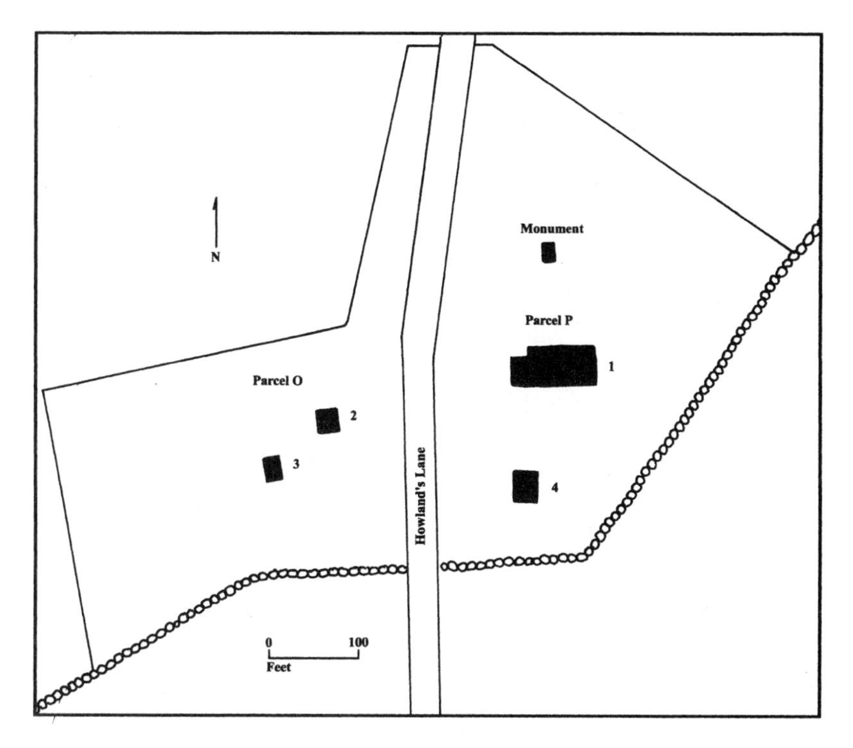

Howland site complex, Kingston, Massachusetts. *Parcel P*: Structure 1—Site of house built by John Howland attributed to his first occupation from 1638, but artifacts excavated indicate it was occupied in the mid- to late seventeenth century. Structure 4—Probably site of house built by James Howland and occupied through the later eighteenth century into the early nineteenth. *Parcel O*: Structure 2—Site of outbuilding adjacent to house attributed to Joseph Howland. Structure 3—Site of house attributed to Joseph Howland, occupied during the late-seventeenth and early eighteenth centuries.

almost a century, providing an important measure of control by eliminating one variable, that of multiple different owners.

The University of Virginia Field School reexposed the foundation attributed to John Howland's initial occupation. For the first time, a detailed plan was made of the foundation. The initial occupation of this structure has not been firmly established as yet, but it is probably no earlier than the mid-seventeenth century. Exactly where on the property John Howland and his family resided prior to that date is uncertain, although it would seem probable that they occupied the house bought from John Jenney. So far it has not been possible to locate the remains of this building. Extensive shovel testing over the entire area where it should have stood failed to produce any evidence of a structure, although a small number of artifacts dating to the first half of the seventeenth century were found. It could well be that the original house was an earthfast structure that would be undetectable in a program of shovel testing where small quantities (a shovelful each) of earth were removed at ten-foot intervals and examined. Occupation of the hall and parlor house built by John Howland (Structure 1) and excavated by Sidney Strickland appears to have lasted through the end of the century, and possibly into the early eighteenth century. Strickland apparently did not screen his back dirt, and when this was done by the Virginia group, a William III penny was recovered, along with a variety of other late-seventeenth-century artifacts. Strickland decided that the house lacked a cellar, but the 1998 excavations revealed one, on the north side of the foundation, possibly having been under a lean-to addition. The cellar was fully excavated during the 1998 and 1999 field seasons. It measured six by seventeen feet with a single row of stones along its top, and was two and a half feet in depth. Seventeen feet is an unusually long dimension for a cellar of the period, and raises problems of access. It may have been under a lean-to with no floor covering, and so could've been entered with either a ladder or simple wooden stairs. A piece of early-eighteenth-century mottled ware, an English ceramic, which postdated the house and therefore was not used by its occupants, was recovered near the top of the fill.

Particularly puzzling are the dates presently assigned to Structure 3, which has until now been identified as the site of Joseph Howland's house. According to a biography of Joseph Howland, John Howland's house (Structure 1) was burned by the Indians during King Philip's War in 1675, and Joseph built a new house nearby in 1676. However, the archaeological evidence contradicts this account; clearly two contemporary structures (Structures 1 and 3) stood on the property during the fourth quarter of the seventeenth century and into the eighteenth. Given this, who then was living in each of the houses? If John Howland's house was not burned in 1676, the most likely scenario is that after John Howland's death in February 1673, his widow, Eliza-

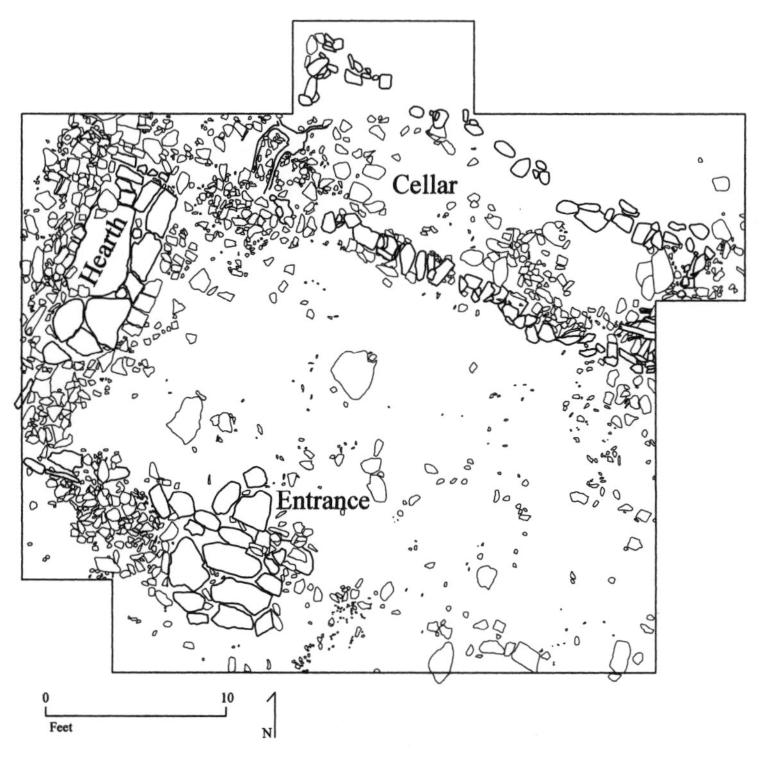

Foundation of John Howland house (Structure 1). Shown above as reexposed in 1998 and 1999 by the University of Virginia. Originally uncovered by Sydney Strickland, 1937–1938.

beth, continued living in the house and that Joseph and his wife, Elizabeth, and their young family occupied a house (Structure 3) across what is now known as Howland's Lane, opposite John Howland's house. Structure 3 is represented by a stone-lined cellar three feet deep and a well-preserved hearth, but broad area excavation was not carried out to determine if footings were present or not. There is a cellar (Structure 2) adjacent to what was probably Joseph Howland's house (Structure 3), but it is not in alignment with it, and at this time can probably be seen to represent a separate building, probably an outbuilding. Again, only the cellar was excavated, and we have no evidence of any other features, including footings, associated with it.

This leaves only Strickland's Structure 4, south of what we believe to have been John Howland's house (Structure 1), unaccounted for.

The building has been assigned a dimension of nine by fourteen feet, but the source of these measurements is unclear. Strickland did not excavate the entire structure but recovered a large collection of materials from its fill. Its date places it beyond the scope of this discussion, but it should be mentioned if only to complete the accounting of the four buildings that stood on the Howland site. Artifactual evidence suggests an occupation through the eighteenth century and into the early nineteenth century. A large number of pipe stems had diameters of five and four sixty-fourths of an inch, supporting these dates. Artifacts include eighteenth-century clay smoking pipes, transfer-printed earthenware, and shell-edged pearlware, a type of earthenware that was most popular in the first decades of the nineteenth century and later. Structure 4 was probably constructed by James Howland, Joseph's son, to whom he had left the property at Rocky Nook, and given the uncertainty concerning its size could well be the remains of his dwelling house. Most significant is the fact that it is remarkably close to the so-called "dew pond." In the summer of 1999 this feature was tested by excavating a five-foot-square test pit in its center. At a depth of two and a half feet, a flat earth floor was exposed, strongly suggesting that the dew pond is the filled cellar of Structure 4.

The most reasonable accounting for the Howland complex is as follows. Upon acquiring the property in 1638, John Howland occupied the house constructed by John Jenney until he built a more elaborate house (Structure 1) about mid-century. There is no evidence that the house burned. Upon his return from military service in 1676, Lieutenant Joseph Howland erected a new house (Structure 3) opposite that which belonged to his father, and where his widowed mother, Elizabeth Tilley Howland, still lived. Elizabeth died at the home of her daughter, Lydia Browne, in Swansea in December 1687. There is no record of when she moved to Swansea, but her will was drawn up in Bristol County in 1686 and she named her son-in-law, James Browne, and her son Jabez Howland as executors, so she had certainly left the Rocky Nook property by that date. Whether Joseph Howland ever occupied the house left to him by his father (Structure 1) we will not know. Structure 2, near Structure 3, could have been the root cellar of an outbuilding associated with the house. Such cellars are known from the period and should this be the case, if Structure 4 is indeed the cellar of a house built by James Howland, all structures are accounted for. Admittedly, there are uncertainties involved in this interpretation, but then in its own way, all of archaeology is detective work of one kind or another. Future excavations in the area around Structures 2 and 3 could go far in clarifying these identifications.

Winslow Site, Marshfield, Massachusetts

Edward Winslow needs no introduction here since he figured prominently in the initial establishment of the colony, including his role in the conclusion of the peace treaty with the local Wampanoags in 1621. As coauthor of *Mourt's Relation*, he left us his classic description of the colony's first harvest festival, transformed today into Thanksgiving. He was one of the most prominent figures in the colony, having served as governor in 1633, 1636, and 1644. In 1646 he traveled to England to represent the colony and to defend it against charges of impropriety in religious practices and of disloyalty. He was asked to remain in England and became involved with the Cromwell government and soon became a member of it, representing the interests of the New England colonies. He was well suited to this position, having been born into the minor gentry and educated at King's School, Worcester, from 1606 to 1611. He died in 1655 off the coast of Hispaniola where he was commissioner of an English invasion force assigned to capture several islands from the Spaniards.

Winslow's rank and social status is evidenced by the fact that he, his son Josiah, and Josiah's wife had their portraits painted by a London artist. His portrait is the only surviving one of any members of the original *Mayflower* company. In 1632 he acquired property in Marshfield, moved there with his family, and presumably constructed a house at that time. He remained resident there until being sent to England in 1646. Upon the occasion of his death, his only surviving son, Josiah, inherited the property and apparently lived there for the remainder of his life. When he died in 1680 he was governor of the colony, having held that position since 1673.

Josiah Winslow fortified the house in 1676 at the time of King Philip's War, and quartered twenty men in anticipation of an attack that never materialized. This additional number of men could well account for the high percentage (59 percent) of pipe stems representing the period between 1650 and 1680. The Harrington histogram for the Winslow site fits these dates quite comfortably, particularly considering the possibility of a significant increase in stem fragments resulting from the troops quartered there. Following Josiah's death, his widow and eldest son, Isaac, continued to live in the Winslow house until 1699 when Isaac built a new house close by, which he named Careswell after the old family estate in Worcestershire. This house still stands today and is exhibited to the public as a historic house museum.

Harry Hornblower excavated the site with the Harvard Excavators' Club in 1940–1943, and again in 1949. In the course of these excavations one of the most charming artifacts ever recovered from an archaeological site in the Old Colony came to light, a small silver whistle, still functional, with the initials E.W. engraved upon it. When blown, it actually produces a seventeenth-century sound.

In spite of Hornblower's meticulous excavation and recording techniques, the plan exposed is the most ambiguous of all of those we have considered. It takes the form of a set of features approximately twenty-seven feet square, and while these are no doubt the remains of a dwelling house, it is very difficult to make logical sense of them. A cellar of indeterminate size was located near the eastern side and in front of it, to the south, was a clearly defined U-shaped trench filled with ash, curiously facing away from the cellar. It has the appearance of a hearth, but its orientation would seem to contradict this. To the south of the cellar is a large six-foot area of brick fragments that probably represents the fireplace base. This would seem to align the house with its long axis oriented east to west, as were all the other foundations of the sites discussed here. There is a logic in this orientation that is seen throughout the colonies until the full establishment of highways, when houses tended to face onto the thoroughfare. In the heat of summer a house oriented east-west receives heat from the sun on its southern exposure, whereas the north side is invariably shaded. It was a simple matter, then, to open windows on both sides, with the rising warm air from the south drawing the cooler northern air through the house, creating a gentle but cooling breeze. As far as the size of the house is concerned, it is impossible to determine on the basis of available evidence.

The Winslow site was the only one excavated in the Old Colony area from which a large sample of animal bone was recovered and curated. It was analyzed by Stanley Olsen of Florida State University, and is informative concerning an aspect of the foodways of the time that is not accessible from documentary sources. We are left with a less than satisfactory body of evidence concerning construction techniques, although there are certain features that might clarify this. These will be discussed in the section on earthfast construction.

Whalers' Tavern, Wellfleet, Massachusetts

The tavern site located on Great Island at Wellfleet on the outermost reaches of Cape Cod is the latest site that we will consider. With a beginning date sometime in the late 1680s or early 1690s, it barely fits within the period of Plymouth Colony's independent history, yet it is a component of such an important part of the economy of early Plymouth that to omit it would be a serious mistake. It appears to have been in use through the first quarter of the eighteenth century.

As early as the mid–seventeenth century, the taking of small whales known as blackfish formed an important part of the economy of the colony. However, it had begun much earlier. A 1613 account describes Englishmen taking a whale by securing the harpoon line to a boat. This is very close to the native method of whale hunting, which was very proficiently done by spearing them with wooden harpoons to which were attached floats to slow the animal's progress. The whale would then be surrounded by canoes and killed with arrows. The similarity of the two methods may well be the result of the colonists adapting and modifying the native practice. The "black thing" mentioned in *Mourt's Relation*, and its subsequent identification as a grampus or whale, show that the local native peoples also probably made extensive use of these small whales. Wellfleet became an important center of what was known as shore whaling, and the industry continued with great success though the first quarter of the eighteenth century. However, by 1727, an article in the *Boston Newsletter*, published on March 20, reads as follows:

> We hear from towns on the Cape that the whale fishery among them has failed much this winter, as it has done for several winters past, but having found out the way of going to sea upon that business, and having had much success in it, they are now fitting out several vessels to sail with all expedition upon that dangerous design this spring, more (it's thought) than have ever been sent out from among them.

It is also significant that in 1690 the colony appointed inspectors of whale, whose duties involved inspecting drift whales—those washed ashore—to be certain that they were properly assigned to whoever had marked them first. Drift whales figure often in the Plymouth court records during the second half of the seventeenth century. The person discovering the whale received a third share of the animal, the colonial government a third, and the town one third. The inspectors of whale could also adjudicate the disposition of whales killed at sea and later washed ashore. Both harpoons and animals were marked according to a complex system developed to provide proper proof of ownership. Although written in 1793, Levi Whitman's description of Cape Whaling is most informative:

> It would be curious indeed to a countryman, who lives at a distance from the sea, to be acquainted with the method of killing blackfish. Their size is from four to five tons weight, when full grown. When they come within our harbors, boats surround them. They are as easily driven to the shore as cattle or sheep are driven on land. The tide leaves them, and they are easily killed. They are a fish of the whale kind, and will average a barrel of oil each. I have seen nearly four hundred at one time lying dead on the shore.

Both eastern Long Island and Cape Cod were leading centers in the development of shore whaling. In fact, a Cape Cod whaler, Ichabod Paddock, was brought by the residents of Nantucket Island to teach them how to kill whales and obtain their oil. Thus, the flowering of whaling on both Nantucket Island and New Bedford had its roots on Cape Cod.

Today, Great Island, the location of the tavern, is part of a long peninsula forming the western side of Wellfleet Harbor. It is connected to the mainland by a long sandbar developed by the constant accumulation of sand that is steadily building up on the western side of the outer Cape. Great Island's shore is made up of a high bluff overlooking a wide beach. It is covered by a thick growth of scrub pine that local informants claim to have been planted to control erosion in the 1830s. A map drawn in 1795 shows the island completely separate from the mainland as well as from Great Beach Hill, which today forms its southern end. An 1831 map of Wellfleet shows Great

Island to be separated from the mainland by the narrow channel of the Herring River, which flowed into the bay at the island's northern end. Today, this channel no longer exists, being blocked by dunes, and the Herring River empties directly into Wellfleet Harbor. The 1831 map shows structures in Wellfleet in great detail, but no building of any kind is shown standing on Great Island.

In 1969, however, archaeologists from Plimoth Plantation and the National Park Service visited Great Island and noticed slight but significant evidence that a structure had once stood there. The local people must have known of its existence, some suggesting that it had been the site of a Dutch trading post. A second tradition told of a tavern that had existed there complete with a sign advertising the hospitality of one Samuel Smith. The site had been visited by unauthorized parties on occasion in the past who had discovered a 1723 English coin, some spoons, and clay-pipe stems. The occasional rock projecting above the surface suggested a buried foundation. During the summer of 1970, an archaeological team under the direction of Eric Ekholm of Plimoth Plantation tested and subsequently excavated the site into the fall of that year. The initial tests produced bits of ivory fans, whale bone fragments, broken drinking containers, clay pipes, and ceramics.

In reality, Great Island is nothing more than an extremely large dune created by drifting sand carried by the waters of the bay to its west. During the eighteenth century Cape Cod was severely deforested, presumably to obtain pine trees, the sap of which was used in the production of tar. This posed a critical problem, as described in an 1802 history of Eastham, of which Wellfleet was once a part:

> Except a tract of oak and pines, adjoining the south line of Wellfleet, and which is about a mile and a half wide, no wood is left in the township. The forests were imprudently cut down many years ago, and no obstacle being opposed to the fury of the winds, it has already covered with barrenness the large tract above described, and threatens the whole township with destruction.

Between the planting of erosion-control scrub pine on Great Island in the 1830s and the abandonment of the site sometime before 1750, ten inches of clean sand covered both the foundation and a six-inch-deep layer of refuse close to it. In all, when the excavation was complete, 24,000 artifacts had been recovered, and were of great help in establishing both the date of the site and the function it served, as no documentary evidence was ever traced to account for the existence of the building. An advantage of the island being formed exclusively of sand lies in the fact that any stone encountered in excavation had to have been brought from elsewhere. The result was the clearest, most unambiguous building plan encountered in all of the archaeology done in the Old Colony. The foundation had been little disturbed by past random digging, and proved to be the ground plan of a very large building, fifty feet in length and thirty feet wide, facing south.

The floor plan was typical of New England construction of the period, with two rooms flanking a large brick chimney, the base of which had survived. The front door was opposite the chimney, with a paved stone walkway leading to it. To the rear of the foundation another paved area was discovered, which probably floored a third room covered by a lean-to facing north. It is possible that this third room extended the full length of the north side of the building. Each of the two main rooms had cellars beneath them, but they differed in the way in which they were constructed. The cellar under the east room was small (eight by ten feet), and lined with rounded cobbles set in mortar, made from sand and clam or oyster shells. Access was through a short stone stairway leading down from the south end of the room above, probably through a trapdoor in the floor of the east front room. Only five feet deep, the cellar had little headroom. The cellar beneath the west room had a sand floor and walls of rectangular stone laid in fairly even courses. It too must have been entered through a trapdoor, but lacking a stone stairway was reached by a ladder or wooden stairs, long since vanished. Both cellars were probably used for food storage and resemble cellars found on other sites in the Old Colony.

Nails were recovered by the thousands, showing beyond doubt that the building had been a wooden-framed structure. It is highly probable that the building was two full stories in height with the upstairs rooms serving as sleeping quarters for what would prove to have been shore whalers who used the building, which provided them with immediate access to the beach. By seventeenth-century standards, rooms of this size could accommodate as many as ten people each, and the downstairs rooms could have been pressed into use as sleeping space as well.

An important source of architectural evidence was provided by fragments of interior wall plaster. The fragments clearly showed impressions of the horizontal laths that held the plaster and that were in turn attached directly to vertical planks. Rust spots on the lath impressions show both the presence and placing of the nails that were attached to the planks. Between the lath impressions one can clearly see the pattern and direction of the grain of the wood to which the laths were nailed. These run at right angles to the direction of the laths, indicating that they were directly attached to the planks. Had the building been stud framed, the majority of lath impressions would have shown no wood-grain marking behind them. Clearly this is an example of the kind of vertical plank siding discussed by Richard Candee in his article on early Plymouth architecture, and extends its presence eastward as far as the outer cape.

Like the other examples cited by Candee, the structure is of a late date. The building then was a large two-story clapboarded structure, with typical late-seventeenth- to early-eighteenth-century casement windows with diamond-shaped panes of thin, greenish glass. The roof was probably shingled, with a massive brick chimney projecting through the center. What purpose, then, would such a large structure in an isolated location on an island in Wellfleet Harbor serve?

The first indication that the building was a shore whalers' tavern was found in the small east cellar. It was filled with clean sand containing no artifacts, but directly on the floor in the southwest corner was a large cervical whale vertebra that had been used as a cutting block, as evidenced by numerous cuts made by knives or cleavers. The bones of a complete whale flipper lay nearby. Whale bone had not been encountered on any other sites in the Plymouth Colony area, but in view of the proximity of the ocean the find was not a total surprise. The vertebra could have been a piece of whale bone picked up on the beach and put to good use. As work proceeded in other areas of the foundation, it became clear that it was not in the cellar by chance. More fragments of whale bone were discovered and when the west cellar was opened, it was found to be rich in whale bone, including several dozen whole pieces and a large number of fragments representing a number of individual animals. The culminating discovery was the foreshaft of a small harpoon or lance.

This accumulated body of evidence seemed to establish a connection between the building that once stood on the site with whales and whale hunting. The conclusion that the building served both as a tavern and housing for shore whalers seemed inescapable. This was further substantiated by a range of artifacts mostly from the refuse area. Large quantities of fragments of dark green glass from squat wine bottles of a type that dates circa 1700 were excavated, and broken wine glasses with baluster stems characteristic of the late seventeenth and early eighteenth centuries were found in great abundance. Finally, a disproportionate number of ceramic mug fragments, the remains of drinking containers, were recovered. Pipe-stem dates are in agreement with those derived from the ceramics, with the majority indicating an occupation from 1710 through 1750, and a smaller sample dating from 1680 to 1710. Taken together, it appears that the most extensive use of the site had been between the late seventeenth century and mid–eighteenth century, which fits perfectly descriptions given above of the time when shore whaling was no longer a profitable venture. Foodstuffs consumed by the patrons of the tavern obviously included whale meat and large quantities of both clams and oysters, the shells of which were abundant in two refuse heaps to the north and south of the building. Were it not for European artifacts scattered in the fill, these features would be mistaken for shell middens more characteristic of native sites. Today Wellfleet oysters are a prime delicacy, and they can still be harvested with ease in the waters just off the beach. This fact was not lost on the members of the excavating crew, who would spend most of their lunch breaks on the beach enjoying freshly shucked oysters. Lemons, cocktail sauce, and Tabasco were as indispensable to their digging kits as were trowels, brushes, and measuring tapes. Of course, only a heathen or some other person lacking in good taste would put any of the three on a fresh raw oyster, which should be consumed unadorned accompanied only by the juices in the shell, in spite of the fact that Eleanor Lightbody in the film *The Road to Wellville* referred to the juices as "oyster piss."

Three other items are worthy of mention. Scrimshaw was a traditional activity of New England whalers to while away the long hours at sea, and the same no doubt applied to what must have been periods of idleness at the tavern between the appearances of whales in the bay. Scrimshaw takes the form of either engraved designs on whale bone or walrus tusks filled with graphite, or full round carvings. In fact, in the Whalers' Museum at New Bedford, there is one entire scrimshaw model ship. What may be the earliest piece of scrimshaw in New England was discovered in the refuse heap just south of the tavern's foundation. It is the carving of a man's head wearing a cap, and measures less than two inches in length. One end has been drilled for attachment in some way or another, but its exact function has not been determined. It is far too small to have been the end of a knife handle, but might have served as an ornament attached to a pipe tamper. A second important discovery was a two-tined fork, the first to be recovered from a New England site. Finally, and most dramatic, was the frontal bone of a middle-aged European male, found in the fill of the west cellar, along with quantities of whale bone, brick, and broken artifacts. It had a series of cuts and slashes over the brows, which could have been what killed the man. There was no trace of his body. Such cuts are consistent with scalping, but this seems very unlikely in view of the virtual lack of a native presence on Cape Cod at that date. It will probably never be possible to account for the grisly discovery, but one possibility offers itself. In 1717, the pirate Bellamy lost his entire fleet on the shore at Wellfleet. He had promised the captain of a snow (a small brig) captured the day before that he could have the vessel if he would only pilot the fleet into Cape Cod Harbor. The captain was suspicious of the offer and instead led the fleet into disaster, as described by Levi Whitman:

> The night being dark, a lantern was hung in the shrouds of the snow, the captain of which, instead of piloting where he was ordered, approached so near the land, that the pirate's large ship which followed him struck on the outer bar: the snow being less, struck much nearer the shore. The fleet was put in confusion; a violent storm arose, and the whole fleet was shipwrecked on the shore. It is said that all in the large ship perished in the waters, except two. . . . After the storm, more than a hundred dead bodies lay along the shore.

It may be that the scarred bone is evidence of this incident in Wellfleet's history. The date of the event is contemporary with the period when the tavern was being used. Or it could be the remains of some unfortunate local individual who became involved in a dispute at the tavern itself.

Great Island today is very different in appearance from what it was in the eighteenth century. But looking across to it from Wellfleet, it is easy to imagine what it might have been like: a grassy eminence with a large building conspicuous on the high bluff, boats drawn up along the shore, and smoke rising from the fires of the whale butchers. When one stands on the site, with the ocean below and seabirds crying, its artifacts and ruins have a special meaning bestowed upon them by the world of which they were, and are, a part.

So what do the excavations of these seven sites tell us concerning the construction of houses in the colony? They confirm that the colonists built houses of varying types, not all of which emerge from documentary sources or can be seen in surviving structures. The Standish byre or long house may have been unique, but the length of the houses at the R. M. and Alden sites also suggest long houses not hinted at elsewhere. But the greatest revelation was the discovery of earthfast construction techniques and the questions it raised.

The Earthfast Question

Earthfast structures are those that are built directly on the ground, without a sill or footings, erected with the posts of the buildings sunk in deep postholes. Cary Carson et al. in a seminal paper, "Impermanent Architecture in the Southern American Colonies," published in 1981, point out that such buildings have their origin in medieval Britain, but by the fourteenth century fully framed structures on stone foundations had become the norm in many places. Despite this, Carson et al. suggest that there had to be a vigorous vernacular earthfast or post-in-ground building tradition that was certainly alive in England in the seventeenth century and may well have continued alongside the more formal building traditions even as late as the nineteenth century. This impermanent building tradition was transplanted to the American colonies in the minds of the settlers and flourished in the Chesapeake region as well as elsewhere in the South in the seventeenth century, persisting through the eighteenth. But did this tradition exist in New England?

In the Chesapeake, impermanent earthfast building has been associated with tobacco monoculture, where profits required a great investment of time in the crop and so less was spent on permanent architecture, as well as with the unstable demographics of the region. In Massachusetts, where there was no tobacco production, and population growth was relatively stable, the implicit assumption has always been that permanent building began at once, as there is a significant number of standing vernacular houses from before 1660, and even more remain from the last four decades of the seventeenth century. Until recently this assumption of a fairly immediate permanent building tradition has extended to Plymouth Colony, but increasingly it is emerging that the reason for the lack of surviving buildings from the first few decades of the settlement may well be due to the presence of an earthfast tradition. With the excavation of the Allerton site in 1972, the first evidence of earthfast construction in New England, let alone Plymouth Colony, was discovered.

The site had been purchased by Allerton in 1628, and presumably built on shortly after that date, so the construction of the house took place early in the history of the colony. In 1973, a year after the Allerton excavation, the remains of a second earthfast structure were excavated in Boston. Since then there has been a slow but steady accumulation of information that would suggest that this type of building was more widespread in New England than formerly believed, and primary documentary sources have been restudied for evidence of such building techniques. A revealing reference in William Bradford's history, *Of Plymouth Plantation*, written in 1635, shows the use of this technique early in the life of the colony. He describes the partial destruction during a hurricane of the trading post, constructed by the colonists in 1627 at Aptucxet in Manomet, some twenty miles south of Plymouth on Buzzards Bay:

> It took off the boarded roof of a house which belonged to this Plantation at Manomet, and floated it to another place, *the posts still standing in the ground* [italics added].

Earthfast building technology would have been known to the early settlers in Plymouth. Carson et al. show that there is evidence for its existence in central Lincolnshire and south Somerset as well as Yorkshire and elsewhere, and Leon Cranmer, in a publication on Cushnoc, the Plymouth Colony trading post on the Kennebec River in Maine, argues that earthfast construction was a technique well known to the Plymouth colonists:

> In the East Midlands and the Southeast, however, where building stone was non-existent, post-in-ground remained a common form of construction. Fully two-thirds of the colonists arriving at Plymouth during the first two years of settlement were from this region, as were a large proportion of settlers arriving in the southern colonies. Thus the emergence of this type of house construction in the American colonies in the 17th century was the continuation of a practice that was still in existence in certain areas of England at that time, rather than the rebirth of a forgotten technology or an adaptation through the development of a new technology.

Of the seven sites in the Old Colony for which we have described construction techniques based on archaeological excavations, only three had stone footings—the Alden site, the John Howland house, and the tavern at Wellfleet. The Standish house is so different in concept and construction that it falls in neither category. This leaves the Allerton house site which we know to have been an earthfast house, and the R. M. and Winslow sites. In both the latter cases a very large area was cleared and carefully examined for archaeological evidence. At the R. M. site an area fifty-two by fifty-eight feet was exposed, and at the Winslow site the area cleared was thirty-five by thirty-two feet. In both cases, since other structural remains were encountered near the center of the area, evidence of footings would have been found had they existed. Even if the stones themselves had been removed in the process of clearing the area for cultivation, one would expect some evidence of the trenches that held such foundations. We are dealing, then, with negative evidence, but of a compelling nature. It must be remembered that in the early 1940s techniques for recognizing post molds and the holes in which the posts were set had not been developed, and given their subtlety of definition could easily have been missed. On both sites there are features, labeled as pits at the R. M. site, and not explicitly identified at the Winslow site, that may well represent post molds and postholes. This is particularly the case in regard to a large feature in the far southeastern corner of the Winslow site plan.

Going further afield, but still within the domain of Plymouth Colony, the trading post established by the Plymouth colonists on the Kennebec River in Maine clearly shows post-in-ground construction.

The colonists needed to establish a trade monopoly in the area in order to secure the fur trade in particular from competitors. A patent to trade was obtained in England by Isaac Allerton in 1628 for exclusive rights for the Plymouth undertakers to trade with the Indians, but it was so poorly worded that a new one had to be obtained. In the meantime, the undertakers went ahead and proceeded to establish a trading post based on the 1628 patent. That same year Bradford wrote:

> Having procured a patent . . . for Kennebec, they now erected a house up above in the river in the most convenient place for trade (as they conceived) and furnished the same with commodities for that end, both winter and summer; not only with corn but also with such other commodities as the fishermen had traded with them, as coats, shirts, rugs and blankets, biscuit, peas, prunes, etc. And what they could not have out of England, they bought of the fishing ships, and so carried on their business as well as they could.

In all likelihood this building, constructed at Cushnoc, located in the modern city of Augusta, Maine, in the vicinity of Fort Western, was largely fabricated at Plymouth and transported to Cushnoc for erection. That such a practice was engaged in by the Plymouth Undertakers is suggested by a passage from Bradford written in 1633 concerning the construction of the colony's trading post at Matianuck, or Windsor, on the Connecticut River:

> But they [the traders from Plymouth Colony with a commission to trade from the Governor of Plymouth] having made a small frame of a house ready and having a great new bark, they stowed their frame in her hold and boards to cover and finish it, having nails and all other provisions fitting for their use. . . . Coming to their place [at Matianuck], they clapped up their house quickly and landed their provisions and left the company appointed, and sent the bark home, and afterwards palisaded their house about and fortified themselves better.

Constructed in 1628, the Cushnoc trading post appears to have lasted until at least 1669, a few years after Plymouth sold its final patent, and was abandoned sometime between then and 1676. Leon Cranmer's 1984 excavation of the trading post revealed an earthfast building forty-four by twenty-five feet with a shallow wood-lined cellar. A second earthfast structure was excavated in 1985 in present day York, Maine, by Emerson Baker, the remains of a manor house built about 1634 for the first governor of Maine, Ferdinando Gorges. The post-hole pattern outlines an earthfast building fifteen by thirty-nine feet. Although Gorges never made the crossing to Maine, his cousin Thomas arrived as deputy governor and described the house:

> I found Sir Ferdinando's house much like your barn, only one pretty handsome room and study without glass windows which I reserve for myself.

The Cushnoc trading post and Gorges manor house evidence does indicate that earthfast construction existed in Maine in the early to mid–seventeenth century, and possibly was more common than has been generally realized.

There is also documentary evidence of its existence in Ipswich, Massachusetts. A careful reading of the building contract drawn up by Samuel Symonds in 1638, cited earlier, shows that he wished to have his two-story house frame to be "strong in timber" and set on posts embedded in the cellar floor of his new house. "Let the frame begin from the bottom of the cellar, and so in the ordinary way upright." Carson et al. comment that while it may have been "ordinary for homestead structures . . . [it] was not ordinary at all for the sorts of seventeenth-century houses that have lasted to the present day in New England." In light of more recent analysis of the archaeological and documentary evidence, however, it would seem that Symonds may well have been referring to what was a common technique in New England at the time, but one that is only now slowly being uncovered by archaeological excavation. It is virtually certain that further archaeological research in New England will produce evidence of additional earthfast buildings, and a far more widespread earthfast tradition than has been believed to have been the case up until now.

The existence of this type of construction in New England in no way contradicts the argument of Carson et al. regarding earthfast construction in the Chesapeake. The common denominator would appear to be expedience. In the Chesapeake limited resources and the cost of tobacco production, a lucrative pursuit, precluded the construction of more permanent buildings, even by those who had the means to do so. In New England, the situation was different and would appear to be due to individual colonists not being sufficiently affluent to erect more permanent buildings until later in the century. This would apply particularly to Plymouth Colony, which never flourished economically, except for a period in the 1630s and 1640s when the colony traded cattle with Massachusetts Bay.

So then, the lack of surviving houses from the first fifty or so years of Plymouth Colony, in addition to the evidence from the Old Colony archaeological sites, the fact that at the colony's trading posts in Aptucxet and on the Kennebec in Maine the buildings were earthfast, as well as the primary documentary sources cited, strongly suggest that an earthfast building tradition could have characterized Plymouth Colony, and existed in Massachusetts Bay Colony as well as Maine, through the first decades of settlement and beyond.

The Artifactual Evidence

By now it should be abundantly clear that archaeology makes a critical contribution to our understanding of the various ways in which houses were constructed in seventeenth-century Plymouth. True, room-by-room probate inventories permit us to suggest the floor plans of houses of the period, but with the exception of the listing of cellars, there is no mention of any subsurface features that, as we have seen, are essential in understanding the variety of house construction methods of the period. Archaeology's second important contribution to our understanding of seventeenth-century life in Plymouth comes from the artifacts recovered themselves. It is beyond the scope of this work even to attempt an exhaustive treatment of this subject, since volumes have been published covering its many aspects. The most notable of these is *A Guide to Artifacts of Colonial America* by Ivor Noël Hume, a superb summary of artifacts encountered on sites of the seventeenth and eighteenth centuries in the American English colonies. Over three hundred pages in length, it covers an amazing variety of subjects, including ceramics, bottles, coins, drinking glasses, firearms, sewing equipment, locks, and tobacco pipes, to cite but a few of the many categories. For the purpose at hand, we will treat only two types of artifacts in detail, ceramics and, curious though it seems, spoons.

There are three broad categories of ceramics found on colonial sites, earthenware, stoneware, and porcelain. Of the three, only the first two occur on sites in seventeenth-century Plymouth, with the first mention of porcelain not appearing until 1735 in a probate inventory. Some definitions are needed before proceeding further.

Earthenware is relatively low-fired (900–1000° C) and is permeable to liquids. For this reason it is covered, at least on the interior, with a glaze produced by applying lead sulfide or lead oxide to the vessel prior to firing, at which time it fuses with the silicon in the clay to form a glassy, impermeable surface. There is no question that lead-glazed pottery fired at such a relatively low temperature released small amounts of lead into the food that was stored, cooked, and consumed in the containers. This, of course, was not known in the seventeenth century, and whether the amount was sufficient to cause death through lead poisoning is not certain. However, an archaeological discovery that does point to a significant presence of lead in European foodways vessels was recorded at Clifts Plantation, Stratford, Virginia, in the late 1970s by archaeologist Frazer D. Neiman. The plantation was occupied between 1670 and 1730, and during the archaeological investigation of the property two closely adjacent cemeteries were excavated, one containing the remains of the European residents of the plantation, and the other those of enslaved African Americans. Analysis of the skeletal material from the European cemetery showed a significantly high level of lead that could have come only from the pottery that they used. In contrast, the lead content of the African American remains was virtually nonexistent, since they used their own handmade unglazed pottery in the processing and consumption of foodstuffs.

In the seventeenth century most lead glazes were yellow, the result of iron impurities in the clays, but were on occasion purposefully colored green by the addition of copper, or purple by adding manganese. Regardless of the color of the glazes, they are transparent and one can see the body of the vessel through them. If tin oxide is added to the glazing material, the result is an opaque white glaze on which a design can be painted, the best known example of which is referred to as delftware, on which designs in blue, and less frequently polychrome (decoration in many colors), are applied. Slipware is the collective name given to another type of earthenware obtained by adding a white slip (a thick solution of the same clay used in making smoking pipes) to the body of a vessel prior to glazing, using various decorative techniques.

Stoneware is a harder, impermeable ceramic fired at a temperature in the vicinity of 1400° C, and is covered with a salt glaze produced by adding sodium chloride to the kiln when it has reached peak heat. The salt condenses on the body of the vessels, producing a surface that has a texture like that of an orange peel, slightly pitted. Stonewares occur in brown, the result of an iron-oxide wash, the most familiar example being that of Bellarmine or Bartmann jugs. Stoneware production was carried out exclusively in Germany until the mid-1680s, when English production, primarily of mugs, began in Fulham, then a suburb of London. A second type of salt-glazed stoneware encountered occasionally on seventeenth-century Plymouth sites is blue and gray, with the addition of purple sometime in the 1660s. The German brown stonewares were mostly produced in Frechen; the blue-gray and later purple in the neighboring Westerwald district. All of the types described above have been recovered from Plymouth sites.

The importance of ceramics in gaining a better understanding of lifeways in seventeenth-century Plymouth can be achieved only from a study of the ceramics excavated from the various sites. Probate listings of ceramics fall woefully short of the kind of specific description needed to understand the role played by pottery in the foodways of the period. Many probate references occur in the form of lumped categories, the most common being "earthenware" or "earthen vessels." There is the occasional reference to vessel forms, such as earthen pots, pans, a platter, or a basin. But such specific descriptions still fall short of providing us with a full understanding of the ceramic inventory of a seventeenth-century colonial household. At this time we lack the capability of determining whether coarse red utility wares, which form the majority of any ceramic assemblage of the period, were produced in America or England, although a number of attempts have been made to analyze the clays using various chemical and isotopic techniques. Local ceramic production began as early as the 1630s in Massachusetts Bay, but it is far from clear whether sufficient pottery could be produced to meet what was certainly a large demand.

Turning now to the ceramics excavated from Plymouth sites, we find that there is a natural division between those that predate circa

1650 to 1660 and those that follow. Prior to mid-century, the vast majority of ceramics recovered are lead glazed, undecorated coarse earthenwares. There is the odd piece of brown Freschen stoneware, borderwares (an English product characterized by a buff body and a glaze ranging in color between yellow and green), the occasional piece of delftware drug or ointment jar, and, at the Allerton site, a pottery known as Merida Red, an unglazed earthenware with micaceous inclusions, which was manufactured in Spain. In addition to these, the occasional fragments of what must have been large polychrome delft plates known as chargers have been recovered.

Ceramics appear to have played a fairly minor role in the preparation and consumption of food. Some borderware fragments are those of pipkins, small three-legged globular cooking pots made of buff earthenware glazed in green or yellow, with a hollow handle in which to insert a rod to move the vessel off and onto a fire. However, these are so scarce that they do not appear to have been central to cooking, which was usually done in metal vessels such as pots, kettles, and skillets. Food was served directly from the cooking pots and eaten on trenchers, small wooden trays with a shallow depression in the center to hold liquids. Beverages seem to have been drunk from containers of leather, pewter, pottery, or, rarely, glass. Trenchers and drinking vessels were communally used, as this was a corporate culture, not one stressing the individual as we know today in modern America.

There was one area, however, in which ceramics played a major role, that of the production and storage of dairy products. Jay Anderson, in his study of the foodways of the Stuart yeoman, shows that milk products were used extensively. Every house would have a dairy, even if it was not a separate outbuilding but an area off the hall or kitchen. Cheese was a more important protein source than was meat, and in fact the folk term for it at the time was "white meat." Contemporary treatises on dairy activities understandably placed great stress on cleanliness, and lead-glazed earthenware was preferred for ease of cleaning. The importance of dairying is clearly indicated in the Plymouth 1627 division of corporately held livestock. Only those animals that produced milk, cows and goats, were carefully divided on a per capita basis. Although we know from contemporary accounts that pigs were very common in the village in 1627, there was no swine division.

Plymouth's role in the production of cattle for trade with the new Massachusetts Bay Colony might also reflect the importance of dairying. The forms of coarse redware ceramics recovered from sites not only in the earlier period but throughout the seventeenth century seem to support this conclusion, since they consist of milk pans (large shallow basins used to skim cream), jars, pitchers, colanders, and crocks, all relating to dairying activities. The probate inventories are helpful in this connection, in that half of those taken on a room-by-room basis show ceramics to be associated with dairy activities in the kitchen or hall. Comparative evidence from contemporary Essex, England, confirms this. Almost all Essex inventories were taken on a room-by-room basis, and in those that mention ceramics, 75 percent of them were specifically located in the dairy.

The odd fragments of large delftware chargers also require comment. It would seem that they played no role in foodways as evidenced by their scarcity and the absence of wear on their interior surfaces. The most likely use that such large plates were put to was as display items prominently located on top of cupboards or hung on the wall. An early eighteenth-century delft plate from an archaeological site at Flowerdew Hundred, Virginia, had two holes pierced in the foot ring that were covered with glaze, indicating that the plate was intentionally produced to be hung. Surviving delftware plates of the period show a distinctive wear on what would be the bottom of the plate, where it rested on a surface for display. Such conspicuous display parallels the common practice of English yeomen to display pewter plates in a similar fashion. Displays of delftware chargers would not only enhance the family's parlor but show other members of the community that they were relatively well-off.

Some time in the mid–seventeenth century, and possibly related to the restoration of Charles II to the throne and his resolve to rein in the English colonies and bring them under greater control of the crown, we see a remarkable increase in the number and variety of finer English ceramics. This probably also relates in part to the Navigation Acts of 1651 and 1660, which required that only English products be shipped to the colonies on British vessels. The coarse redwares and very occasional bits of other ceramic types were now joined by a range of colored slipwares produced by various techniques, as well as a significant increase in the number of ceramic drinking vessels. The majority of stoneware mugs found in such large quantities at the Wellfleet tavern were of the Fulham brown stoneware type. Food appears to have continued to be taken from trenchers, for they are equally common in probate inventories through the end of the seventeenth century. Although by 1700 matched sets of delftware dinner plates were commonly used in England, this pattern appears not to have been transferred to Plymouth. This probably reflects both the conservative nature of the culture of the colonists as well as the relatively low level of affluence of the majority of Plymouth's citizens.

Only through archaeology can the insights concerning foodways and trade offered above be arrived at. Other classes of artifacts may not be as effective in this regard, but they have their uses nonetheless. One important value is a purely emotional one. It is one thing to read an entry in an inventory that states, "Item, one felling axe," and quite another to excavate such an axe and holding it, rusted, with particles of dirt still adhering to it, realize that you are the first person to have handled this axe head in three or more centuries. Artifacts add a texture and dimensionality to the informative but matter-of-fact entries in probate inventories, valuable as they may be. They can also bring one into direct contact with items that are not mentioned in inventories, the "small things forgotten," such as the pipe tongs discovered when excavating the Joseph Howland site. It was a small common object, so much a part of everyday life that it was never recorded. How often would Joseph himself or one of his sons, or even his wife, Elizabeth, have used that simple little gadget to pick up a hot coal from the fire to light a smoking pipe? And for more than two hundred and fifty years it lay buried in the ground, defying recognition when discovered, initially misidentified by a local physician as a surgical clamp. It was part of the evidence from the site, finally correctly identified, its significance part of the complex investigation of past life, which is what archaeology is all about. Artifacts and their place on a site are evidence critical to the solving of the mysteries of the past, working out constructions of what may have been taking place, in many ways a parallel to a homicide or other crime scene investigation. Unearthing and analyzing artifactual evidence is part of what drives archaeologists, which only their painstaking uncovering of the shrouding soil can open to our understanding.

But archaeology makes other valuable contributions; were it not for the hundreds of thousands of smoking pipe stem fragments with different bores that have been recovered, there could be no Harrington dating method. There are other artifacts that undergo fairly rapid change in form, providing yet another method of assigning a date to a site. Recalling the nearly restorable wine bottle found in the Allerton site cellar, it was its shape that led to an awareness that the cellar was of a much later date than originally thought. The earliest English wine bottles were produced in the mid–seventeenth century and are essentially spherical with long necks. By 1700 they had become very squat with wide bases and short necks, almost as if, like the Wicked Witch of the West, they were melting into themselves. In the course of the eighteenth century they became taller once again, but with shorter necks, so that by the third quarter of the eighteenth century they were not that different from wine bottles as we know them today. One of the most intriguing finds of wine bottle fragments came from the cellar of Structure 3 at the Howland site. Several fragments of what had been a globular wine bottle, which looked for all the world like the pieces of a modern interlocking jigsaw puzzle, were recovered. This was certainly not the result of breakage in the usual way. The excavators puzzled over these, and it seemed inconceivable that someone would purposely make a three-dimensional puzzle of a wine bottle. The pieces were sent to Robert Brill at the laboratory of the Corning Glass Company in New York, who gave us a perfectly reasonable explanation of what created such a curious set of bottle fragments. He explained that when two dissimilar lots of glass are mixed, but not thoroughly, breakage occurs along the lines separating the two types. The break would be clean, and the result precisely that which was recovered from the cellar.

One type of item often encountered on Plymouth sites deserves special mention, that being ox shoes, which to those who have never encountered them before are extremely puzzling objects. They resemble wide iron inverted commas, and were no doubt forged by the same smiths who produced horseshoes. The reason for their distinctive shape is that eight were needed to shoe an ox, since oxen have cloven hooves. The wide portion of the shoe fit the "heel" of the hoof, and the narrower curve at the top covered the "toe."

Of the large variety of artifact types recovered from Plymouth sites, we will conclude this description of artifactual evidence by considering spoons. To paraphrase a paraphrase of George Harrison's response to a reporter interviewing him about Ringo Starr in the film *A Hard Day's Night*, "Drums loom large in his life." In like manner, spoons loomed much larger in the lives of seventeenth-century people than they do in ours today. Granted, we have spoons, and in a considerable variety—teaspoons, tablespoons, dessert spoons, serving spoons, iced-tea spoons, and demitasse spoons, but these are part of a much larger complex of dinnerwares, including various types of knives and forks. Such was not the case in seventeenth-century Plymouth, the only fork having been recovered from the Wellfleet Tavern site, dating at the earliest to the end of the seventeenth century. Food was consumed using spoons, knives, and fingers, and given the nature of seventeenth century foodways, spoons were probably more important in the consumption of the various pottages and stews that were standard fare.

Probate inventories are of limited use in determining precisely what types of spoons were owned and used at the time, the most explicit mention being latten and alchemy spoons. Latten is an alloy of copper and zinc, similar to fine brass, and alchemy is a mixed metal, harder than pewter. The cellar of Structure 3 produced no fewer than five different types of spoons, no two alike (the concept of matching sets did not make its appearance until the mid- to late-eighteenth century). Included in their number is the end of a handle with the knopf (knob at the end of the handle opposite the bowl) in the form of a figure of St. Philip. Such a spoon is known as an Apostle spoon, and as one would expect, they normally came in sets of twelve, and modern sets of Apostle spoons are still obtainable today. Whether the other eleven were ever owned by the household we have no way of knowing. In contemporary England they were usually made of silver, but the one from the Howland site is of latten. A complete pied-de-biche spoon was recovered; these were produced in England from around 1663 until 1700. Handles could be flat, rectangular, or six-sided. The Howland specimen has a flat handle with a trifid terminal produced by flattening the handle end and cutting two notches into it, hence the name. On the basis of Percival Raymond's typology, this spoon can be dated somewhere in the middle of the period of production, around 1680. Spoon bowls and handles of latten spoons were also recovered. The knopf of the latter is of the so-called long seal type rather than a simple sphere located beneath the seal such as was found on the sealtop spoons from the R. M. and Allerton sites. The Howland specimens have a lozenge-shaped section below the seal that is sometimes decorated with acanthus leaves. Spoons of this type have a characteristic fig-shaped bowl, and the Howland examples show evidence of their once having been plated with tin, a technique thought to have begun about the middle of the seventeenth century. By contrast, the shape of the trifid spoon bowl more closely resembles those of modern spoons, supporting Raymond's generalization that the narrower the front of the bowl, the later the spoon. A single "slipped-in-the-stalk," or slip top, spoon was also recovered. In this form there is no knopf whatsoever, but the handle is obliquely truncated as though planed off from front to back. The handle is hexagonal, the material latten, and the form of the bowl unknown.

The most spectacular spoon to be recovered from the Howland site was a complete so-called rat-tail spoon, named for the projection of the handle onto the back of the bowl, which, with some imagination, might be seen to resemble the tail of a rat. Two things made this spoon special. First, being made of pewter, it is remarkable in its state of preservation since pewter does not survive well in the ground, oftentimes reduced to a gray smear with no recognizable form. Pewter spoons and other small pewter objects were produced at home using molds, since pewter melts at a low enough temperature to permit such domestic casting. But the truly remarkable aspect of this spoon takes the form of a coincidence that is at first almost impossible to accept.

Early in 1968, Lothrop Withington visited the research department at Plimoth Plantation to compare a late-seventeenth-century latten spoon with one excavated at the Bradford site in Kingston. This spoon is a part of a collection of spoons and spoon molds that he and his son, Ellis Brewster Withington, had been assembling over the years. He also brought a number of other spoons for the staff to inspect, including a striking seventeenth-century mold for a pewter spoon and some new casts made from it. This mold produced a rather ornate spoon, with elaborate scrollwork on the back of the bowl, and a portrait of an individual who is probably King William III on the end of the handle. But the most interesting thing about the cast was its apparent similarity to the spoon excavated at the Joseph Howland site. The excavated spoon was produced from the collections, and upon comparing it with the one cast from the Withington mold, it became obvious that indeed they might not be merely similar, but possibly identical. Further comparison of the two spoons was simple. A series of measurements were made of the distances between various points in the decorative scrolls on the bowl, and it was found that the two were identical in that respect. The numbers of dots, beads, and lines in the design were compared, and also found to be identical. Furthermore, alignments between different portions of the design were the same. The extremely unlikely possibility that the only pewter rat-tail spoon ever recovered archaeologically in Plymouth actually was made from the mold in the Withington collection seemed in fact to be true. Further support of this remarkable match came through comparing the two spoons with three others of a similar type in the Withington collection. While the same general decorative ideas are evident on all four spoons, they vary quite considerably in detail, and none of the others even remotely matches the Howland spoon in the degree of specific resemblance shown by the spoon cast from the mold. The Howland spoon had been excavated in 1959, and in all the years it had been at the plantation, no one had ever closely examined the handle for any design. With the fresh cast from the mold in hand, it was scrutinized minutely, and found still to bear the extremely faint traces of the same portrait as was on the cast. The mold has a minute flaw in the handle, which produced a bump on the face of the portrait. Under low microscopic examination and proper lighting, a possible remnant of this bump can be seen on the Howland spoon handle.

The mold owned by Lothrop Withington was obtained in Connecticut, but it had been in a New Bedford antique shop before that, and so was in the Old Colony area until quite recently. What its history was prior to that is not known. However, the simple fact that after two centuries a mold and one of the spoons that it had produced could be brought together again is truly amazing, and doubly so in view of the fact that the Howland spoon is the only complete one of its type ever excavated from an Old Colony site.

Archaeology plays a vital role in recovering the physical form of a world long vanished. Court records, probate inventories, and other documents form a very important component in this process, but only archaeology provides the actual evidence of the nature of possessions and buildings once in use but now visible only to the trained eye of an excavator. Archaeology also has the immediacy, dimensionality, and direct connection with people in the past that enables us to construct something of the material world in which they lived. Wherever archaeologists uncover the material evidence of the past, connections are made with the people who ate pottages using the spoons that have been excavated, who downed wine from squat green bottles, skimmed cream in cool dairies, gave birth, and were laid out in death in parlors still discernible as traces in the ground. These are connections that enable one to make the leap between a name on an inventory page and some real yet intangible sense of knowing the individual.

At the same time that the archaeology of early Plymouth was taking place in the 1960s and 1970s, a radical experiment in living history was being developed at Plimoth Plantation. Connections were being made between the way in which the early English settlers in Plymouth Colony lived and the types of houses they built, and the manner in which it was possible to create a living experience in which visitors to Plimoth Plantation's re-creation of the earliest village of the settlement could participate. The archaeology of early Plymouth was a part of this experiment, the story of which is now to be told.

Postscript

Patricia Scott Deetz

The Allerton site was to be the last seventeenth-century Plymouth site to be excavated under the immediate direction of James Deetz. Between 1972 and 1978, when he accepted a position at the University of California, Berkeley, he spent the bulk of his time developing the living history, first-person interpretative program for the 1627 village exhibit at Plimoth Plantation, one that would eventually become a model for others in the United States. He was involved in one other excavation in the Old Colony in the field seasons of 1975 and 1976, when he supervised the uncovering of a tiny settlement on the boundaries of Plymouth and Kingston. Known as Parting Ways, it had been occupied by three free African American families between 1794 and the turn of the twentieth century. The results were extremely important in the context of African American history of the area as well as in their broader implications, and have been published in Deetz's well-known book *In Small Things Forgotten: An Archaeology of Early American Life.*

Deetz's field archaeology days, however, were far from over. In his first season at Berkeley he directed the excavation of Somersville, an abandoned late-nineteenth-century coal mining town in the hills some forty miles west of San Francisco. In 1980 he initiated an extensive program of archaeology in Tidewater Virginia that would occupy the next fifteen years of his life. The results of this work are reported in *Flowerdew Hundred: The Archaeology of a Virginia Plantation, 1619–1864.* By 1995 circumstances changed, and he effectively hung up his trowel after at least thirty-five summers in the field. His involvement with Plymouth, however, was not over. By now he was on the faculty of the University of Virginia. He returned to a combination of documentary and field research, continuing the rigorous examination of the ethnohistory of the colony that had driven his work and shaped the living history program at Plimoth Plantation. In the spring of 1996 he added a seminar on the historical ethnography of Plymouth Colony to his teaching commitments, and two of his graduate students directed field schools at the Howland site in Plymouth in the summers of 1998 and 1999. And so his interest in and commitment to the archaeology of early Plymouth has come full circle.

Looking back over some forty or more years of archaeology, Deetz admits to a feeling of immense satisfaction, and would not change one small part of the experience. What might have been the beginning of an interest in things buried took place when he was six years old. He and his two cousins had heard a rumor that a dead baby and some gold coins were buried somewhere in the yard of his home in Cumberland, Maryland. He reasoned, with something of the same type of outrageous logic that would characterize his archaeological analysis in later years, that the only place people had not looked was the privy still in use at the rear of the yard. So, equipped with coal scuttle, shovel, and a length of rope, the excavation was under way. Deetz was lowered into the privy—it was his idea after all—and loaded up the scuttle for his cousins to sift through, but no skeleton, no gold. Before they could finish, however, looking up, Deetz saw his mother's face framed in the circle of the privy seat hole, a sight that will remain burned in his memory for life, as she yelled: "Jimmy, you come up here this minute!" It was an isolated incident and probably bore no relationship to what he would become when he "grew up," but who knows? He went on to attend Harvard—the first hillbilly affirmative action case, he maintains—where he intended to become a doctor, but failed Chemistry I. "So much for premed," he muses, "and who knows how many saved lives." Then, in his second semester he took a course in introductory archaeology and was immediately hooked. Until then he had no conscious interest in archaeology, having been preoccupied with astronomy, birds, butterflies (he was nicknamed "Bugs" Deetz at school), jazz, amateur dramatics, and stamp collecting.

And so began a career that ranged from digging Arikara Indian sites on the Missouri River to La Purisima, a Franciscan mission in Southern California, and a small Chinese community in Paradise, Nevada. English settler sites became his specialty, from those on the eastern seaboard of America in Plymouth and Virginia, to those of British settlers many thousands of miles away on the eastern Cape frontier of South Africa. As Jimmy Stewart said, "It's been a wonderful life." But perhaps the greatest time of that life was the years at Plimoth Plantation, where not only did he and his wife and family of nine children have the time of their lives (they would not trade their Plymouth years for anything), but the first-person interpretive program that emerged reflects for us today a closer understanding of the men, women, and children who experienced a time unlike anything else in their lives and left us the richer for it.

Chapter 7

The Time of Their Lives

Plimoth Plantation

At Plimoth Plantation in 1997, traveling to another age has become as natural as traveling to another town. Once we stroll by a sign marking the line between the 20th and 17th centuries, the past comes alive with a vibrant clarity. Every sense is engaged. The earthy tones of the little village, set off against the deep blue of the ocean, bewitch the eye. Each fence post we touch, or fabric of a bed curtain we feel deepens our awareness of the year 1627 . . . [telling] us that we have left the twentieth century. The sense of the past is heightened by the sounds of bygone colonial life: the bite of the ax, the report of a musket, the lowing, clucking and bleating of farm animals. It is very human with all the authentic smells of the pen and cow-yard which, combined with the pervasive tang of wood smoke and the scents of the ripening fields and gardens, evoke the tangible presence of another era. . . . This is a

> magnificent and intriguing way to learn about the past, but it took many years to achieve.
>
> The Plimoth Plantation of the 1960s was a successful and well-regarded historical exhibit. However, despite superficial similarities, a visit to Plimoth Plantation in 1997 is worlds apart from a visit in 1967. Dramatic changes occurred at Plimoth Plantation in the late 1960s and after which totally transformed the museum. Innovations in "living history" and "first-person interpretation" propelled what had been a modest regional institution to one with international recognition and acclaim. The Plantation's program has become a standard by which other open air museums are judged. How this came about is part of the engrossing history of Plimoth Plantation.
>
> —James W. Baker, *Plimoth Plantation: Fifty Years of Living History*, 1997

The village of 1627 at Plimoth Plantation as it is represented now in the twenty-first century is the culmination of a vision long held by Harry Hornblower. In 1967 he wrote, "What you see here today is the outgrowth of my teenage dream to create an exhibit which would show the visitor to Plymouth the life and times of the early Plymouth settlers. . . ." How it came into being and the changes it underwent form the basis of the following account.

The village would not exist today if it were not for the generosity of the Hornblower family. In 1945 Harry convinced his father, Ralph Hornblower, to make a gift of $20,000 to the Pilgrim Society, based in Plymouth, to initiate planning and building an open-air museum that would be a construction of the first settlement in the colony. Within two years the Pilgrim Society had realized that the village project was too ambitious for their resources, and in 1947 an independent corporation was established, known as Plimoth Plantation Inc. The articles of incorporation explicitly state that it should serve as a "memorial to the Pilgrim Fathers [to further] the historical education of the public with respect to the struggles of the early settlers in the Town of Plymouth, with the expansion of that settlement and the influence of the Pilgrim Fathers throughout the world. . . ." No small mission, this, the second part of which contains echoes of claims made by some writers concerning the importance of the Mayflower Compact.

A board of governors composed of prominent local businessmen was formed and the first projects undertaken. From its inception, every effort was made to base the interpretation of the museum on historical and archaeological research. A small house that was thought to represent the first type of dwelling in the original village was constructed on the Plymouth waterfront adjacent to Plymouth Rock. It opened to the public in May 1949. In 1953 a second building was erected on the waterfront, the "fort/meeting house," similar in form to that described by Isaack de Rasieres in his classic account of Plymouth in 1627. Both buildings were designed by the architectural firm of Strickland and Strickland (Sidney Strickland and his son, Charles). In 1955 a third structure was added, the so-called "1627 House" intended to illustrate improvements in architecture over the first seven years of the settlement. While the "First House" was thatched and had simple, vertical plank siding, the 1627 house was clapboarded, had casement windows with diamond lead-set panes, and a massive stone fireplace and chimney. It also had a shingled roof rather than thatch. Both houses would serve as models for the first structures erected in the full village established on Plimoth Plantation land. The 1627 house served in part as a gift shop, a function that it continues today as it still stands on the Plymouth waterfront.

In 1955 Harry's grandmother, Hattie F. Hornblower, left a tract of 140 acres two and a half miles south of Plymouth on the north bank of the Eel River to Plimoth Plantation Inc. Topographically, it was very similar to the site of the original settlement, with the Eel River occupying the same situation as did Town Brook to the south of the town, the land rising above the stream, overlooking the ocean. The tract of land was more than adequate to accommodate the projected settlement, and plans could now go forward for the construction of a complete village. The date chosen to be represented was 1627 since that was the last year in which all of the settlers lived together in a single fortified community, and because of the cattle division we have a reasonably accurate "census" of the people living there at the time. In addition, the best eyewitness description was written by Isaack de Rasieres in 1627.

Ground was broken on the new site in 1957 by Harry Hornblower, seated on a large bulldozer. The same year, another dimension to the open-air museum became a reality. Initially Plimoth Plantation had plans for a waterline full-size representation (non-seaworthy) of the *Mayflower* to be moored in the Eel River. The plans were based on meticulous designs by William Baker, an eminent naval architect. However, in the meantime, a corporation quite independent of Plimoth Plantation was formed in England, Project Mayflower Ltd., which constructed a fully seagoing replica of the *Mayflower* using Baker's plans (with his permission), and successfully sailed her across the Atlantic in 1957, arriving safely in Plymouth Harbor in June of that year. So visitors to the waterfront who may have come just to view Plymouth Rock now had four other attractions, but soon there were to be more at the plantation itself.

By 1959, five houses had been constructed (in street order, Bradford, Brewster, Howland, Fuller, and Warren). Nineteen house lots were staked out for the construction of future houses. The entire village was enclosed by a long rectangular lightweight palisade that, on the sides facing away from the village, carried metal signs that read "Walpole Fence Co." By then, the fort/meeting house had been moved from the waterfront to the village site where it stood for a brief period before being moved to the head of the street closer to the position suggested in the de Rasieres description.

A visitor entering the village in 1959 was confronted by a postcard-perfect picture. Indeed, far too perfect to begin to approximate the true nature of the early settlement, but it was right up to date with what was then known concerning the establishment of period open-air museums. Grass was kept neatly mowed, houses were surrounded by tidy herb gardens in raised beds, and there was no evidence anywhere of cultivated foodstuffs. Each house had a large wooden sign to the left of the entrance with the names of the occupants printed on it. Guides and hostesses (a curious gender distinction) were present in the village in small numbers to give a very simplified account of the community. They were attired in spotless clothing of polyester modeled more on nineteenth-century paintings of the "Pilgrims" than on contemporary knowledge of early seventeenth-century clothes. A certain air of quaintness pervaded the scene. Two of the most popular artifacts exhibited were a bed key, used to tighten the ropes of bedsteads, and a device known as a "niddy-noddy," which consisted of two horizontal rods at right angles to each other and attached to opposite ends of a rod of equal length. These were used to wind yarn, but one cannot help but think that this particular artifact was chosen because of its quaint name rather than any evidence that it was in use at the time.

Limited demonstrations were engaged in, the most frequently shown being candle dipping, which is all too often still a part of historic house and open-air museum tours. This comes as no surprise, since three of the museums that served as models for the village in its formative years were Colonial Williamsburg, Old Sturbridge Village, and the Jamestown Festival Park, at the time the leading open-air museums in America, all of which were committed to demonstrations of one or another type of craft activity. By far the commonest "human" figures encountered in the village were mannequins, including a full church congregation in the fort/meeting house, Myles Standish negotiating trade in furs with a local Wampanoag, the conclusion of the peace treaty in the William Bradford house, and in "Dr." Samuel Fuller's house two figures showing Fuller about to lance a particularly large, ugly carbuncle on the shin of a patient.

The houses were furnished with seventeenth-century antiques acquired in England by Harry Hornblower and Charles Strickland, who were told at one point by a dealer that there were probably not enough genuine seventeenth-century pieces in the country to furnish the number of houses that had been planned. Visitors were separated from the house interiors by velvet ropes. The chimneys of the houses were massive, built of field stone, based on what was later recognized to be a misinterpretation of the evidence uncovered by Sidney Strickland at the John Howland house site of the Howland complex. Houses interpreted as having been built before 1627 had thatched roofs; later ones were roofed with shingles, incorrectly as it turned out, the result of a misreading of the regulations passed by the court concerning the covering of houses. Based on the precedent set by the 1627 house on the waterfront, the windows of the Warren house, located at the end of the street and interpreted as one of the most recently built at the time, had casement windows with diamond leaded panes. The glass was in subtly different colors, producing an effect evocative of a stained glass window, which was probably intentional, designed to reflect the reverence in which the early settlers were held. The village retained its "squeaky clean" and tidy appearance until 1960, which was to become a year of radical change in all aspects of the village exhibit.

In 1959 a new position, that of assistant director, was created and James Deetz was appointed to the post. He divided his time evenly between teaching anthropology at Brown University in Providence, Rhode Island, and overseeing all aspects of exhibits, interpretation, and research, including archaeology. The split appointment had definite advantages in that a number of Brown students became involved in various aspects of the plantation's programs. Prior to 1960 the research staff had consisted of a single person, but in the months following the Deetz appointment it was increased initially to three, and by year's end to five. Given such an expanded research capacity, attention was first directed toward the village exhibit, to exploring the ways in which it might be changed and improved. The basic premise that underlay all of the proposed changes was that if visitors were made to believe that they were entering the world of 1627, there should be nothing present that would not have been there at that time. This was radical and innovative thinking, new to existing concepts of outdoor museum presentations, and resulted in the removal of a large number of items that were a part of the exhibit. The large signs listing the inhabitants of the houses were taken down and small labels identifying various herbs in the house gardens removed, as were all printed explanatory texts in the houses. Save for the fort/meeting house where they would remain for a few more years, all mannequins were taken out. This particular removal was what prompted an irate letter from a Boston schoolteacher to the plantation, who every year promised her students prior to their field trip to the plantation that they would see "the boil" in the Fuller house, and since it was no longer present she felt betrayed—so much for some educational field trips.

Careful analysis of a set of the earliest probate inventories available showed that the village houses were overfurnished, and their contents were reduced to conform as closely as possible to the probate data. Prior to this, every house had a bedstead, but these were reduced in number, as the probates showed that beds (meaning mattresses in seventeenth-century usage), placed directly on the floor, were more common. A house of 1627 would normally have had a thick layer of rushes strewn over the floor, but this was impractical, given the nature of the amount of foot traffic that was anticipated in the houses once all the changes had been made. So one house, that of John Howland, was given a rush floor covering so that visitors could see it as an example of this practice. This was the only exception, and guides and hostesses, by now referred to as interpreters, were on hand to provide appropriate explanations for this and to respond to all other questions.

The other major change was the removal of all the antique furniture from the houses to be replaced by reproductions of the highest quality. There were three reasons for taking this step. First, the environment in which the antiques were located underwent great changes in humidity and temperature in no way consistent with responsible curatorial practice. Second, the visitor should be confronted with furniture that looks new rather than shows the effects of three centuries of wear, since in 1627 such antiques would have to have been made in the 1300s. Many of these antiques were sold at a controversial auction held in 1972 that was deplored by traditionalists who in part had come to believe that they were "authentic" artifacts somehow associated with the early settlement, but of course such was not the case, and many of them were stylistically of a later period than the year represented. Third, the use of reproductions permitted the interpretive staff to make full use of all furnishings and equipment in a house. The barriers were removed and the public had free access to the interiors and was encouraged to touch and handle everything that was present.

Coincidental with this, a foodways program was implemented under the direction of Jay Anderson, a specialist on Stuart yeoman foodways. Anderson instructed the interpretive staff in the proper manner of preparing various dishes in seventeenth-century style, and cooking became standard practice in all of the houses, save for the Howland house. Prior to this, cooking would be done on rare occasions using an open fire, but more typical was an exhibit in one of the houses that showed plastic fish on a grill over a bed of cold, charred logs with bits of red metal foil tucked between them to simulate flames.

These innovations did not all occur in the first year, but took place steadily over the next decade. The changes in the period between 1960 and 1969 were based on research obtained in the process of creating a major database on early Plymouth material culture and lifeways, which incorporated both documentary and pictorial material. The changes implemented radically altered the nature of the village exhibit, but others were yet to come. In 1969 livestock were introduced, with the first animals being sheep, soon followed by cattle, goats, and chickens. They were allowed to wander freely on the street and highway, and, in the case of chickens, in the yards. The old split rail fences around the houses were replaced by animal-proof palisade-type fences such as that excavated at the Allerton site. Pigs were also introduced, but for safety reasons were confined to a large pen with posts set deep in the ground to prevent them from rooting their way out. The livestock and foodways programs contributed a set of sounds and smells never encountered in the earlier village, but that would certainly have been typical of 1627 Plymouth.

At the same time research was under way to create accurate attire for the interpretive staff. Research staff visits to the Victoria and Albert Museum in London and consultations with Janet Arnold, a well-known designer of costumes for dramatic productions, provided the necessary professional training. Arnold had in fact published a book of patterns that was invaluable to what had by this time become a fully developed costume department. Polyester was discarded, and for good reason in addition to the historical, since it burns easily and is transformed into something resembling napalm; one woman interpreter suffered severe burns when her dress ignited while she was working by the fire. Linen and wool, which were the fabrics newly put to use, do not present this problem. Should an ember fall on a woolen garment, it will simply burn a hole through it rather than ignite it. In their new attire, carrying on an increasing number of daily chores, the interpreters soiled their dress, just as the colonists would have at the time.

James Baker is correct in stating that constructions such as those that had emerged at the plantation in the 1960s are partly a function of the time in which they took place. The emergence of the "new" village by the late 1960s coincided with the height of the development of the hippie movement and counterculture, and the similarities were more than passing. The village was populated by people dressed in brightly colored clothes, many of the male interpreters with long hair appropriate for the time, the settlers represented as sleeping, as often as not, directly on the floor. It should come as no surprise that what had certain outward resemblances to a hippie commune, even though it was supported by solid research, met with great disapproval by traditionalists. Three quotes from letters received reflect this. One, from a *Mayflower* descendant, deplored the "removal of the furniture and mannequins from the houses. Everything looks run down and shabby including the hostesses and guides." Another referred to it as "a hippie village of today," and, in a widely circulated letter by another *Mayflower* descendant, the demand was made to "get rid of the realism, so-called, and give people some ideals to live up to. Clear out the radicals in command and get some 100% Americans."

In spite of this response, the governing board, President Harry Hornblower, and the director, David B. Freeman, remained steadfast in their support of the changes. The plantation had moved into a phase referred to today as "living history," in which visitors experience all of the aspects of day-to-day life and craft demonstrations had been abandoned. What visitors saw was what might've been happening on a given day, whether building a house, planting or harvesting crops, tending livestock, or musket drills. While it was not possible to create an area outside the village as large as that cultivated in 1627, two acres were planted in corn and wheat. An unplanned but very positive result of involving the interpreters directly in all of the daily routines of life was that they increasingly made use of first person rather than third. If one had cooked a meal or worked on a house, or planted crops, it no longer made sense to use third person past plural—"we do" this or that, rather than "they did," became the norm.

This particular change received great impetus in 1973 when Henry Glassie of Indiana University, a vernacular architecture specialist, was employed to assist in research on the architecture of the period. After spending only two weeks at the plantation assessing the research that had already been done, he informed the staff that what they knew was probably all they were going to find out. Having been employed for the summer, Glassie was in need of something else to engage his time. One night, or to be more precise, very early one morning —about 3 AM—Glassie and Deetz, over a bottle of Virginia Gentleman bourbon, decided that it might be a very worthwhile effort to construct a house in the village based on the excavated Allerton site plan, using period techniques and tools, and the interpretive staff to build it. Prior to this time, the houses had been built by contractors who used modern tools and simulated seventeenth-century features using, for example, colored concrete to resemble a daubed chimney. This was expensive, for the work could only be done during the hours when the village was closed, which in some cases required overtime compensation. In comparison, building a house with the assistance of the interpretive staff would involve little more than the cost of the materials. Neither Glassie nor Deetz were experienced carpenters. Glassie had built a set of shelves, and Deetz's only carpentry effort was a total disaster, consisting of a box in which he planned to mount his butterfly collection, made of two-by-fours and a one-inch-thick bottom. The end result was crooked, far too heavy, and had to be abandoned. This in no way served as a deterrent, since Glassie was intimately familiar with how mortises and tenons were cut and how a house was raised and assembled. Plans proceeded apace, and sufficient oak was obtained from a local lumber company to construct the house. The roof was to be thatched by the plantation master thatcher, Peter Slevin, who amongst other contracts had thatched the roof of Anne Hathaway's cottage in England, and numerous other British and American houses. The lumber arrived and construction began.

The lumberyard had sawed all of the pieces that would be needed in framing the entire house, following specifications furnished by the plantation. The only remaining work to be done on the framing pieces was to smooth them with an adze, removing the circular saw marks that were clearly visible. In the seventeenth century, dressing such timbers with an adze would also have been done, but in that case to remove cut marks made by the broad axe used to shape the piece. Once this had been accomplished, all mortises and tenons were cut so that the house could be fit together. The location chosen for the erection of the house was that occupied in 1627 by John Billington, its position being shown on Bradford's sketch plan of "the meersteads and garden plots of those who came first layd out 1620." It was on the south side of the street, across the highway from the house of William Brewster. The erection of Billington's house would provide the fourth and last of the houses located at the intersection of the two byways. Further incentive to represent John Billington's house was the fact that, given the then current policy for house construction, it was highly unlikely that it would ever have been built. Billington was the first person executed in the colony, for his murder of John Newcomen in 1630, and at this time funds for the construction of houses were provided by the various associations of descendants of the person indicated as living there. Except for four dollars contributed by a Plimoth Plantation staff member, there were no funds in the Billington house account.

Once all of the framing pieces had been adzed, and all of the joints cut, it was time to raise the frame. Following seventeenth-century practices, this was a community effort in its initial phase. Work on raising the house began at seven in the morning, with some thirty interpreters and volunteers in period dress on hand to assemble the very heavy primary parts of the frame. The four corner posts were inserted, two each in the two major horizontal framing pieces, or plates. This produced two large U-shaped sections, known as bents, one of which would form the rear wall of the house, and the second, the front. The building was earthfast, so they were placed adjacent to deep postholes excavated to receive them. Since each section weighed well over a ton, most of those present were needed to raise the house frame. Ropes were attached to the plates, and while some people pulled on these, others pushed from behind with their hands or with poles. Slowly the bents rose from lying horizontal on the ground to their proper vertical position, firmly seated in the holes dug to accommodate the corner posts. The next step was to lift two massive, horizontal beams, known as girts, which tied the front and rear bents together. At this point a rectangular frame stood in place and the postholes were filled and solidly tamped. The basic frame was now quite stable, permitting further framing to take place. This last step involved raising two large pairs of rafters (principal rafters) on either end. Each pair of rafters was connected by a horizontal piece of timber known as a collar beam, which served to keep the rafters from spreading under the weight of the roof. The principal rafter pairs were connected by two horizontal timbers known as purlins, one on either side, on which smaller rafters (common rafters) would rest with their lower ends mortised into the plate. The entire process of raising the basic house frame was completed by 7 PM of the same day. Further work, such as setting the studs, attaching the common rafters, and building the chimney would need only a few workers in comparison to the major effort required in raising the frame.

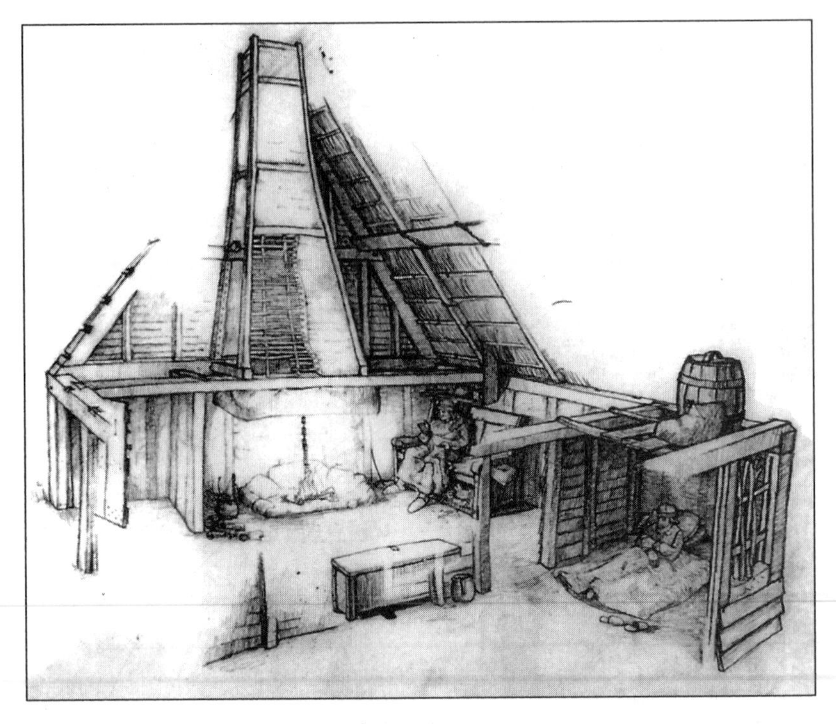

Billington house interior showing construction details. Drawn by Henry Glassie, 1973, based on data from the Allerton site.

In the ensuing months, the wooden chimney was framed and wattle was used to fill the space between the studs and also in the chimney frame. Thatch poles of peeled cedar were attached to the common rafters and these in turn would have the roof thatching bound to them with heavy twine. The walls and chimney were given a thick coat of daub, covered with hand-riven clapboards on the exterior, and whitewashed on the inside. There was a dirt floor, which in the seventeenth century would have been covered with rushes. Shutters were installed in the windows and a hearth of cobbles, similar to that excavated at the Allerton site, was placed beneath the chimney.

The only step in the process that was omitted was providing protection against entry by witches, which was in fact done at about the same time in the Hopkins house. A broken Bellarmine jug was suspended from the trammel bar in the chimney, which routinely would be used for pot hooks from which to suspend kettles over the fire, and beneath the threshold stone an intact Bellarmine containing common pins and human urine was buried. A small strip of red felt was attached over each window. These procedures were all intended to prevent access to the house by a witch, whether through the chimney, door, or windows. These precautionary steps appear to have been successful, since no witch made an appearance during the ten years the house stood. Of course, we cannot be certain of this, since to our knowledge none of the other houses not protected in this way were visited by witches either. But it made the point that in seventeenth-century Plymouth a set of folk beliefs were firmly in place to prevent such access.

The visiting public showed great interest in the process of building the Billington house, and there was always a large number of people watching the craftsmen at work. When complete, the house was by far the most comfortable of those in the village due to its thick walls of daub and thatched roof, relatively cool in hot weather, and snug and warm in cold. The construction of the Billington house initiated an important new phase in village house building. Professional building contractors were no longer employed, and all work was done by the plantation artisans, assisted by interpreters. A second

Billington house exterior. Drawn by Charles Cann, 1977.

house, the Allerton house, just west of the Billington house, was framed entirely in place, but later houses were fabricated in the plantation workshops and assembled on site, much as would have been done at the Plymouth trading posts in Connecticut and Maine.

In 1972, a year before the construction of the Billington house, the plantation made a long overdue and very important decision to involve members of the Native American community in the program and to expand greatly the Native American exhibit. Prior to this time, the native presence was acknowledged solely by a winter-type bark-covered longhouse located inside the 1627 village. Except for a brief period in the early 1960s when it was staffed by a nonnative person who was skilled in flaking arrowheads, which he demonstrated, the exhibit was completely static, could not be entered by the public, and was interpreted through labels and textual material. The new plan would involve members of the native community both as directors and interpretive staff members, all funded by the plantation.

In the spring of 1972, a Native American clearing house conference was convened in Concord, Massachusetts. The purpose of the conference was to provide information on various job opportunities available to members of the Native American community. It was attended by two plantation representatives, exhibits director Tom Young and James Deetz, who, in retrospect, naively believed that the plantation's proposed plan would be received with enthusiasm. What actually occurred was precisely the opposite, with the majority of those present expressing a range of negative reactions. These centered around three themes, two of which had great merit, and the third, in the estimation of Young and Deetz, somewhat less so. First, great concern was voiced regarding affiliating with a museum that represented the very people whose arrival in 1620 led to the eventual destruction of the native culture. Second, it was thought that members of the interpretive staff would be viewed by the visitors more like animals in a zoo than in their own right. Third, some of those present stated that their culture was their own, which they had every right to keep private and not communicate to the public at large. All told, it was a very unsettling experience, marked on occasion by moments approaching open hostility.

As Art Zimiga, a Lakota Sioux, put it in the style of punning for which Native Americans are known, the entire episode could have been called "Face the Nations." However, after the meeting formally adjourned, three Narragansetts from Rhode Island, Eric and Ella Thomas, and Eric's sister, Mary Gorman, approached the plantation representatives and said that they would be interested in becoming involved. Thus it was that the new Native American Studies Program at Plimoth Plantation had its beginnings. Along with Steve Figueroa, an Apache from Los Angeles, they were soon joined by several Wampanoag Indians from the town of Mashpee on Cape Cod, a community that to this day is composed largely of Wampanoags.

The site for the new exhibit was located on the Eel River at the foot of the hill below the fort and had in fact been occupied by the Wampanoag until the period of initial European contact. Three years previously the longhouse had been removed from the village and a summer-type hemispherical wigwam was constructed on this site, and while a step in the right direction, it was still a static exhibit with no interpreters present. All of this would change with the initiation of the new program. Eric and Ella Thomas were sent by the plantation to visit a number of museums in the northeast, most notably the Smithsonian Institution, the American Museum of Natural History, and the Peabody Museum at Harvard University. The purpose of their trip was to collect information on native material culture of the period. This was scantly represented compared to material from further west acquired in the late eighteenth and nineteenth centuries, but there was still more than sufficient material, both archaeological and ethnographic, to provide a firm foundation for the careful reproduction of the material world of the Wampanoag in the early seventeenth century.

A summer encampment was chosen as the most reasonable to represent, with a single wigwam surrounded by an acre of cultivated land. Short stumps of trees with their ends cut to simulate their having been felled, using either stone or metal axes, and charred with a blow torch, were planted in the field. This represented a field cleared by the so-called "slash and burn" technique where trees were felled, allowed to dry, and then burned, returning nutrients to the soil. The field was planted with white flint corn, appropriate to the period, the seed for which was obtained from the Rhode Island Agriculture Experimental Station, as well as squash and beans. Meat was also an important part of the foodways of the Native Americans, and once the program was open to the public, this was obtained by buying bears at auctions of roadkills held by the state of New Hampshire, as well as through the initiative of some of the interpreters themselves. Each night, as they drove home to Mashpee, they would scan the road carefully. On their return to work in the morning, they would stop and collect fresh roadkills of raccoon, squirrel, and other small wild animals that would have formed part of the diet of the Wampanoag in the seventeenth century. The animals would be skinned and used both for their hides and meat.

The wigwam was covered on the outside with sewn cattail mats, and the interior lined with woven mats of cedar bark or bulrush. All of these mats were made by hand and required many hours of work to complete, but the accuracy and quality of the artifacts that the staff produced during the months that followed were remarkable. Contemporary clothing was made using buckskin decorated with quillwork in various colors, and included leggings and breech clouts for the male interpreters, and buckskin dresses for the women. As the crops grew and food, including the roadkills, was prepared both inside and outside the house, the final effect was dramatic and convincing.

The exhibit featured prominently in the pilot program for a television series hosted by Bill Moyers for the bicentennial in 1975–1976. Filmed entirely at Plimoth Plantation, the film's title was *The Peach Gang*, and it dealt with the trial of three men, Arthur Peach, Thomas Jackson, and Richard Stinnings, for the murder of a Nipmuck Indian in 1638, for which the three were found guilty and executed. It is probably the only film ever made in which the native actors spoke Algonquian. To make this possible, the film company obtained the services of Ives Goddard, the foremost authority on various Algonquian languages. The only exception to this was Chief Dan George, who won an Academy Award for his role in *Little Big Man*, and portrayed Canonicus, chief sachem of the Narragansett, in *The Peach Gang*. He refused to deliver his lines in Wampanoag, insisting that he use his native Salish, but it is highly unlikely that many viewers would have been aware of the difference.

First-person interpretation was not possible in this context, and the interpreters were given considerable latitude in what subjects they could discuss with the public, including, when they arose, contemporary Native American issues. Although the concerns expressed at the Concord conference regarding the interpreters being viewed as exhibits themselves were shown to be largely unfounded, there were occasional problems with visitor reactions. At times they would speak of the interpreters in their presence as if they did not understand English. One visitor asked Steve Figueroa if he was allowed to go home at night, to which Steve replied, "Yes. I own a car. I go home, have a beer and watch TV." On another occasion, a plantation staff member found a missionary from India attempting to convert Ella and Eric Thomas to Christianity, believing that he had found, finally, some unreached tribal Indians, suitable subjects for conversion. He was promptly and firmly told not to continue and to leave. While incidents such as these were witness to the fact that the interpretive program was a highly convincing one, and did not happen too frequently, they were sufficient to generate a significant amount of stress on the part of the interpreters, and for the first couple of years that the program was in place, they regularly visited the home of the assistant director of the plantation after work to discuss the day and release some of the tension that had built up. Nevertheless, the program succeeded beyond expectations and gradually members of the local Native American community became convinced of its value. Five years after its implementation it was given a citation by the Boston Indian Council for its important role in contributing to greater public understanding of local native culture. Unfortunately, under the leadership of a new director, the program was discontinued in 1978, but was reopened in 1984 and then replaced in 1989 by a homesite representing the residence of Hobomok, a Wampanoag known to have lived just outside the village in 1627. This is the exhibit that is in place today, still staffed and directed by Native Americans.

By the mid-1970s, interpretation in the 1627 village had naturally evolved into exclusively first-person presentation. After all, it was the interpreters who helped in house construction, planting and harvesting crops, cooking over open fires, and all of the other tasks that occupied people of the time. So it would be unusual to refer to the former inhabitants using third person, past tense. This fact was not lost on members of the plantation staff, and a final change was implemented in the form of presentation that visitors encounter in the village today. For the first time, interpreters adopted the role of a particular persona who had lived in the village, people such as Myles and Barbara Standish, William Brewster, William and Alice Bradford, John and Elinor Billington, Stephen Hopkins and his wife, Elizabeth, as well as brought in children to play relevant roles. The interpreters were allowed to use a part of their salaried time to research the individuals they represented in as much detail as possible. This change to the program met with immediate success, and even *Mayflower* descendants who had separated themselves from the museum since the changes in the late 1960s and early 1970s, were won over, by being able to converse directly with those representing their forebears. This change led to two others, both critical to the ultimate success of the program. Interpreters now spoke in the dialect of the region from which the person they were representing came, so that visitors would appreciate the varied backgrounds of the residents of the village. Of course, once these changes were implemented, it was imperative that the interpreters' representations of the settlers have no knowledge of any events that occurred after 1627. It takes considerable skill and initiative on the part of interpreters to acquire the correct dialect and handle questions that arise concerning events and inventions of which they can have no knowledge, but it can be done with remarkable results. The later history of the colony was, and still is, dealt with in exhibits installed in the visitors' center.

The change to first-person interpretation brought about a profound and positive change in the way in which visitors encountered life as it was represented in the 1627 village. What has been described as a remarkable cultural performance, a type of theater, emerged as interpreters became not just actors, but part of a simulation of life in which all senses are involved, feeling, thinking, and acting in an environment as close to reality as research could make it. It is this type of performance that caused novelist James Carroll to write of Plimoth Village in *The New York Times* in 1984 that: "The actors' perfect balance between earnestness and playfulness is what enables us to suspend our disbelief, if only in flashes, that we've stumbled on another time."

Like other outdoor living history museums such as Colonial Williamsburg and Old Sturbridge Village, to name but two, Plimoth Plantation is a construction of a past world done as carefully and thoroughly as research permits. However, it does not, and cannot, constitute a *re*-creation of the village as it was in the early seventeenth century. One hopes, however, that should the original inhabitants reappear, while they would not recognize the little community in all of the details that it possessed in their time, they would at least feel

somewhat at home in their surroundings. The plantation's goal is to provide its visitors with some sense of what life was like in Plymouth almost four centuries in the past. If it has done this, then the 1627 village exhibit continues to play a vital and unique role in bringing the birthplace of the Old Colony into the twenty-first century. One can hope to do little more, but if the product of years of careful research and program development is as successful in the future as it's been in the past, it is the closest thing to time travel that will ever be accomplished, as visitors encounter a very close approximation of the reality of life in Plymouth in the early seventeenth century.

What of the myths that have surrounded the "Pilgrim Story" for many generations? Already, Baker suggests, these have been supplemented by the radical changes brought about at Plimoth Plantation in the 1960s and 1970s: "Pilgrims as 'real people' are today recognizable images alongside the Rock, Priscilla and John, and Thanksgiving." For centuries myths have shaped ideals and driven aspirations of people of all nations, and the "Pilgrim" image is no exception in the American ethos. When one strips away the misconceptions, and glimpses instead something of the reality of the historical facts, and the social and material world of the Plymouth settlers, a stronger, more powerful content can be given to those mythic images, which all Americans can respect and new immigrants identify. In this way "the light here kindled" can burn with a small but steady flame of integrity—not "just the facts," but the facts and something of the vision and courage that not only thrust these people into the new world out of the old, but sustained them in it.

Sources and Notes

CHAPTER 1 Partakers of Our Plenty

1 **Our harvest being gotten in . . .**
Edward Winslow, December 11, 1621. "A letter sent from New England to a friend in these parts, setting forth a brief and true declaration of the worth of that plantation; as also certain useful directions for such as intend a voyage into those parts." In Dwight B. Heath (Ed.), *Mourt's Relation: A Journal of the Pilgrims at Plymouth* (Applewood Books, 1963), p. 82. *Mourt's Relation* was first published in London in 1622, printed for John Bellamie, to be sold in his shop at the two Greyhounds in Cornhill near the Royal Exchange.

3 **They began now to gather in the small harvest . . .**
William Bradford, *Of Plymouth Plantation 1620–1647*, Samuel Eliot Morison (Ed.) (Knopf, 1952), p. 90. At times in the text and sources this will be referred to as Bradford's history. Morison's edition of Bradford's history is generally accepted as the best available in print.

Winslow's letter, written on December 11, 1621 . . .
Dates are those used in the original texts and are from the Julian calendar, which was in use by England and her colonies until 1752. Ten days must be added to the date to bring it into conformity with the modern Gregorian calendar.

The winter of 1620–1621 was an extremely difficult one . . .
This is discussed in Chapter 2, "I Will Harry Them Out of the Land!"

4 **That which was so sad and lamentable was . . .**
Bradford, p. 77

5 **. . . in 1618 King James I even issued a book encouraging "lawful sports" . . .**
Book of Sports, 1618; reissued in 1633 by Charles I under the title, *The King's Majesty's declaration to his subjects concerning lawful sports to be used*, which is in the "tenor" of the 1618 declaration published by James I. The 1633 text is reprinted in Samuel Gardiner, *The Constitutional Documents of the Puritan Revolution 1625–1660* (Clarendon Press, 1906), pp. 99–103.

... Governor William Bradford objected to a new group of settlers playing ...
Bradford, p. 97. These new settlers were among those who arrived on the *Fortune* in November 1621.

... leading Henry VIII to berate the farmers ...
Cited in James Deetz and Jay Anderson, "The Ethnogastronomy of Thanksgiving," *Saturday Review of Science*, November 25, 1972, p. 35.

6 **In harvest time, harvest folk ...**
Thomas Tusser, *Five Hundred Pointes of Good Husbandrie ...*, W. Payne and Sidney J. Herrtage (Eds.) (London: Trübner, 1878), p. 132, chap. 46, "Augusts husbandrie," verse 26. The first edition of Tusser's work was published in 1557. It went through thirteen editions in his lifetime and contains a considerable amount of valuable information concerning feast days and customs.

Ye ask, What eat our merry Band ...
Couplet by Ebenezer Cook, published in *The Sot-Weed Factor*, by John Barth (Anchor Books/Doubleday, 1967), p. 211.

7 **... alkermes berries ...**
Emmanuel Altham to Sir Edward Altham, September 1623. In Sydney V. James Jr. (Ed.), *Three Visitors to Early Plymouth: Letters About the Pilgrim Settlement in New England During Its First Seven Years* (Applewood Books, 1963), p. 26.

For fish and fowl ...
Winslow, December 11, 1621. In *Mourt's Relation*, p. 84.

8 **... our victuals being much spent, *especially our beer* ...**
"A Relation or Journal of the beginning and proceedings of the English Plantation settled at Plymouth in New England," *Mourt's Relation*, p. 40. Italics added.

After salutations, our governor kissing his hand ...
Ibid., p. 56.

10 **... the early Plymouth settlers were first referred to as "pilgrims" in a sermon delivered in Plymouth by the Reverend Chandler Robbins ...**
The sermon was preached in 1793, and although Robbins referred to Bradford's description, he himself used the terms "our fathers" or "our forefathers" to describe the settlers. The sermon was published in 1796 in Stockbridge, Massachusetts. James Baker, however, in a personal communication, has drawn attention to the fact that by 1793 a Pilgrim Society had been established in Concord, Massachusetts.

... but they knew they were pilgrims ...
Bradford, p. 47. The quotation is from Hebrews, chapter 11, verse 13,

and would have been used by Bradford in a biblical sense, referring to troubles in this world that they were just passing through as pilgrims on their way to heaven. It had nothing to do with their arrival in the New World.

11 **Pilgrim Hall today is much different . . .**
The Pilgrim Hall Museum, America's Museum of Pilgrim Possessions, is online at http://www.pilgrimhall.org.

14 **. . . the great American orator Daniel Webster gave an address . . .**
Daniel Webster, *A Discourse Delivered at Plymouth, December 22, 1820 in the Commemoration of the First Settlement of New England* (Wells and Lilly, 1821).

. . . Saints . . . Strangers.
Bradford, p. 3, note 1, p. 44, note 9.

. . . Old Comers . . .
Morison notes that this was a term first used by Bradford to refer to the *Mayflower* passengers (Bradford, p. 178, note 8). Eugene Aubrey Stratton, in his *Plymouth Colony: Its History & People 1620–1691* (Ancestry Publishing, 1986), p. 27, comments that "Old Comers" came to be used to refer to all those who were living in Plymouth by 1627. However, both Morison and Stratton use Old Comers interchangeably with the term "Purchasers" (Bradford, p. 429, note 7 and Stratton, p. 76), who were the "Planters" from Plymouth who in 1626 purchased the shares of the London adventurers who had underwritten the settlement at Plymouth. There were fifty-three Plymouth settlers who were Purchasers, and their names are listed in *The Records of the Colony of New Plymouth in New England*, Nathaniel B. Shurtleff and David Pulsifer (Eds.) (William White, 1855–1861; AMS Press, 1968), vol. 2, p. 177 (cited hereafter as *PCR* [Plymouth Colony Records], volume and pagination from the AMS Press 1968 reprint).

. . . Old Planters, or simply Planters . . .
Bradford used the term "Old Planters" to describe those who arrived before the "new ships" in 1623, essentially the surviving *Mayflower* passengers. He wrote: "The Old Planters were afraid that their corn, when it was ripe, should be imparted to the newcomers, whose provisions which they brought with them they feared would fall short before the year went about, as indeed it did" (Bradford, p. 132). Like "Old Comers," the term came to be used to refer to those who were resident in Plymouth before 1627, and was also used interchangeably with "Purchasers" in the 1641 "Surrender of the Patent to the Body of Freemen" (Ibid., p. 430).

. . . Old Colony Club.
George F. Willison, *Saints and Strangers* (Cornwall Press, 1945), pp. 409–413.

. . . the designation of December 22 as the date to celebrate the landing . . .
A group of passengers from the *Mayflower* first set foot on the Plymouth mainland on December 11, 1620, according to the account given in *Mourt's Relation*, p. 38. As mentioned above, ten days must be added to the Julian calendar used by the English until 1752, to bring it into conformity with our own Gregorian calendar. This would make it December 21, 1620, but there was some confusion concerning the dates and number of days to be added to the calendar from the eighteenth century, which was compounded in the nineteenth when much information in circulation was not based on recently consulted primary sources.

15 **National Monument to the Forefathers.**
Designed by Hammatt Billings in 1855. See John Seelye in *Memory's Nation: The Place of Plymouth Rock* (The University of North Carolina Press, 1998), pp. 444–448, 495, 528, the most comprehensive and recent work on Plymouth Rock.

James Baker writes, "Although it was dedicated to the Pilgrims . . ."
"Haunted by the Pilgrims," p. 348. In Anne Elizabeth Yentsch and Mary C. Beaudry (Eds.), *The Art and Mystery of Historical Archaeology: Essays in Honor of James Deetz* (CRC Press, 1992), pp. 343–358. Cited as "Haunted by the Pilgrims."

16 **For a century and a half . . . the rock lay unmarked . . .**
Willison, pp. 1–2.

New Melo Drama . . . THE PILGRIMS . . .
Ibid., pp. 484–485.

17 **Thomas Faunce**
The account of Faunce's identification of the rock was first published in James Thacher's *History of the Town of Plymouth, from Its First Settlement in 1620, to the Year 1832* (Marsh, Capen & Lyon, 1832), pp. 29–30. He based his account on that of "the late venerable Deacon [Ephraim] Spooner, who at the age of fifteen years, was present on the interesting occasion." The question of the chair in which Faunce was taken to the rock is worth noting. Thacher's words are: "A chair was procured and the venerable man conveyed to the shore . . ." The implication is not that Faunce was taken there in his wheelchair, as recounted by George F. Willison in *Saints and Strangers* (Cornwell Press, 1945), p. 411, but that in the late eighteenth century in Plymouth sedan chairs were available and that it was in this manner that Faunce reached the harbor shore.

We know from the account in *Mourt's Relation* . . .
The details are recorded in "A Relation of Journal of the

Proceedings of the Plantation settled at Plymouth in New England," one of the records published in *Mourt's Relation*, pp. 32–39, and Bradford, pp. 68–72.

18 **. . . occasioned partly by the discontented and mutinous speeches . . .**
Bradford, p. 75.

In the name of God, Amen.
The Mayflower Compact. *Mourt's Relation*, pp. 17–18. Bradford's transcript in his *Of Plymouth Plantation* (pp. 75–76) is very similar to that which was published in Nathaniel Morton's *New England Memorial* (Cambridge, 1669), John Davis (Ed.) (Crocker and Brewster, 1826), pp. 37–38. The differences between the three versions lie mainly in punctuation, capitalization, and the use or omission of definite articles. The sense is exactly the same. The names of the signatories, of which there were forty-one, were first published in Morton's *New England Memorial*, p. 38. Morison comments in his edition of Bradford's history that he believes that Morton had access to the original document with the signatures (Bradford, p. 441, note 1). The names given in Morton are listed below in alphabetical order. Those known to have embarked from Leiden, and so presumably who were members of the Scrooby congregation, are marked with an asterisk. They comprise sixteen of the forty-one signatories. Where known, estimated age and/or occupation at the time of emigration are stated in parentheses, based on information from Robert Charles Anderson's three-volume *The Great Migration Begins: Immigrants to New England 1620–1633* (New England Historic Genealogical Society, 1995), Caleb Johnson's Web site, http://members.aol.com/calebj/passengers2.html, or, for additional occupational information, *The Mayflower Descendant* CD ROM (Search & ReSearch Publishing Corporation). Anderson and Johnson have been used only for estimated ages. Where there is a difference, two figures are given, and Anderson precedes Johnson.

> John Alden (21, cooper), Isaac Allerton* (34, tailor, merchant), John Allerton (29, seaman), John Billington (38), William Bradford* (30, fustian maker), William Brewster* (54, printer), Richard Britteridge (39), Peter Browne (20), John Carver* (35), James Chilton* (64, tailor), Richard Clarke, Francis Cooke* (37, wool comber), John Crackstone* (45), Edward Doty (21/18–25, servant), Francis Eaton (25, house carpenter), Thomas English (master of the shallop), Moses Fletcher* (55, smith), Edward Fuller* (45), Samuel Fuller* (40, say-weaver, surgeon), Richard Gardiner (38, seaman), John Goodman, Stephen Hopkins

(38/42, tanner), John Howland (21, servant), Edward Lester (18-25, servant), Edmund Margesson (34), Christopher Martin (38, merchant), William Mullins (52/48, boot and shoe dealer), Degory Priest* (41, hatter), John Rigsdale, Thomas Rogers* (48), George Soule (18/18–25, servant), Myles Standish (27/36, military captain), Edward Tilley* (32), John Tilley (49), Thomas Tinker* (wood sawyer), John Turner* (30, merchant), Richard Warren (42, merchant), William White* (30, wool comber), Thomas Williams* (38), Edward Winslow* (25, printer), Gilbert Winslow (20).

The *Mayflower* passengers consisted of fifty men, nineteen women, three of whom were pregnant, fourteen young adults, and nineteen children, a total of 102. One baby was born on the voyage across the Atlantic (Oceanus Hopkins), one after the ship's arrival in Cape Cod Harbor (Peregrine White), and on December 22, 1620, Mary (Norris) Allerton had a stillborn son while the *Mayflower* was anchored in Plymouth Harbor (Bradford, p. 442, *Mourt's Relation*, p. 41).

19 **One of these remarkable incidents . . .**
John Quincy Adams, *An Oration, Delivered at Plymouth, December 22, 1802 at the Anniversary Commemoration of the First Landing of Our Ancestors, at That Place* (Russell and Cutler, 1802), pp. 17–18.

20 **They drew up one of those familiar church or sea compacts . . .**
Henry Steele Commager, *Godspell Visits Plimoth Plantation with Henry Steele Commager for Thanksgiving* (Boston, Mass.: WGBH Educational TV, 1972).

. . . only four of the ten adult servants aboard the Mayflower signed [the Compact]. . . .
The four were John Howland, George Soule, Edward Doty, and Edward Lester. The remaining six were Roger Wilder (25), Elias Story (18–25), John Langmore, Robert Carter, William Holbeck, and Edward Thompson. In an article entitled "Why Did Only Forty-one Passengers Sign the Compact?" George Ernest Bowman suggests that the reason the other adult male passengers did not sign was because they were servants and in terms of their contracts were not free agents, even if they were twenty-one years old; the remaining two who were not servants, William Trevore and Mr. Ely, were seamen whose contracts would have likewise bound them. His conclusion is that all free male passengers of legal age signed the compact. See *The Mayflower Descendant*, 1920, vol. 22, no. 2, pp. 58–59.

This is not, however, to decry the fact . . .
George L. Haskins, in "The Legal Heritage of Plymouth Colony," discusses the contribution of the Plymouth legal system, including

the Mayflower Compact, in Chapter V of *Essays in the History of Early American Law,* David H. Flaherty (Ed.) (University of North Carolina Press, 1969), pp. 121–134.

21 ***The Midnight Ride of Paul Revere***
First published in Henry Wadsworth Longfellow's *Tales of a Wayside Inn* in 1863.

Myles Standish, who was in his late twenties or early thirties . . . Priscilla Mullins . . . was seventeen.
Anderson (p. 1743), gives a birth date for Standish as about 1593, based on the date of his first marriage. Caleb Johnson simply gives ca. 1584 (http://members.aol.com/calebj/standish.html, February 20, 1999). Johnson gives no birth date for Priscilla Mullins, but Anderson suggests about 1603 (p. 1316).

Peter Gomes . . . comments . . .
Peter Gomes, "The Darlings of Heaven," *Harvard Magazine*, November 1976, p. 33.

22 **James Baker tells us . . .**
Personal communication.

. . . James Baker offers an explanation . . .
Baker, "Haunted by the Pilgrims," pp. 350–351. A reproduction of the print from *Frank Leslie's Illustrated* appears on p. 351.

It was only after the turn of the century . . .
Ibid., pp. 351–352.

23 **I am Pocahontas . . .**
Quoted from the film *Addams Family Values* (Paramount Pictures, 1993).

In 1970, Thanksgiving was declared a National Day of Mourning by Native Americans . . .
Herald Traveler (Boston), Nov. 27, 1970, p. 1.

24 **. . . the second execution in the colony, involving three Englishmen . . .**
Bradford, pp. 299–301. The first execution was that of John Billington in 1630 for the murder of John Newcomen (Bradford, p. 234).

26 **Sam Shepard's . . . *Rolling Thunder Log Book* . . .**
Viking Press, 1977.

Plymouth is a donut of a town . . .
Ibid., p. 24.

"What concert?" . . .
Ibid.

27 **Ginsberg is already chanting . . .**
Ibid., p. 37.

Scarlet Rivera . . . "the mysterious dark lady of the fiddle . . . "
Ibid., p. 19.

28 **. . . sarcophagus containing a small number of bones . . . discovered in 1855**

. . . identification supported by Oliver Wendell Holmes . . .
William Russell, registrar of deeds and keeper of the Plymouth Colony records in the mid–nineteenth century, discusses in some detail Holmes's identification in his *Pilgrim Memorials and Guide to Plymouth* (Boston: Crosby and Nichols, 1864), pp. 83–85. Seelye (pp. 293 and 529) also refers to this incident.

On this hill . . .
Thacher, p. 355.

29 **The bones that repose in the sarcophagus today were first placed in a canopy that was erected over Plymouth Rock . . .**
The first canopy for the rock was designed by Hammatt Billings in 1855, but not completed until 1867 (Seelye, pp. 495, 539, and 542). It was demolished in preparation for the tercentennial celebrations of 1920 (Ibid., pp. 590–591). The sarcophagus was built for the 1920 tercentennial (Ibid., p. 592).

As one small candle may light a thousand . . .
Bradford, p. 236.

Chapter 2 I Will Harry Them Out of the Land!

32 **. . . three well-known Separatist ministers were executed.**
These were Henry Barrow, John Penry, and John Greenwood, hanged in 1593. David S. Lovejoy, *Religious Enthusiasm in the New World* (Harvard University Press, 1985), p. 24.

. . . "I will make them conform . . ."
James I (1603-1625), at the Hampton Court Conference in 1604. Cited in William Bradford, *Of Plymouth Plantation 1620–1647*, Samuel Eliot Morison (Ed.) (Knopf, 1952), p. 9, note 4.

So after long waiting and large expenses . . .
Bradford, pp. 11-12.

33 **. . . the master espied a great company . . .**
Ibid., p. 13.

Being thus apprehended . . .
Ibid., p. 14.

34 **Many were engaged in the production of cloth . . .**
Ibid., p. 17, note 4.

The place they had thoughts on . . .
Ibid., p. 25.

But that which was more lamentable . . .
This and the immediately following quotation are also from Bradford, p. 25.

35 **. . . they managed to form a joint stock company . . .**
Ibid., pp. 29–46.

36 **There were 102 passengers . . . 180 tons . . .**
William Bradford compiled a list of passengers on the *Mayflower.* On the left of each group listed is a numeral showing the number of passengers in each. These total 104. He includes in this the two children born before the final arrival at Plymouth, Oceanus Hopkins and Peregrine White. Bradford, pp. 441–443.

Bradford states that the ship they sailed in "was hired at London, of burthen about nine scoure [score] . . ." (Bradford, p. 47). Dimensions given here are those estimated by Caleb Johnson in his Web page, "The *Mayflower*: Dimension and Images," http://members.aol.com/calebj/mayflower_dimensions.html, February 20, 1999.

. . . not only were the sixty-nine adult passengers mainly in their thirties . . .
The information given concerning ages of the *Mayflower* passengers is based on the estimated or known ages provided in Robert Charles Anderson, *The Great Migration Begins: Immigrants to New England 1620–1633* (New England Historic Genealogical Society, 1995) and Johnson's Web site, op. cit.

. . . the adult members of the Scrooby congregation from Leiden . . . comprised 42 percent of those on board.
There were twenty-nine congregational members out of the sixty-nine adults who sailed on the *Mayflower*. They were John and Katherine Carver, William and Mary Brewster, Edward and Elizabeth Winslow, William and Dorothy Bradford, Isaac and Mary Allerton, Samuel Fuller Sr., John Crackstone Sr., William and Susanna White, William Holbeck, Edward and Ann Tilley, Francis Cooke, Thomas Rogers, Thomas Tinker and his wife, James and Susanna Chilton, Edward Fuller and his wife, John Turner, Moses Fletcher, Degory Priest, and Thomas Williams.

. . . the majority being the younger, more vigorous members of the congregation . . .
Edward Winslow, in his "Brief Narration of the true grounds or

cause of the first Planting of New England," published in 1646 as part of his *Hypocrisie Unmasked*, mentions specifically that the church at Leiden determined that "the youngest and strongest" part of the congregation should be the first to establish the settlement in New England. If it was successful, they could then help to bring over "such as were poor and ancient and will to come." See Alexander Young, *Chronicles of the Pilgrim Fathers of the Colony of Plymouth from 1602 to 1625* (Little and Brown, 1841), p. 383.

In addition to being friends, some were related, by both blood and marriage.
Copies of the Leiden marriage records of six of the passengers on the *Mayflower* were published in *The Mayflower Descendant* in a series entitled "The Mayflower Marriage Records at Leyden and Amsterdam." They appear in the original Dutch with English translations. The records show relationships among the passengers that enrich our understanding of some of the dynamics at work between them. The six marriage records published in *The Mayflower Descendant (MD)* are as follows:

Degory Priest to the widow Sarah (Allerton) Vincent, sister of Isaac Allerton who was married to Mary Norris, a double marriage on November 4, 1611. *MD*, 1905, vol. 6, no. 3, pp. 129–130.

William White to Ann (Susanna) Fuller, February 11, 1612. *MD*, 1905, vol. 7, no. 4, pp. 193–194.

Francis Cooke to Jennie (Hester) Mahieu, June 30, 1603. *MD*, 1906, vol. 8, no. 1, pp. 48–50.

Edward Winslow to Elizabeth Barker, May 12, 1618. *MD*, 1906, vol. 8, no. 2, p. 100.

Samuel Fuller, widower to (1) Agnes Carpenter, April 24, 1613—he was accompanied by his brother-in-law William White. Agnes Carpenter was accompanied by her sister Alice Carpenter, who ten days later married Edward Southworth, on May 4, 1613. After Southworth's death, Alice married William Bradford in Plymouth on August 14, 1623. *MD*, 1906, vol. 8, no. 3, p. 129.

Samuel Fuller, widower to (2) Bridget Lee, on May 27, 1617. *MD*, 1906, vol. 8, no. 3, p. 130.

37 **Elizabeth Tilley**
Baptized in 1607, Elizabeth Tilley was the youngest daughter of John Tilley and his wife, Joan (Hurst) Rogers. They joined the *Mayflower* in London, to travel with his younger brother Edward from Leiden, who was accompanied by his wife, Agnes, or Ann (Cooper) Tilley

and little Humility Cooper. Henry Samson, son of James and Martha (Cooper) Samson, also joined the ship in London. Elizabeth's marriage to John Howland had taken place by 1624, the year it is estimated that their first daughter, Desire Howland, was born. (Anderson, pp. 1022, 1623, 1819, 1822). Elizabeth (Tilley) Howland died in 1687, fourteen years after her husband, at the age of eighty at the home of her daughter Lydia, wife of James Brown, in Swansea. Elizabeth's will and the inventory of her estate have been published in C. H. Simmons (Ed.), *Plymouth Colony Wills and Inventories, Vol. 1: 1633–1669* (Picton Press, 1996), pp. 13–14.

Humility Cooper

Anderson, and Eugene Aubrey Stratton, in his *Plymouth Colony: Its History & People 1620-1691* (Ancestry Publishing, 1986), draw attention to the work of Robert Leigh Ward, whose research has provided evidence that Edward Tilley's wife was Agnes Cooper, and that Humility, born in Holland, was the daughter of Agnes Cooper Tilley's brother, Robert Cooper. Bradford comments that "the girl Humility . . . was sent for into England and died there" (Bradford, p. 446). She was baptized in London at the age of nineteen in March 1639 and so would have been less than a year old when she sailed to America as a passenger on the *Mayflower* (Anderson, p. 479; Stratton, p. 273).

Henry Samson

Henry Samson (or Sampson) was baptized on January 15, 1604, the son of James and Martha (Cooper) Samson. Martha was sister to Ann (Cooper) Tilley, wife of Edward Tilley. Henry was sixteen when he made the crossing to America. In February 1636, sixteen years after his arrival in Plymouth, he married Anne Plummer (*Plymouth Colony Records (PCR)*, vol. 1, p. 36), with whom he had nine children. Anne was the sister or cousin of Mary Plummer, who married John Barnes of Plymouth in 1633 (*PCR* 1:16). Samson was one of the Plymouth Purchasers, settled in Duxbury, and died sometime between December 1684 and March 1685, some eighty years of age (Anderson, pp. 1621–1624; Stratton, pp. 347–348).

John Howland

John Howland was born in 1593 or 1594, as he was "above eighty years," or "in his eightieth year" when he died in February 1673 (*PCR* 8:34). This means that he was about fourteen years older than Elizabeth Tilley when he married her ca. 1623, based on the estimated birth date of their first child, Desire Howland, which was ca. 1624. He would have been twenty-six or twenty-seven when he sailed to America. It has been argued by Johnson that Howland was more likely to have been born in 1599, but based on the evidence of his age at the time of his death even the traditional date of 1592 would seem to be a year or two too early (Anderson, pp. 1020–1024; Stratton, pp. 311–312, and Johnson, http://members.aol.com/calebj/howland.html, February 20, 1999).

Myles Standish

Myles Standish was hired in Leiden to serve as military captain in preference to the experienced Captain John Smith, who had offered his services to the members of the former Scrooby congregation planning to emigrate to America (see reference below to page 71, under "... the 'Brownists of England, Amsterdam and Leyden ...' "). For biographical details on Standish see Anderson, pp. 1741–1747; Johnson, http://members.aol.com/calebj/standish.html, February 20, 1999, and Stratton, pp. 357–359).

John Alden

John Alden was working as a cooper at Southampton when he was hired as such by the *Mayflower* emigrants. He was free to return to England at the end of his contract, but married Priscilla Mullins and settled in the colony. He was evidently well educated, judging by the fact that he was involved in public service for over fifty years, including acting as deputy governor on two occasions, serving as colony treasurer, on the council of war, on committees to revise the laws of the colony, as deputy for Duxbury in the 1640s, as well as serving as an assistant to the governor from 1650 to 1686. He and Priscilla moved to Duxbury from Plymouth in 1632, where they brought up their six daughters and four sons. Although Alden was initially a cooper, he became actively involved in the fur trade, particularly through the Plymouth Kennebec trading post in Maine. He and Myles Standish were closely associated, both moving from Plymouth to Duxbury in 1632, where they were neighbors, and his daughter Sarah had married Standish's son Alexander by 1660. Alden died in his eighty-ninth year on September 12, 1687. For a biographical summary see Anderson, pp. 21–26.

Mary Chilton

John Seelye in *Memory's Nation: The Place of Plymouth Rock* (The University of North Carolina Press, 1998), pp. 384, 651, gives John Davis, *A Discourse before the Massachusetts Historical Society, Boston, December 2, 1813. At their anniversary commemoration of the first landing of our ancestors at Plymouth, in 1620* (John Eliot, 1814), as the first published source of the story concerning Mary Chilton being the first to set foot on Plymouth Rock. Susanna Latham, eldest daughter of John and Mary (Chilton) Winslow, was involved with her husband, Robert Latham, in a murder trial concerning the death of their servant John Walker in January 1655 (*PCR* 3: 71–73, 82, 143).

. . . the More children, Jasper, Richard, Mary, and Ellen . . .
This information is based on Anderson, pp. 1282–1287, and Johnson, http://members.aol.com/calebj/more.html, February 21, 1999.

John Billington Sr.
In September 1630, Billington was executed for killing a young man, John Newcomen, whom he waylaid about a former quarrel and shot. The case caused great distress to the colonists in Plymouth as he was the first person found guilty of willful murder, and Billington, although a troublesome, problematic person, was one of the survivors among the first comers on the *Mayflower*. Bradford provides a brief account of the event in his history, pp. 234 and 446. For further details on Billington's life see Anderson, pp. 173–174.

William and Dorothy Bradford
William and Dorothy Bradford's son, John, left behind in Leiden, was born ca. 1617. His parents were married in Leiden on December 10, 1613, and in marriage intentions recorded in Amsterdam the previous month, Dorothy May gave her age as sixteen. She would have been twenty-three when she left Holland on the *Mayflower*. While it is probable that her death by drowning after she fell overboard, while the ship was anchored off Cape Cod, was accidental, there is not a consensus on this matter. In 1952, Samuel Morison, in the introduction to his definitive edition of Bradford's history, emphasized his belief that it was suicide (Bradford, p. xxiv). Much earlier, in 1869, a fictitious account of her having committed suicide was published by a Mrs. Jane (Goodwin) Austin, the year before the two hundred fiftieth anniversary of the landing of the Pilgrims, which cannot be unconnected to Morison's theory. It led to a popular belief that this was what happened, but George Bowman, in an article published in *The Mayflower Descendant* in 1931, provided cogent reasons why he believed that this could not have been the case, and then in a further note in the same journal, published Mrs. Austin's own admission that this was "what might have been" rather "than what is known to have been" (*The Mayflower Descendant*, 1931, vol. 29, no. 3, pp. 97–102, and 1933, vol. 31, no. 3, p. 105). It is important to note, however, that the first account of Dorothy Bradford's death by drowning was only published eighty-two years after her death by Cotton Mather in his *Magnalia Christi Americana, or The Ecclesiastical History of New England* (Hartford: Silas Andrus and Son, 1855), p. 111. The work was first published in London in 1702. William Bradford gives no details in his history, simply the bare statement of her death: "William Bradford his wife died soon after their arrival . . ." (Bradford, p. 444). Nathaniel

Morton, in his 1669 volume, *New England's Memorial*, makes no reference to her death. Their son, John, later joined his father in Plymouth, married Martha Bourne, but left no issue. For brief biographical details on William Bradford see Anderson, pp. 207–209. Samuel Morison has a short biographical sketch of William Bradford at the beginning of his edition of Bradford's history, pp. xxii–xxvii. A longer, possibly less reliable account is provided by Mather in his *Magnalia* (1855 ed., pp. 109–114).

38 **Neither *Mourt's Relation* nor Bradford's history . . . even mentions the *Mayflower* by name.**
The first reference to the *Mayflower* in the Plymouth primary sources is in the *Plymouth Colony Records* in a manuscript volume entitled "Plimouth's Great Book of Deeds." The handwriting is that of William Bradford, and the reference is in a section on the allotment of lands made in 1623 to the passengers of the three first ships to arrive, the *May-Floure*, the *Fortune*, and the *Anne* (*PCR* 12:4). In his history Bradford simply refers to the *Mayflower* as the "bigger ship" and the *Speedwell* as the "lesser" or "small" ship (Bradford, pp. 52 and 47). It is Nathaniel Morton, in his 1669 *New-England's Memorial*, (Cambridge, 1669), John Davis (Ed.) (Crocker and Brewster, 1826), p. 23, who first mentions the names of both ships in a published account.

They were encountered many times with cross winds . . .
This and the following brief quotation concerning the "great iron screw" are from Bradford, pp. 58–59.

39 **. . . a lusty [lively] young man called John Howland . . .**
Ibid., p. 59.

John Howland . . . died in 1673 at an age of "above eighty years."
In the Plymouth Register of Births, Marriages, and Deaths, it is recorded that John Howland Sr. died on February 23, 1672, and that he lived until "he attained above eighty years in the world" (*PCR* 8:34). On March 4, 1672/73, Elizabeth (Tilley) Howland was granted letters of administration to administer the estate of Mr. John Howland Sr. of Plymouth, "late deceased" (*PCR* 5:110). John Howland in fact died in February 1673. The Plymouth settlers in their official records used the Julian calendar, which was in use in England and her colonies until 1752. March 25, not January 1, was regarded as the start of the civil, legal, and ecclesiastical year. So March was the first month and February the twelfth month of the year. John Howland therefore died in February 1672 according to the Julian calendar, but in the historical year of 1673. The main body of published primary source data on Plymouth Colony is the

court records, and the editors follow the convention of double dating for events occurring between January 1 and March 24 to indicate the legal and historical years in which an event took place. For example, February 23, 1672/73, indicates that Howland died on February 23, 1672, in the legal year of the Julian calendar, but his death actually took place in the historical year 1673. Throughout *The Times of Their Lives*, however, we have used the historical year to avoid confusion but retained the day of the month according to the Julian calendar, as this is what appears in the primary sources. The Julian calendar dating is also known as old style dating as opposed to the new style that follows the Gregorian calendar, which is still in modern use, and is ten days ahead of the Julian calendar.

proud and very profane young man
This and the following quotation are from Bradford, p. 58.

William Butten [or Button],
Bradford notes that "In all this voyage there died but one of the passengers, which was William Butten, a youth, servant to Samuel Fuller, when they drew near the coast." Ibid., p. 59.

Four more would perish during the time the *Mayflower* lay at anchor . . .
The four were Edward Thompson, Jasper More, Dorothy Bradford, and James Chilton.

Edward Thompson, servant to William White, died on December 4, 1620, "the first that dies since their arrival," according to Thomas Prince in his five-volume *Chronological History of New England in the Form of Annals* (Boston, 1736; Edinburgh, private printing, 1887–1888), vol. 3, p. 8. Prince lists at intervals extracts from "A Register of Governor Bradford's in his own hand, recording some of the first deaths, marriages and punishments at Plymouth." According to Anderson (p. 1809), this register has subsequently been lost. Bradford in his history simply comments that "Mr. White and his servants died soon after landing" (Bradford, p. 445).

Jasper More's death took place the day before that of Dorothy Bradford. Bradford, in his register, noted: "Dec. 6, Dies Jasper, a boy of Master Carver's" (Prince, vol. 3, p. 12).

As noted above, William Bradford's only mention of the death of his wife, Dorothy (May) Bradford is "William Bradford his wife died soon after their arrival . . ." (Bradford, p. 444). Thomas Prince, citing Bradford's register, gives the date of Dorothy Bradford's death as December 7 (Prince, vol. 3, p. 12). Bradford was not on the *Mayflower* when her death occurred. It was the day after the final expedition to search for a suitable harbor had left, and when he returned to the ship with the group that set out on December 6 with the good news that Plymouth Harbor and the land adjacent to it were suitable for planting their colony, he learned of his wife's death.

James Chilton died on December 8, 1620, off Cape Cod (Prince, vol. 3, p. 12). Bradford says that James Chilton and his wife died in the first infection (Bradford, p. 446).

It seems clear that when Bradford compiled his record of "decreasings and increasings" (published in his history, pp. 443–448), he did not consult his register, apparently kept as a log while at sea and after landing, to which Thomas Prince later had access. There are too many inconsistencies. What is clear is that Bradford notes in his register that there were forty-four deaths between December 1620 and the end of March 1621. After this he adds that Governor John Carver died in the spring, and his wife, Katherine (White) (Leggatt) Carver, some five or six weeks later in the summer. As the register has all the appearance of being the earliest record, it should be preferred to the later list of decreasings.

41 **And for the season it was winter . . .**

Bradford, p. 62.

On November 11 . . .

The accounts that follow of the explorations along the coast in search of a settlement and the initial building of houses at Plymouth through mid-January 1621 are from "A Relation or Journal of the Proceedings of the Plantation settled at Plymouth in New England," published in *Mourt's Relation* in 1622, specifically from pp. 19–48. The authorship of the *Relation* is not known, but Dwight Heath, editor of the edition of *Mourt's Relation* that is most readily available (*Mourt's Relation: A Journal of the Pilgrims at Plymouth*, Applewood Books, 1963), states that: "It is almost certain the principal author was Edward Winslow, although it is generally believed that William Bradford also had a hand in the effort." Bradford's history, *Of Plymouth Plantation*, includes passages from *Mourt's Relation*, together with material that does not appear in it. It is probable that the manuscript of the *Relation* was taken to England from Plymouth on the *Fortune* in December 1621, by Robert Cushman. (*Mourt's Relation*, pp. xi–xiii).

48 **Thomas Jefferson . . . gives a description of his excavation of an Indian burial mound . . .**

Notes on the State of Virginia, by Thomas Jefferson, William Peden (Ed.) (Norton, 1972), pp. 97–100.

55 **Samuel de Champlain**

Champlain's chart of Port du Cap. St. Louis (Plymouth) was first published in his *Les Voyages du Sieur de Champlain* (Paris, 1613). It has been reproduced in *Mourt's Relation* with an accompanying note by the editor, on pp. [xviii–xxi]. Samuel Eliot Morison in his *Samuel de Champlain: Father of New France* (Little, Brown, 1972) has also reprinted the chart including Champlain's key, which is missing from *Mourt's Relation* (Morison, p. 65).

58 **. . . but by December 11, 1621, Edward Winslow . . . stated, "We have built seven dwelling houses . . ."**
Mourt's Relation, p. 81.

59 **In the time of most distress . . .**
This and the immediate quotation following are from Bradford, p. 78. He describes the winter in a section that Samuel Morison has captioned "The Starving Time," pp. 77–79.

The passengers included Richard Britteridge . . .
Prince, vol. 3, p. 17.

. . . Rose Standish, who died on January 29, after which entry Bradford writes, "This month, eight of our number die."
Ibid., p. 31.

Mary (Norris) Allerton . . .
Ibid., p. 32.

Edward Winslow's wife, Elizabeth (Barker) Winslow died on March 24 . . .
Ibid., p. 38.

60 **This month, thirteen of our number die . . .**
Ibid., pp. 38–39.

Of the fifteen men who came out on their own, it seems that only eight survived the winter.
The survivors appear to have been John Alden, Peter Browne, Mr. Ely, Richard Gardiner, John Goodman, William Trevore, Thomas Williams, and Gilbert Winslow. There is some question about the date of John Goodman's death. Bradford includes him in those who died soon after their arrival in the general sickness (Bradford, p. 447), but Goodman was still alive in mid-January 1621 (*Mourt's Relation*, pp. 45–48), although not in good physical shape. He is listed as one of those who received land in 1623 (*PCR* 12:4). He is not listed among those who were part of the cattle division of 1627, so he must have died by then. Bradford, in his list of "decreasings and increasings," notes that Thomas Williams and six others "died soon after their arrival in the general sickness that befel" (Bradford, p. 447). Williams, however, appears to have been alive in the spring, as on March 22, 1621, the day the peace treaty with Massasoit was

signed, reference is made to a "Master Williamson," who with Captain Standish met Massasoit at the brook (*Mourt's Relation*, p. 56). As there was no Williamson among the passengers, the editor, Dwight Heath, suggests that this was in fact Thomas Williams. Thomas Williams is not mentioned in the land division of 1623, so he could have died by then, or simply left the colony.

There were also some who died in the spring and summer . . .
Bradford, p. 86.

Alice Mullins and her adult son, Joseph . . .
Although Bradford mentions that "Mr. Mullins, his wife, his son and his servant died the first winter" (Ibid., p. 445), both were mentioned in William Mullins's will, dated April 2, 1621. Mullins's will was published in *The Mayflower Descendant*, vol. 1, no. 4, 1899, pp. 230–232. In notes accompanying the published will, George Bowman draws attention to the fact that both Alice and Joseph Mullins must have been alive at the time the will was drawn up.

. . . John Pory, visiting Plymouth in August 1622, wrote . . .
Sydney V. James Jr. (Ed.), *Three Visitors to Early Plymouth: Letters About the Pilgrim Settlement in New England During Its First Seven Years* (Applewood Books, 1963), p. 7. Cited as *Three Visitors to Early Plymouth.*

. . . Emmanuel Altham, wrote to his brother in England . . .
Ibid., p. 24.

The wind was south, the morning misty . . .
Mourt's Relation, p. 50.

61 **The spring now approaching . . .**
Bradford, p. 84.

. . . the Indians came skulking about them . . .
Ibid., p. 79.

Samoset, an Abenaki Indian from Pemaquid Point, Maine.
The account that follows concerning Samoset and Squanto is based on a number of sources. There is some contradiction, and it is not always clear just what the sequence of events was, but the main points concerning what is known of them have been extracted from the accounts in Bradford's *Of Plymouth Plantation*, pp. 79–84, and *Mourt's Relation*, pp. 50–52, 55. The description of Samoset and Squanto's subsequent visits to the village, and of the treaty made with the Massasoit, is based on *Mourt's Relation*, pp. 51–59.

62 **. . . the Wampanoag did not have musical instruments . . .**
Catherine Marten, "The Wampanoags in the Seventeenth Century:

An Ethnohistorical Study," *Occasional Papers in Old Colony Studies*, no. 2, December 1970, pp. 1–40. Marten's fine study is meticulously documented from historical sources and it is the only extensive ethnography available of the Wampanoag. The statement concerning the lack of musical instruments among them is from her section on entertainment, pp. 17–18.

They had brought three or four skins, but no trading took place . . .
It was a Sunday, and the Separatist leaders of the colony would not have transacted business on the Sabbath even though fur trading in beaver skins was a crucial part of the potential economy of the settlement.

Burial Hill
Known as Fort Hill, then Burying Hill, Burial Hill stands in the center of downtown Plymouth. According to a map of Plymouth published in James Thacher's *History of the Town of Plymouth, from Its First Settlement in 1620, to the Year 1832* (Marsh, Capen & Lyon, 1832), it rises 165 feet above sea level, and covers some eight acres (p. 351). Later the colonists erected a fort on the top of the hill and used it as a meeting house, and so it is not surprising that the hill naturally evolved into the cemetery, it being common English practice to bury the dead in proximity to their place of worship. The earliest grave marker still standing is that of Edward Gray, who died "the last of June, 1681." It is commonly supposed that the initial grave markers would have been made of wood, and so didn't survive. Wooden grave markers, usually called "grave rails," were known to have been used in England at this time, and mention is made of them in some probate inventories. Grave rails resembled wooden headboards of a bedstead, and were set in the ground parallel to the body of the deceased, a wooden post being placed at the top and bottom of the grave to demarcate the head and foot. For a discussion of gravestones in New England, including Plymouth, see James Deetz, *In Small Things Forgotten: An Archaeology of Early American Life* (Anchor/Doubleday, 1996), pp. 89–124.

63 **. . . they drew up a peace agreement . . .**
The text of the agreement and subsequent quotation are from *Mourt's Relation*, pp. 56–58.

64 **. . . Metacom, known by the English as King Philip . . .**
The most recent works on King Philip's War are *King Philip's War: The History and Legacy of America's Forgotten Conflict*, by Eric B. Schultz and Michael J. Tougias (Norton, 1999) and *The Name of War: King Philip's War and the Origins of American Identity*, by Jill Lepore (Knopf, 1998).

Squanto . . . instructed them in how to plant their corn to ensure a good crop.
The method used is described by Bradford (p. 85) and in *Mourt's Relation*, pp. 81–82.

. . . they began to see that Squanto sought his own ends . . .
Bradford, p. 99. Edward Winslow also gives details of Squanto's duplicity toward both the English and the Wampanoag in his *Good Newes from New England: or a true Relation of things very remarkable at the Plantation of Plimoth in New-England* (London, 1624), published in Alexander Young, *Chronicles of the Pilgrim Fathers of the Colony of Plymouth from 1602 to 1625* (Little and Brown, 1841), under the title "Winslow's Relation," pp. 285–290. Cited as *Good Newes from New England* (Young).

65 **. . . a ship sent to New England by Thomas Weston . . .**
One of the main leaders of the London merchant adventurers who sponsored the colonization of Plymouth, Thomas Weston decided to form a colony in New England near to Plymouth, for his own gain and to the detriment of the settlers who needed all the colonists and supplies that they could get. In 1622 he sent a fishing vessel, the *Sparrow*, to Massachusetts Bay, with a small party of seven men to find the most suitable place for a colony and prepare for the arrival of a large group of single men whom he proposed to send out. The ship anchored at the Damaris Cove Islands off the coast of Maine, and a group of ten, including some crew from the *Sparrow*, sailed down to Plymouth in a shallop, arriving there on May 31, 1622, just as Massasoit's men were demanding that Squanto be handed over to them for execution. They brought letters to the governor from Weston, but no provisions for which the settlement was in desperate need. Phineas Pratt was one of Weston's settlers, and he and his six companions were given hospitality in Plymouth until the *Charity* and the *Swan* arrived with the main party of Weston's settlers at the end of July or early August 1622. For information on Thomas Weston see Samuel Morison's comments in Bradford, p. 37, note 2, and Anderson, pp. 1967–1970.

[Squanto] died in the fall of 1622 . . .
His death took place at Manamoyick Bay, the modern Pleasant Bay, during an expedition to obtain corn for the settlers at Weston's plantation at Wessagusset, "shortly after harvest" in 1622 (Bradford, pp. 113–114; *Good Newes from New England* (Young), pp. 299–301).

. . . a plan made by Bradford entitled, "The meersteads & garden plots of [those] which came first layd out 1620."
This record is in the handwriting of William Bradford. A facsimile

appears in the *Plymouth Colony Records* (*PCR* 12:[2]). A meerstead has the same meaning as "messuage," which is a portion of land intended as a site for a dwelling house with its outbuildings and adjacent land (*The Shorter Oxford English Dictionary on Historical Principles*, 3rd rev. ed., 1947). Although Bradford's plan shows seven lots, and these coincidentally agree with Winslow's mention of seven houses being in place before December 11, 1621, James Baker has commented that "the two houses actually mentioned in *Mourt's Relation* are those of the Governor and Stephen Hopkins. It seems unlikely that Goodman, for example, would get a house (as opposed to a lot) before Bradford, Fuller, Standish, &c." Baker, personal communication.

. . . sent a messenger unto them with a bundle of arrows tied about with a great snakeskin . . .
Bradford, p. 96.

66 **In the meantime, knowing our own weakness . . .**
Good Newes from New England (Young), p. 284.

67 **At the end of May 1622, word reached Plymouth of the Indian uprising in Virginia that had taken place the previous March 22 . . .**
Opechancanough (1607–1644), was the younger brother of Powhatan, chief of the Pamunkey, and assumed his authority a short while after his death in 1618. Powhatan was the dominant Indian leader in Tidewater Virginia, occupying territory from the James River to the Potomac. The English settlers spread out along the James River, ignoring the rights of the indigenous occupants, which was deeply resented by them. Without any warning, in an effort to rid himself once and for all of the English invaders, Opechancanough planned a carefully coordinated surprise attack on the various small settlements, and on March 22, 1622, the Indians attacked. There were 347 English men, women, and children killed that day, and the Virginia Company's policy of restraint and integration toward the Indians completely changed to one of destruction and enslavement. See Edmund S. Morgan, *American Slavery, American Freedom: The Ordeal of Colonial Virginia* (Norton, 1975), pp. 97–101 for a brief account of the uprising.

The Indians began again to cast forth many insulting speeches . . .
Good Newes from New England (Young), p. 295.

68 **This summer they built a fort with good timber . . .**
Bradford, p. 111.

We have a hill called the Mount, enclosed within our pale . . .
Good Newes from New England (Young), p. 295. Work on the fortification continued through the summer and harvest, with

considerable difficulty considering the small size of the colony, and the initial eagerness of some having evaporated. In March 1623 Winslow remarks:

> Now was our fort made fit for service, and some ordnance mounted; and though it may seem long work, it being ten months since it begun with such small means, a little time cannot bring [it] to perfection . . . As amongst us divers seeing the work prove tedious, would have dissuade from proceeding, flattering themselves with peace and security, and accounting it rather a work of superfluity and vainglory, than simple necessity." [Ibid., p. 335]

69 **As well appeareth by their building . . .**
John Pory's letter of January 13, 1623, and later, to the Earl of Southampton, who was treasurer of the Virginia Company in London, was more the nature of a report. Pory had stopped at Plymouth in part on company business. It was first published in 1918, edited by Champlin Burrage, as *John Pory's Lost Description of Plymouth in the Earliest Days of the Pilgrim Fathers*, and has been reprinted together with a second letter of Pory's to Sir Francis Wyatt, in the *Three Visitors to Early Plymouth*, pp. 1–18. The passage cited in the text is from the *Three Visitors*, p. 11.

Emmanuel Altham
There are three letters that Altham wrote while with the *Little James* in New England waters, two to his brother, Sir Edward Altham, and one to James Sherley, treasurer of the New Plymouth adventurers in London. Altham returned to England in 1625 after a year on the *Little James*. He made a second voyage to New England later that year, hoping to find employment in the colony at Plymouth, but without success. A fourth letter, written to his brother in June 1625, has been published with the other three in the *Three Visitors to Early Plymouth*, pp. 19–59. Altham later joined the East India Company, where he had a brief but successful career, dying in India in January 1636. His letters were published for the first time in 1963 in the *Three Visitors to Early Plymouth*. The passage cited in the text is from the *Three Visitors*, p. 24.

Captain John Smith
The most recent scholarly edition of the writings of Smith is that of Philip L. Barbour, *The Complete Works of Captain John Smith (1580–1631)*. It was published in three volumes in 1986 for the Institute of Early American History and Culture, Williamsburg, Virginia, by University of North Carolina Press, Chapel Hill and London.

. . . a map of New England . . .
Dated 1616, Smith's map is considered to be "the most detailed of the early ones that survive, for the area covered in it: the coast of Maine W. of Mount Desert Island, the narrow bit of New Hampshire, and E. Massachusetts to the underside of Cape Cod" (Barbour, vol. 1, verso of reproduction of the map, opposite p. 323). The map went through a number of reprints during Smith's life, with a few changes being made but the date 1616 retained. One of the amendments was the addition of "New" above "Plimouth." It has been reprinted in *Mourt's Relatio*n, with notes by the editor, Dwight B. Heath (pp. xxiii–[xxv]).

. . . "humbly entreating his Highness he would please to change their barbarous names for such English . . ."
Captain John Smith, in Book Six of his *The Generall Historie of Virginia, New-England and The Summer Isles* . . . (London, 1624; reprinted in Barbour, vol. 2, pp. 401–402). Book Six, the New England section of *The Generall Historie*, incorporates much of the material published by Smith in his 1616 edition of *A Description of New England* and 1622 edition of *New England's Trials*, with some additional material that includes his description of Plymouth in 1624.

71 **. . . the "Brownists of England, Amsterdam and Leyden, [who] went to New Plimouth . . ."**
The True Travels, Adventures and Observations of Captaine John Smith (Barbour, vol. 3, p. 221). Barbour points out that *True Travels*, published in London in 1630, is the only truly autobiographical work that Smith wrote, and that autobiographies, with which we are very familiar, were so rare in the early seventeenth century that Smith would not have had a model to follow even if he had searched for one (Barbour, vol. 3, p. [125]). It was Smith's greatest ambition to return to New England, and the fact that the Separatists preferred to use his published map and books rather than pay him was evidently something he never forgot. He not only refers to the matter in this passage, giving it a marginal note, "The effect of niggarlinesse," but also twice in his last work, *Advertisements For the unexperienced Planters of New-England, or any where*, published in 1631 (Barbour, vol. 3, pp. 283, 285–286). Smith did not have any sympathy for the "Brownists" [Separatists] at this stage of his life, but if he had gone with them to Plymouth, perhaps that would have been different. It is sobering to consider, though, that if Smith had accompanied the *Mayflower* passengers to New England instead of Standish, many lives might well not have been lost over the long time taken exploring the coastline of Cape Cod, with the resultant illness of passengers and crew alike.

. . . there is an echo of the Pocahontas episode in Smith's account of his capture by Turks . . .
The details are to be found in Smith's *True Travels* (Barbour, vol. 3, pp. 184–189, 200–203).

Charatza Trabigzanda
Smith uses "Tragabigzanda" or "Tragabigsanda" (Barbour, vol. 3, p. 186, and vol. 2, p. 402, respectively), but Barbour prefers "Trabigzanda," which is "a distortion of a Greek phrase meaning 'girl from Trebizond'" (Barbour, vol. 3, p. 186, note 2).

. . . Smith gives a description of New Plymouth as it was in that year [1624].
First published in his *Generall Historie* (1624), the source Smith used is not known. Barbour has suggested that Smith's description of "The present estate of the plantation at New-Plimouth. 1624" was perhaps based on a communication or report made by Edward Winslow after he arrived in London late in 1623 (Barbour, vol. 2, p. 472, note 1; Smith's description is on pp. 472–473). We would suggest that whoever gave Smith the details had to have knowledge of the fire in Plymouth that burned some of the houses on November 5, 1623, almost two months after Winslow had left for England on the *Anne* on September 10, 1623 (Bradford, pp. 136-137). Winslow does not mention this event in his *Good Newes from New England*, which he had published in London in 1624. *Good Newes* covers events in Plymouth Colony from December 1621 through September 1623. In addition, Smith refers to the saltworks that had been built. This took place after Winslow's return to Plymouth in March 1624. Bradford refers to carpenters sent to their fishing grounds at Cape Anne "to rear a great frame for a large house to receive the salt," which was burnt the following year. Bradford does not give specific dates, but the context is 1624 and 1625 (Bradford, pp. 146–147). Smith reprinted his 1624 account of Plymouth in chapter 8 of his *Advertisements* (1631) (Barbour, vol. 3, pp. 282–284). He expanded it slightly, and included a reference to his having been in touch with the colonists before they sailed. He also mentions the correct distance between Plymouth and Cape Cod, nine leagues, or twenty-seven miles, another instance of the accuracy of his informant (Ibid., p. 283, note 8), or of his own knowledge of the area when he surveyed the coast in 1614. He wrote:

> At the first landing at Cape Cod, being an hundred passengers, besides twenty they had left behind at Plimoth [England] for want of good take heed, thinking to find all things better than I advised them, spent six or seven weeks in wandering up and

> down in frost and snow, wind and rain, among the woods, cricks, and swamps, forty of them died, and three score were left in most miserable estate at New-Plimoth, where their Ship left them, and but nine leagues by Sea from where they landed, whose misery and variable opinions, for want of experience, occasioned much faction until necessity agreed them.

72 **. . . he [Smith] only made two voyages to New England . . .**
The first was made in 1614, five years after Smith returned to England from Jamestown. He sailed in March as captain of a small fleet, returning in late August after a reasonably profitable venture in fishing and fur trading (Ibid., p. 301). The title of admiral, which appears on his map of New England, was presumably granted to him as a result of the excellent coastal survey he conducted (see Barbour, vol. 1, p. lxviii). His second voyage began in March 1615, but he had to turn back for his ship to be repaired. He set out again three months later, but was intercepted by a French fleet. He was taken on board the flagship to give account for his ship when the French saw another ship and set sail to capture it. Smith's crew took advantage of this to escape, leaving him on board the privateer. He had to accompany his captors until eventually they returned to La Rochelle in France, where Smith escaped, eventually making his way back to England (*Generall Historie*, Barbour, vol. 2, pp. 427–435). Still determined to return to New England, Smith obtained backing for another voyage to New England in 1617, but his ships were delayed in England in Plymouth Harbour for three months by strong winds, and he had to abandon his plans (Barbour, vol. 1, p. 302). He returned to London, but evidently did not give up hope of returning to New England, as we know that he was in touch with the Separatists in Holland, hoping to accompany them to America, but without success. He never did return, but spent the rest of his life writing and promoting New England in every way he could, feeling a strong personal involvement in the colonization of the territory that he had explored, mapped, and named.

At New-Plimoth there is about 180 persons . . .
Barbour, vol. 2, p. 472.

Isaack de Rasieres
The original letter written by de Rasieres to Samuel Blommaert, an Amsterdam merchant and a director of the Dutch West India Company, was in Dutch, and was written after de Rasieres returned to Holland. He had visited Plymouth in October 1627, but his letter is undated, and has some missing pages. It has been suggested that it was written after his return to Holland in 1628 or 1629. The

translation published in the *Three Visitors to Early Plymouth*, pp. 65–80, is that by William I. Hull, originally published in 1909 in *Narratives of New Netherland, 1609–1664*, edited by J. F. Jameson, but has been amended and checked by the editor, Sydney V. James Jr. The passage cited is from the *Three Visitors*, pp. 75, 76–77.

It is important to note that in the description, where de Rasieres describes the site of the town of Plymouth as having a street 800 feet long leading down the hill on which the fort stood, and a cross street in the middle extending "northwards to the rivulet and southwards to the land," he has reversed the bearings. The rivulet was Town Brook, which lies to the south of the town.

74 **In a footnote by J. F. Jameson . . . he says, "the street first mentioned was longer, 1,150 feet."**
Three Visitors to Early Plymouth, p. 76. See *Narratives of New Netherland*, p. 112, note 1. Jameson does not provide any details as to the source from which he obtained this measurement.

. . . apparently a stockade of this type was a standard part of town planning on the English frontier at the time.
See Anthony N. B. Garvan, *Architecture and Town Planning in Colonial Connecticut* (Yale University Press, 1951), pp. 27–40 for a discussion that includes the development of the Ulster plantations, which turned out to be an experimental ground for colonial town planning. For their use in riot control see Garvan pp. 28–29.

75 **The *Fortune* arrived in November 1621 . . .**
Bradford, pp. 90–92. The names of the passengers on the *Fortune* are listed in the 1623 division of land in the *Plymouth Colony Records* (*PCR* 12: 5). They include Thomas Prence, who was to become one of Plymouth's governors; Jonathan Brewster, eldest son of William Brewster, whose sister, Patience, came out on the *Anne* in 1623, and who married Thomas Prence the following year, the first of his four wives (she died in 1634); Thomas Cushman, eldest son of Robert Cushman, traveling with his father—the latter returned to England, and John Winslow, brother of Edward Winslow, who by 1627 had married Mary Chilton of the *Mayflower.*

So they were all landed . . .
Ibid., p. 92.

Early in June of 1622, Edward Winslow traveled to Maine . . .
Bradford, pp. 110–111, *Good Newes from New England* (Young), pp. 292–294.

. . . in late June two other ships arrived, the *Charity* and the *Swan* . . .
Good Newes from New England (Young), p. 296. These ships carried settlers for Thomas Weston's Wessagusset settlement.

77 **In 1623, two more ships, the *Anne* and the *Little James*, arrived . . .**
Bradford, p. 127. The ships arrived at the end of July and beginning of August (*Good Newes from New England* [Young], p. 351). The names of the passengers are listed in the 1623 division of land in the *Plymouth Colony Records* (*PCR* 12:5–6). In addition to Alice Southworth, whose marriage to William Bradford took place on August 14, 1623, they include Fear and Patience Brewster, George Morton and his twelve-year-old son Nathaniel Morton, who became secretary to the colony and wrote *New England's Memorial*, published in London in 1669. George Morton's wife, Juliana Carpenter, was the sister of Alice Southworth, so when George Morton died less than a year after his arrival in Plymouth, his son was left in the care of his aunt and uncle by marriage, Alice and William Bradford. By 1624 Myles Standish had married a Barbara, surname unknown, who came over on the *Anne*.

Alice Southworth
As mentioned above, Alice (Carpenter) Southworth was the widow of Edward Southworth. Their two sons, Constant and Thomas, joined William and Alice Bradford in Plymouth in 1628, when they were about sixteen and eleven, respectively (Anderson, pp. 1709–1715). Alice also had close family ties to Samuel Fuller Sr., who had married her sister Anna (Agnes) Carpenter in 1613, in addition to George Morton being her brother-in-law and Nathaniel her nephew. Alice Bradford left a will and her estate was inventoried and probated after her death in 1670 (both documents are published in *The Mayflower Descendant*, 1901, vol. 3, no. 3, pp. 144–149). She and William had two sons and a daughter (Anderson, p. 208).

In 1623, land outside the palisade was allotted to individuals for their private use . . .
Plymouth Colony Records (*PCR* 12:4–6).

In 1624, the *Charity* arrived again with "some passengers and goods" . . .
Edward Winslow returned to Plymouth on the *Charity* in March 1624 from a six-month visit to England, bringing cattle and letters (Bradford, p. 141).

. . . the most important part of the cargo may well have been the "three heifers and a bull" . . .
Bradford describes them as "the first beginning of any cattle of that kind in the land" (Ibid.).

Altham's account is the first to mention livestock.
Three Visitors to Early Plymouth, p. 24.

78 **The cattle . . . were evenly divided between twelve "companies" . . .**
Known as the Cattle Division of 1627. The details are given in the *Plymouth Colony Records*, vol. 12, pp. 9–13. Eleven cows, one bull, two steers, two calves, and one unborn calf are listed. Apart from the three heifers and a bull that Winslow brought on the *Charity*, the *PCR* records "a great black cow," "the lesser of the black cows," and "a great white back cow" which came in the *Ann* [sic] (*PCR* 12:9, 11–12). As the *Anne* arrived in July 1623, before the *Charity*, which Bradford says brought the first cattle to the colony, we can only assume that the *Ann* mentioned here came over later. Four black heifers were sent out on the *Jacob* (*PCR* 12:9), and in a letter to Bradford and the others in leadership, from James Sherley and others, dated December 18, 1625, mention is made of cattle being sent to the colony. Morison points out in a footnote that these included a heifer, a gift from Sherley, "to begin a stock for the poor, a bull, and three or four jades [horses]" citing Bradford's *Letter-book*, p. 35 (Bradford, p. 174, and note 7). This was presumably the "Red Cow which belongeth to the poore of the Colony," mentioned in the cattle division (*PCR* 12:10). It is unlikely that the horses survived the journey, as none are mentioned in the colony until well after the Massachusetts Bay Colony began importing on a regular basis (James Baker, personal communication).

In 1627, after considerable debate as to how the debt owed the adventurers could be liquidated . . .
Bradford, pp. 184–188, 194–196. George D. Langdon Jr., in his excellent *Pilgrim Colony: A History of New Plymouth 1620–1691* (Yale University Press, 1966), pp. 26–37, covers the liquidation of the joint-stock company and the economic development of the colony in its early years.

79 **By 1628, people were moving . . . first to Duxbury, whose residents complained . . . "they could not long bring their wives and children to the public worship . . ."**
Bradford, p. 253.

. . . after 1627, between 1636 and 1687, twenty towns were established . . .
The first of the towns to be incorporated as such was Scituate, established by settlers from Kent in 1636. Duxbury, adjacent to Plymouth, was incorporated as a town by the general court in 1637. Ten families were granted permission to settle at Sandwich on Cape Cod the same year, but the town was only incorporated in 1639.

Barnstable was also established in 1639 by a group that left Scituate under the minister John Lothrop. Taunton, first known as Cohannett, also dates its incorporation to 1639, although the first constable was appointed on December 4, 1638. Yarmouth, adjacent to Barnstable on the cape, was incorporated in 1639 as well. Marshfield was first referred to as a town by the general court on March 2, 1641, and was initially known as Green's Harbor, then Rexham. The township of Rehoboth was organized in 1643 when about sixty families established a village on the edge of an open plain at the head of Narragansett Bay. The area was known as Seekonk, and had been recently abandoned by the Wampanoag. By 1645, when Rehoboth was first listed as a town at the general court of June 4, a village had been established around a commonage 200 acres in size, named the Ring of the Green. A meeting house and cemetery were centered on the Green, and house lots radiated out from it ("Landscape, Communities, and 'the Community at palmers River': Settling a River System, 1665–1737, Rehoboth, Massachusetts," Leslie C. Abernathy III, D.Phil. dissertation, Brown University, 1981, pp. 13–14). Nauset, on Cape Cod, was incorporated as a township in 1646. Its name was changed to Eastham in 1651. Bridgewater, in the center of the colony, northwest of Plymouth, was established in 1656. It was eight years before the next town, Dartmouth, was officially recognized as such in 1664, although its first constable, Samuel Jenney, had been appointed in June 1662. As the colony continued to expand, the demand for new land increased, and there was a steady trade in Indian lands. A particularly controversial action by the general court insofar as the Wampanoag were concerned, was the establishment of Swansea, or Wannamoisett, as a township in October 1667. Wannamoisett was on the border of Rhode Island, and Swansea actually encroached on the birthplace, at Sowams, of the new Wampanoag sachem, Massasoit's son Metacom, or Philip, as he was increasingly known to the English. It was no coincidence that King Philip's War began with an attack on Swansea by the Wampanoag in June 1675. An area known as Namassakett, to the west of Plymouth, received township status in June 1669, and was given the name Middleborough. The last seven towns to gain that status in the colony were Edgartown (1671), Tisbury (1671), Little Compton, home of Benjamin Church (1682), Freetown (1683), Rochester and Falmouth (1686), and Nantucket (1697).

For now as their stocks increased . . .
Bradford, p. 253.

Chapter 3 There Be Witches Too Many

82 **"The colony's controversies were relatively minor . . ."**
Darrett B. Rutman, *Husbandmen of Plymouth: Farms and Villages in the Old Colony, 1620–1692* (Beacon Press for Plimoth Plantation, 1967), p. 64.

Rutman suggests . . .
Ibid., p. 65.

It was cool that day, September 16, the year of Our Lord 1633.
The last will and testament of William Wright, as well as the inventory of his possessions, has been published in C. H. Simmons (Ed.), *Plymouth Colony Wills and Inventories, Vol. 1: 1633–1669* (Picton Press, 1996), pp. 45–50. This construction is based on details contained in these documents as well as some biographical information concerning both Priscilla Carpenter and Will Wright in Robert Charles Anderson's *The Great Migration Begins: Immigrants to New England 1620–1633* (New England Historic Genealogical Society, 1995), pp. 313–314, 2075–2076.

83 **. . . he had married a woman who bore the name "Carpenter," Priscilla Carpenter . . .**
Priscilla Carpenter was the daughter of Alexander Carpenter, who had emigrated from England to Holland with his family, possibly by 1600. He had five daughters of whom Priscilla was the youngest. Four of the sisters settled in Plymouth, Alice marrying William Bradford shortly after her arrival on the *Anne* in 1623. Anderson (p. 2076) suggests that Priscilla and Will Wright were married sometime between 1629 and 1633.

84 **. . . on November 27, 1634, Priscilla married John Cooper.**
Plymouth Colony Records (*PCR*), vol. 1, p. 32.

It is recorded that she [Priscilla (Carpenter) (Wright) Cooper] died on December 28, 1689, in her ninety-second year.
Vital Records of Plymouth, Massachusetts, to the year 1850, Lee D. van Antwerp (Comp.) and Ruth Wilder Sherman (Ed.) (Picton Press, 1993), p. 133.

85 **And always present was the constant fear of the unknown . . .**
Bears, wolves, rattlesnakes, and other fauna and flora new to the colonists are described in *New-Englands Rarities Discovered*, by John Josselyn (London, 1672). It has been reprinted by Applewood Books (Bedford, Mass.).

86 **"There are none that beg in this Country, but there be Witches too many . . ."**
John Josselyn, Colonial Traveler: A Critical Edition of Two Voyages to

New-England (London, 1674), Paul J. Lindholdt (Ed.) (University Press of New England, 1988), pp. 126–127.

From the time that Christianity was brought to England in the sixth century by St. Augustine . . .
One of the most thorough treatments of systems of popular belief current in sixteenth- and seventeenth-century England is *Religion and the Decline of Magic* by Keith Thomas (Weidenfeld & Nicolson, 1971; Oxford University Press, 1997). Cited as Thomas, 1997.

87 **. . . popular magical practices of inquiry such as conjuration, divination, and necromancy.**
Conjuration was the art of getting in touch with angels and demons, and was practiced by wizards or cunning men and women. There were secret formulas for such ceremonies, involving rituals of fasting and prayer, drawing chalk circles on the ground, incantations, and the use of holy water, candles, wands, and other devices. It was believed, for example, that conjurers could be used to locate missing property and identify and track down criminals, a process of divination not unlike those used in modern African rituals (Thomas, 1997, pp. 212–219). It was a capital crime, and so practiced in great secrecy. Necromancy was an ancient form of sorcery, involving magic using the skull or body parts of a dead person.

In 1644 Governor John Winthrop . . . wrote of the thirty "monstrous births or thereabouts, at once," delivered from a pregnant Anne Hutchinson in 1638 . . .
John Winthrop, *A Short Story of the Rise, Reign, and Ruine of the Antinomians, Familists & Libertines that Infected the Churches of New England* (Ralph Smith, 1644). Reprinted in Charles Francis Adams (Ed.), *Antinomianism in the Colony of Massachusetts Bay, 1636–1638* (The Prince Society, 1894). The quotation cited is from page 88 of the reprint.

88 **Richard Greenham, a Puritan theologian, whose works were owned by Will Wright . . .**
Greenham's works were published in 1599. Wright was evidently literate and well read. In his bedchamber he had "one great Bible & a little bible. 1 Greenham's works. 1 [p]salme book with 17 other smale books" (Simmons, p. 49). The references to Greenham's views are based on a selection from entries in Thomas, 1997, pp. 61, 493, and 610.

89 **. . . in 1692, Increase Mather stated, "I am abundantly satisfied that there have been, and still are most cursed Witches in the Land."**
Cited from the unpaged postscript to Increase Mather's *Cases of Conscience Concerning Witchrafts*, written in October 1692, but only

published the following year in Boston under the title *Cases of Conscience Concerning Evil Spirits* (Benjamin Harris, 1693).

What did people in the seventeenth century understand by witchcraft?
The section that follows draws on a number of sources. For those who would like to read further, we recommend in addition to Thomas, 1997, the following works. Specific cases are reprinted in a volume edited by David D. Hall, *Witch-hunting in Seventeenth-century New England: A Documentary History, 1638–1692* (Northeastern University Press, 1991). John Putnam Demos's *Entertaining Satan: Witchcraft and the Culture of Early New England* (Oxford University Press, 1982) is a highly readable, scholarly analysis of this aspect of early American society. And for those who wish to discover more about the Salem witch trials, the interpretation of events by Paul Boyer and Stephen Nissenbaum in *Salem Possessed: The Social Origins of Witchcraft* (Harvard University Press, 1974) cannot be bettered.

91 **People lived in a small, face-to-face society . . .**
John Demos, in *A Little Commonwealth: Family Life in Plymouth Colony* (Oxford University Press, 1970), develops the argument that repression of hostility in the family circle could lead to frequent violent conflicts outside it. See in particular pp. 49–51.

A most Horrible witchcraft or Possession of Devils which had broke in upon several Towns . . .
Letter from Governor William Phips to the home government in England, written from Boston, October 2, 1692. George Lincoln Burr (Ed.), *Narratives of the Witchcraft Cases: 1648–1706* (Scribner, 1914), p. 196.

A number of the settlers came from East Anglia, and customs that survived there . . . would certainly have been familiar practices.
George Ewart Evans, in his *The Pattern under the Plough: Aspects of the Folk-Life of East Anglia* (Faber & Faber, 1966) has made a detailed study of these customs, and this section draws on some of his examples, particularly those found on pp. 52–60 and 74–82.

92 **There are three explicit references to witches and witchcraft in the Plymouth Colony records.**
The list of capital crimes including that of witchcraft is to be found in *PCR* 11:12. The case of Holmes vs. Sylvester is in *PCR* 3:205–207, 211. The indictment of Mary Ingham as a witch is recorded in *PCR* 5:223–224.

. . . Dinah Sylvester, nineteen-year-old daughter of Richard and Naomi Sylvester of Marshfield . . .
Anderson's biographical sketch of Richard Sylvester includes a list of his children (Anderson, pp. 1677–1681).

The suit was brought by William Holmes and his wife . . .
PCR 3:205–207, 211.

93 **Dinah and her eldest brother, John, faced criminal charges . . .**
PCR 7:134.

Naomi's arrest followed a court case against her . . .
Ibid., pp. 132–133.

. . . the family's relationship with John Palmer Jr. . . .
Ibid., pp. 101, 115–116, 119–120.

94 **Dinah later appeared in court on a charge of fornication . . .**
PCR 4:141,162. Fornication was a serious crime in seventeenth-century Plymouth. It involved sex outside of marriage in the case of two single persons, whether or not a contract to marry was in existence, or between a married man and a single woman. Married men who had sexual intercourse with unmarried women were guilty of fornication, not adultery. Fornication was punishable by a fine, a whipping, or a sentence to sit in the stocks, depending on whether or not it was a first offense, although the latter two sentences were more likely in the earlier years of the colony's history when the court tended to enforce the laws more strictly.

. . . Elkanah Johnson of Marshfield was bound to appear . . . "in reference to a child laid unto him by Dinah Sylvester."
PCR 5:22.

. . . Naomi Sylvester's situation after the death of her husband in 1663.
PCR 4:46.

In 1638, Francis Bauer of Scituate was presented . . . "for offering to lie with the wife of William Holmes, & to abuse her body with uncleanness."
PCR 1:98.

95 **In 1641 Gowen White . . . was fined for assaulting William Holmes . . .**
PCR 2:24.

By 1658 Holmes and his wife had moved from Scituate to Marshfield . . .
PCR 8:190, 20.

. . . Holmes had one term of office as a constable . . .
William Holmes served as constable for only one year (*PCR* 4:38), but considering the run-in that one of his successors had with the

Sylvesters two years later (*PCR* 7:134), it is not inconceivable that he had difficulties with them himself, considering that he had won his case against Dinah Sylvester a short two years before.

. . . William Holmes died in 1678, aged eighty-six years.
"Vital Records of Marshfield, Massachusetts," *The Mayflower Descendant,* 1900, vol. 2, no. 3, p. 181.

At the time he drew up his will in March 1678 . . .
Transcripts of William Holmes's will and probate inventory are contained in vol. 4, pt. 1, pp.15–18 of an unpublished volume of Plymouth Colony wills and inventories, cited as *Plymouth W&I.* The four original volumes of wills and inventories, 1633-1685, and a set of transcripts are held in the office of the Plymouth County Court Commissioner. Volumes 1 and 2 have been published as *Plymouth Colony Wills and Inventories, Vol. 1: 1633–1669,* edited by C. H. Simmons (Picton Press, 1996). Plimoth Plantation holds two sets of transcripts of the unpublished Plymouth Colony wills and inventories, volumes 3 and 4, 1670 through 1685.

. . . the Holmeses do not appear to have been as wealthy as the Sylvesters (Richard Sylvester's estate amounted to £244, not including lands and housing).
It is not always easy to gain a sense of just where people stood in relation to their neighbors in terms of wealth, but probate inventories do provide some guidelines. In the 1660s the sumptuary laws of Massachusetts Bay Colony permitted members of families with an aggregate wealth of above £200 to wear silks adorned with gold and silver lace (Demos, *Entertaining Satan*, pp. 22, 127, 446, note 138), and although these laws did not apply in Plymouth Colony, they do provide a guide as to the position in society that Richard Sylvester and his family held.

. . . the case of Mary Ingham of Scituate . . .
The indictment of Mary Ingham as a witch is recorded in *PCR* 5:223–224, but no other information concerning her has been traced. The construction in the text is shaped around the Plymouth court records for March 6, 1677 (*PCR* 5:217–229).

96 Ordinaries were open to the public coming in to town, beer and cider easily available . . .
In June 1674, a number of new acts were passed by the court in Plymouth. Initially only strangers coming into a town were allowed to use the ordinaries (inns or taverns), not any of the residents (*PCR* 1:38). The new enactment of 1674 implicitly recognized that this had changed, and required that ordinary keepers did not serve any kind of drink to inhabitants of the town on the Lord's Day, and

also that the ordinaries had to be closed, except to lodgers, by sundown (*PCR* 11:236).

. . . forced searches of her body by the authorities to find evidence of her crime.
The detection of "witch marks" on the body was often an important part of the evidence against an accused person. In the case of a woman, a committee of women could be appointed to examine the body of the defendant for such marks. See Demos, *Entertaining Satan*, pp. 180–181 and 459, note 117, and also Thomas 1997, pp. 445–446 and 551.

Like Margaret Jones of Charlestown . . .
On June 15, 1648, Margaret Jones of Charlestown, Boston, was executed by hanging after her conviction as a witch the previous month. John Winthrop, governor of the Massachusetts Bay Colony at the time, described the case in his journal, including the forced body search and the "very great tempest at Connecticut" that took place "the same day and hour she was executed." *Winthrop's Journals*, James Kendall Hosmer (Ed.) (New York, 1908), vol. 2, pp. 344–345.

She was splay footed . . .
Although there is no evidence, as stated in the construction, that Mary Ingham had any physical deformity, Robert St. George, in *Conversing by Signs: Poetics of Implication in Colonial New England Culture* (University of North Carolina Press, 1998) draws attention to the sensitivity to bodily deformities in seventeenth-century English culture, and its association with Satan. See in particular pp. 163–173.

. . . Mehitable Woodworth, the daughter of Walter Woodworth, of Scituate . . .
At the time Mary Ingham was tried, Mehitable Woodworth was about fifteen. Her father, Walter, had emigrated to Plymouth in 1633, and she was probably his ninth child. When he drew up his will in November 1685 she was unmarried, aged about twenty-three. Woodworth's estate was valued at £355 10s, the bulk of which was in real estate. Nothing emerges from the records to indicate why Mehitable was considered to be a victim of witchcraft. Biographical details concerning Walter Woodworth may be found in Anderson, pp. 2064–2067.

97 **In New England, women were in general the ones accused of witchcraft, as in Old England, although in Europe it seems that the gender bias was not as clear.**
David D. Hall includes a short section on gender and witch-hunting, as well as some selected bibliographic references, in his *Witch-hunting in Seventeenth-century New England*, pp. 6–7, 319.

John Putnam Demos, in *Entertaining Satan*, has a detailed analysis of the sex of witches in New England, and found that women outnumbered men by four to one (pp. 60–64). Keith Thomas, in *Religion and the Decline of Magic* (1997), also discusses the sex of witches (pp. 436, 520, 562, 568–569). Robin Briggs in *Witches and Neighbors: The Social and Cultural Context of European Witchcraft* (Viking, 1996) finds, however, that in Europe a greater number of men than is realized were executed as witches, about 25 percent of all cases, although he stresses that it remains true that more women than men were accused as witches. See his chapter "Men against women: the gendering of witchcraft," pp. 259–286.

. . . Malleus Maleficarum. . .
Translated with an introduction, bibliography, and notes by the Reverend Montague Summers (Blom, 1928). The *Malleus Maleficarum* was drawn up in response to an Apostolic Bull *Summis desiderantes affectibus* issued on December 9, 1484, by Pope Innocent VIII to Fathers Henry Kramer and James Sprenger, both officers of the Inquisition. It is a theological and legal work, written in great detail, and became the standard text on witchcraft for close to three hundred years. It was published around 1486, and it not only actively encouraged witch-hunting in Europe through the sixteenth and seventeenth centuries, but also shaped and sustained much credulity in its supernatural dimensions.

This detailed work . . . stresses the credulity of women . . .
Malleus Maleficarum, 1928, p. 44. The emphasis of the *Malleus Maleficarum* on women's unbridled sexuality as the cause of witchcraft did more than anything else to promote a general belief that women rather than men were to blame for any strange, unaccountable, and therefore supernatural occurrences in society.

98 **. . . the ancient medieval practice of trial by ordeal.**
This is discussed in Thomas, 1997, pp. 219–220, in the context of conjuration and divination. It was a test used by judges and jurors in cases of homicide. It is one of several circumstances to be noted during the examination of a person accused of homicide set out in an early seventeenth-century law manual by Michael Dalton. Entitled *The Countrey Justice, Containing the Practise of the Justices of the Peace out of their Sessions . . .* (London, 1619), the manual was certainly in use in New England as it was the model used in the drafting of the first Rhode Island legal code according to Bradley Chapin's *Criminal Justice in Colonial America, 1606–1660* (University of Georgia Press, 1983), pp. 6, 46. The greatest efficacy of the demand by the jurors that the defendant touch the body to see if it "bled anew" lay more in its deterrent effect, as on its own it was not considered to be sufficient to

convict. The belief that guilt could be determined by supernatural means could be forceful enough to prevent a crime from being committed. On another level, but no less powerful, the cunning men's powers of divination were a form of social control. In addition to the two Plymouth cases discussed, a third appears to contain an oblique reference to trial by ordeal. In January 1654, the jury appointed to view the dead body of John Walker, servant to Robert and Susanna Latham, and determine the cause of death, submitted their report to the court of assistants. They ended it with the cryptic remark (as it appears to us reading the text well over three centuries later), "and also, upon the second review, the dead corpse did bleed at the nose" (*PCR* 3:72). These appear to be the only cases of trial by ordeal in the *Plymouth Colony Records*, but Chapin refers to one in Massachusetts and another in Maryland (Chapin, pp. 46, 159, note 94).

John Sassamon

The details given concerning the murder of Sassamon have been drawn from the following sources: the official court records, *PCR* 5:159, 167–168. Contemporary accounts by John Easton (1675), "N. S." (Nathaniel Saltonstall) (London, 1675 and 1676), reprinted in *Narratives of the Indian Wars 1675–1699*, Charles H. Lincoln (Ed.) (Scribner, 1913), pp. 7–8, 24–25, 54–55; William Hubbard's *A Narrative of the Troubles with the Indians in New England* (Boston, 1677), reprinted in *The History of the Indian Wars in New England*, Samuel G. Drake (Ed.) (Franklin, 1971 reprint), vol. 1, pp. 60–64. Cotton Mather's *Magnalia Christi Americana* (Hartford, Conn., 1702) contains the only account of the supernatural encounter between Tobias and John Sassamon's corpse. It may be found in the two-volume edition of the *Magnalia* edited by Thomas Robbins (Andrus, 1853), vol. 2, pp. 559–560. His description is followed by mention of the fact that "the total eclipse of the moon in Capricorn" could be interpreted by troopers as they marched out of Boston on June 26 as "ominous of ensuing disasters" (p. 561). It is accounts such as these by one of the leading Puritans that do underline not only that they were very familiar with the mindset of the greater part of the population, but that their own beliefs and teaching were a part of it.

99 **. . . the cause of death of Anne Batson's child . . .**
PCR 5:261–262.

100 **. . . the case of the guilty canoe.**
PCR 3:208–209.

102 **The widow was to receive a third of her husband's land during her life, as she would have in England, as well as a third of his goods that she could dispose of as she liked.**
Relevant sections of the Plymouth laws are to be found in *PCR*

11:13, 133. See also *PCR* 11:52 for the recognition of a wife's rights concerning the disposal of her husband's property.

103 **Captain Thomas Southworth**
A biographical outline of Thomas Southworth has been published in Anderson, pp. 1712–1715. See also *Plymouth W&I*, vol. 3, pt. 1, pp. 1–2.

104 **John Howland, however, went further, in leaving his wife, Elizabeth Tilley . . .**
Howland's will, dated March 5, 1673, is recorded in *Plymouth W&I*, vol. 3, pt. 1, pp. 49–50.

. . . Agatha Vause of Middlesex County in Virginia . . .
Cited by Darrett B. Rutman and Anita H. Rutman in *A Place in Time: Middlesex County, Virginia 1650–1750* (Norton, 1984), p. 119.

Notwithstanding their marriage and improvement of their stock together . . .
The agreement cited between Joanna and Dolar Davis of Barnstable appears as testimony taken on July 2, 1673. *The Mayflower Descendant*, vol. 24, no. 2, 1922, pp. 71–73.

John Demos . . . states that women in seventeenth-century England did not have the right to make contracts except under exceptional circumstances.
Demos, *A Little Commonwealth*, pp. 85–86.

105 **. . . Elizabeth Poole of Taunton. . .**
Details concerning Elizabeth Poole have been obtained from the *Plymouth Colony Records*, and from her will and inventory published in Simmons, pp. 299–303.

106 **Women Before the Court**
The section that follows draws on a thorough survey and analysis of references to women in Plymouth Colony in the court records from 1633 to 1668 prepared by Anna Neuzil of the University of Virginia as an independent study project under James Deetz in the fall of 1998.

In 1668 William Tubbs was granted a divorce from his wife, Marcy . . .
PCR 4:42, 46–47, 66, 104, 187.

107 **. . . Captain William Hedges of Yarmouth drew up his will . . .**
Plymouth W&I, vol. 3, pt. 1, pp. 20–21.

John Barnes of Plymouth . . .
John Barnes's will, drawn up in 1668, and his 1671 probate inventory are recorded in *Plymouth W&I*, vol. 3, pt. 1, pp. 31–36.

108 **One of Plymouth's most turbulent wives was Hester Rickard . . .**
Sources for the account of Hester Rickard are as follows, in the

order discussed: *PCR* 8:13, 17; 7:96; 3:210; 4:50, 8:109; 4:111; 5:87. John Rickard's probate inventory is recorded in *Plymouth W&I*, vol. 3, pt. 2, pp. 122–123.

109 **Janet A. Thompson . . . comments . . .**
Wives, Widows, Witches and Bitches: Women in Seventeenth-Century Devon (Lang, 1993), chap. 3, "Defamation Cases and Sexual Slander in Seventeenth-Century Devon," pp. 81–99.

110 **Only a small proportion of women left wills . . .**
For the period 1633 through 1685, there are 289 Plymouth Colony wills. Of these only twenty-three (8 percent) are those of women, twenty-one widows and two spinsters.

. . . John Demos shows that one can use the deviant cases found in the court records to infer at least three basic obligations in marriage that the law sought to protect.
Demos, *A Little Commonwealth*, pp. 91–99.

. . . Henry Howland's sensitivity to his wife's need for privacy . . .
Plymouth W&I, vol. 3, pt. 1, p. 26.

111 **. . . Josiah Cooke's gift to his wife, Elizabeth, of his servant and apprentice, Judah, an Indian . . .**
Plymouth W&I, vol. 3, pt. 1, p. 90.

Bound by Indentures—Servitude in Plymouth Colony
This section draws on a thorough extraction and analysis of references to servants in the *Plymouth Colony Records (PCR)* prepared by anthropology graduate student Jillian E. Galle as a paper for Anthropology 509, Historical Ethnography, at the University of Virginia, in the spring of 1998. Demos's *A Little Commonwealth* contains a chapter on masters and servants, pp. 107–117. Stratton's *Plymouth Colony* also contains a useful section on the relationships between man and master, pp. 179–190, and refers to a shipping list that gives details concerning Poole and Deane's servants, who traveled with them on the *Speedwell* in 1637 (p. 67).

. . . a contract . . . between Edward Doty and Richard Derby . . .
PCR 12:21, 1:94.

112 **. . . the first duel on New England soil . . .**
Bradford describes the event as:

> The first duel fought in New England upon a challenge at single combat with sword and dagger, between Edward Doty and Edward Leister, servants to Master Hopkins, both being wounded, the one on the hand, the other in the thigh. They are

> adjudged by the whole company to have their head and feet tied together, and so to lie for twenty-four hours, without meat or drink; which is begun to be inflicted, but within an hour, because of their great pains, at their own and their master's humble request, upon promise of better carriage, they are released by the Governor.

Thomas Prince in *A Chronological History of New England in the Form of Annals* (Boston, 1736; Edinburgh, 1887), vol. 3, p. 40. Prince extracted the information from William Bradford's lost register. There is a biographical outline of Edward Doty in Anderson, pp. 573–577.

113 **What of laws relating explicitly to servants?**
In the order discussed, these may be found in *PCR* 11:113; 29; 47,48; 66.

114 **. . . a law was passed that "no Indians that are servants to the English shall be permitted to use guns for fowling or other exercises . . .**
PCR 11:242.

. . . the case of Thomas Wappatucke.
PCR 6:153.

. . . Margaret, "an Indian slave" . . .
Plymouth W&I, vol. 3, pt. 2, pp. 110–112.

As early as 1653, John Barnes had a "neager maid servant," . . .
PCR 3:27.

115 **Captain Thomas Willett's negro servant, Jethro . . .**
PCR 5:216.

. . . in a will drawn up by Walter Briggs . . .
PCR 6:135.

. . . John Dicksey of Swansea . . . and John Gorum . . .
The inventories are to be found in *Plymouth W&I*, vol. 3, pt. 1, pp. 106–108 and 162–163, respectively.

Captain Willett's 1674 estate included eight Negroes valued at £200, and in his will, drawn up three years earlier . . .
Thomas Willett's will and estate inventory are to be found in *Plymouth W&I*, vol. 3, pt. 1, pp. 114–128.

John Demos comments that the pattern with white servants . . .
Demos, *A Little Commonwealth*, p. 110.

Only three of the indenture agreements in the court records relate to women . . .
The case of Alice Grinder: *PCR* 1:21, and Anderson, p. 823; Mary Moorecock: *PCR* 1:129; Elizabeth Billington: *PCR* 2:38.

116 **The Billington case raises the . . . question of the "putting out" of children . . .**
This is discussed by Edmund S. Morgan in *The Puritan Family: Religion and Domestic Relations in Seventeenth-Century New England* (Harper, 1966), pp. 75–79, and briefly by Demos in *A Little Commonwealth*, pp. 71–72. The case of Zachariah Eddy is to be found in *PCR* 2:112–113; John Eddy, ibid., p. 82; Joseph Billington, ibid., p. 58. Anderson has a biographical outline of Francis Eaton that contains details concerning his marriage to Christian Penn and their children (pp. 608–610).

118 **In 1637 Edith Pitts accused her master, John Emerson, of abuse . . .**
PCR 1:48–49.

. . . Jane Powell, servant to William Swift . . .
PCR 3:91.

119 **Fourteen-year-old John Walker was the servant of Robert and Susanna Latham of Marshfield.**
The Latham case is to be found in *PCR* 3:71–72, 82, and 143. The important connection between Robert and Susanna Latham and the Winslow family was researched by Teresa Smith, a University of Virginia undergraduate, in a first-year university seminar on Plymouth Colony in the fall of 1998.

120 **. . . the ancient English custom of "benefit of clergy."**
This is discussed by Bradley Chapin, pp. 48–50.

Bradley Chapin . . . comments that benefit of clergy was "a highly irrational and erratic means . . ."
Ibid., p. 48.

. . . desired the benefit of the law . . .
PCR 3:78.

121 **They held it as a basic truth . . .**
Chapin., p. 49.

122 **. . . an extension of the policy of providing for the poor . . .**
PCR 3:118, 132, 134.

To "Drink Drunk"—The Regulation of Alcohol
This section draws on a thorough extraction and analysis of references to alcohol consumption and ordinaries in the *Plymouth Colony Records (PCR)* prepared by Allison Devers as a paper for Anthropology 509, Historical Ethnography, at the University of Virginia, in the spring of 1998.

123 **The first case documented for alcohol abuse . . .**
PCR 1:12.

As Holmes . . . was presented again in 1639 . . .
PCR 1:118.

124 **For being drunk at Mr. Hopkins his house . . .**
PCR 1:75.

"to the oppressing & impoverishing of the colony" . . .
PCR 1:87.

. . . "drink drunk" . . .
The earliest law against drunkenness in Plymouth Colony is that in the 1636 codification of the laws: "That such as either drink drunk in their persons or suffer any to drink drunk in their houses be enquired into amongst other misdemeanors and accordingly punished or fined or both by the discretion of the bench."
PCR 11:17.

And by drunkenness is understood . . .
PCR 11:50.

125 **. . . there is mention as early as 1638 of the widow Palmer being present at Stephen Hopkins's house . . .**
PCR 1:75.

. . . Ann Hoskins accused Hester Rickard of being "as drunk as a bitch . . ."
PCR 4:111.

And in 1669 James Clark, Philip Dotterich, Mary Ryder, and Hester Wormall . . .
PCR 5:15.

John Demos suggests . . .
Demos, *A Little Commonwealth*, p. 90.

126 **. . . in 1678 Mary Williamson of Marshfield was granted a license . . .**
PCR 5:271–272.

. . . in 1684 Mistress Mary Combe was licensed to keep a tavern . . .
PCR 6:141. References to Francis Combe being licensed to keep an ordinary and serving as a selectman for Middleborough may be found in *PCR* 5:273 and 6:59, 84.

. . . "understanding that James Leonard, of Taunton . . ."
PCR 4:54.

One of the few cases brought against a woman is that of Ann Savory . . .
PCR 3:212.

127 **It is also strongly implied that Hester Rickard was drunk . . .**
PCR 5:87.

In March 1664 . . . the court . . . called in the license of Edward Sturgis Sr. due to "much abuse of liquors . . ."
PCR 4:54

. . . Thomas Lucas was required . . . to answer "for his abusing of his wife . . ."
PCR 4:55.

The selling of liquor to Indians was prohibited . . .
The 1646 law is set out in *PCR* 11:54.

Whereas there has been great abuse by trading wine . . .
PCR 3:60–61.

128 **In 1662 the court ruled that if any Indian be found drunk . . .**
PCR 11:140.

The government in Plymouth . . . in 1677 reiterated the need for "prevention of the growing intolerable abuse by wine . . ."
PCR 11:244.

Chapter 4 In an Uncivil Manner

131 **In February, rest not for taking thine ease . . .**
Thomas Tusser, *Five Hundred Pointes of Good Husbandrie. . . .* W. Payne and Sidney J. Herrtage (Eds.) (Trübner, 1878), pp. 228–232.

132 **Population figures are difficult to obtain with accuracy . . .**
In 1643 a list of all those able to bear arms between the ages of 16 and 60 years was published, and includes some 600 names *(Plymouth Colony Records (PCR)*, vol. 8, pp. 187–196). The estimate of 1,500 to 2,000 is an extrapolation that seems reasonable. See Eugene Aubrey Stratton in his *Plymouth Colony: Its History and People 1620–1691* (Ancestry Publishing, 1986), p. 70, and note 33 on p. 72; for an assessment of the population from 1660, see D. C. Parnes, *Plymouth and the Common Law (1620–1775): A Legal History* (The Pilgrim Society, 1971), pp. 37–38.

Bradley Chapin . . . suggests . . .
Bradley Chapin, *Criminal Justice in Colonial America, 1606–1660* (University of Georgia Press, 1983), p. 139.

134 **Sex-Related Capital Crimes**
This section, and the one that follows dealing with noncapital crimes, draw on a thorough extraction and analysis of references to sexual misconduct in the *Plymouth Colony Records (PCR)* prepared by anthropology graduate student Lisa M. Lauria as a paper for Anthropology 509, Historical Ethnography, at the University of Virginia in the spring of 1998.

The winter's day was shrewdly cold . . .
The construction that follows is based on facts available in the court records. The account that John Holmes presented to the court may be found in *PCR* 2:51. Details concerning Holmes's estimated age and family are based on the biographical profile in Robert Charles Anderson, *The Great Migration Begins: Immigrants to New England 1620–1633* (New England Historic Genealogical Society, 1995), pp. 977–978. Holmes was about thirty-one when he executed Granger, and Love Brewster around thirty-nine (Anderson, p. 229).

135 **And after the time of writing of these things befell . . .**
William Bradford, *Of Plymouth Plantation 1620–1647*, Samuel Eliot Morison (Ed.) (Knopf, 1952), pp. 320-321.

136 **. . . it is described very factually, there are no bizarre details or circumstances as there were in one case in particular in the Bay Colony.**
The case in question is that of George Spencer of New Haven, accused and convicted of buggery with a sow in 1641. Spencer was deformed, having only one eye, and when a sow had a "monster" one-eyed piglet, described in horrific detail, Goodwife Wakeman, who brought the charge against him, said her spirit was impressed by "some hand of God, " and that George Spencer had "been actor in unnatural and abominable filthiness with the sow." He was pressured to confess before being tried, and did so on eleven separate occasions, but denied it at his trial, having misunderstood the grounds of the mercy that confession was supposed to bring. The case is discussed by Chapin, pp. 38–39, 128–129. It may be found in the *New Haven Colonial Records*, vol. 1, p. 62.

137 **Less than a year later . . . the odd case of John Walker . . .**
PCR 2:57.

138 **The next instance of buggery to come before the court . . .**
PCR 4:116.

Thomas Saddler, you are indicted . . .
PCR 6:74–75.

In a letter to Richard Bellingham . . .
Bradford, pp. 318–319.

139 **Edward Michell, who propositioned Lydia Hatch . . .**
PCR 2:35–36.

Perhaps the greatest single reform in early colonial America . . .
For a discussion of the changes to the law in early colonial America in so far as the death penalty was concerned, see Chapin, pp. 8–9.

140 **John Alexander & Thomas Roberts were both examined . . .**
PCR 1:64.

141 **There are only two rape cases in the Plymouth Colony court records.**
The case of Fish vs. Fish may be found in *PCR* 5:245–246, and that of Freeman vs. Sam the Indian in *PCR* 6:98. Stratton's account of the case may be found on pp. 198–199 of his *Plymouth Colony.*

A strange twist to the story . . .
The Tupper incident is to be found in *PCR* 6:152. The survey of constables in Duxbury between 1642 and 1665 formed the basis of a paper prepared by Benjamin Chadwick for Anthropology 509, Historical Ethnography, at the University of Virginia in the spring of 1999.

142 **In 1682 Philip's severed head was still on display in Plymouth . . .**
See Jill Lepore, *The Name of War: King Philip's War and the Origins of American Identity* (Knopf, 1998), pp. 174–175, 309, notes 7, 8.

Eugene Stratton sees this case . . .
Stratton, p. 107.

Laurel Thatcher Ulrich comments . . .
Good Wives: Image and Reality in the Lives of Women in Northern New England 1650–1750 (Vintage Books, 1991), pp. 97, 174.

143 **Historian Keith Thomas has shown . . .**
Keith Thomas, "The Double Standard," *Journal of the History of Ideas*, 1959, vol. 20, no. 2, p. 213.

. . . whosoever shall commit Adultery . . .
PCR 11:95.

144 **The first case of adultery . . . involved Mary Mendame . . . and Tinsin, an Indian.**
PCR 1:132.

In 1670, Samuel Halloway of Taunton . . .
PCR 5:29, 31–32, 41–42.

145 **On July 4, 1673, John Williams of Barnstable . . .**
PCR 5:127.

In June 1686, John Glover, also of Barnstable . . .
PCR 6:190.

146 **In 1661 Elizabeth Burgis petitioned for and was granted a divorce . . .**
PCR 3:221.

The first case of incest involving fornication . . .
PCR 3:197–200.

147 **Nine years later, in 1669, Christopher Winter and his daughter, Martha, came before the court to answer charges of incest . . .**
PCR 5:13–14, 21.

. . . thirty years earlier Winter and his wife, Jane Cooper, were found guilty of fornication before marriage.
PCR 1:132.

148 **Between 1633 and 1691, sixty-nine cases of fornication were presented.**
The statistics and assessment discussed in this section draw on those given by Lauria in her paper "Sexual Misconduct in the Colony of New Plymouth" prepared for the authors' University of Virginia Anthropology 509 seminar, spring semester 1998.

149 **. . . Jane Powell . . . who admitted to fornication with another servant . . .**
PCR 3:91.

Typical entries in the court records for cases of fornication . . .
References for the eight cases cited are as follows: *PCR* 2:138; 3:5; 3:6 (Launders and Davis); 3:11; 4:34; 4:42; 4:47.

150 **Hannah Bonny was convicted of fornication . . . "with Nimrod, [a] negro . . ."**
PCR 6:177.

151 **A typical case . . . was that of John Peck of Rehoboth . . .**
PCR 3:75.

A slightly more unusual case was that of Nathaniel Hall of Yarmouth . . .
PCR 5:169.

In October 1668, Samuel Worden laid a charge against Edward Crowell and James Maker . . .
PCR 5:8.

152 **In March 1656 John Gorum was fined forty shillings . . .**
PCR 3:97.

In 1642 Lydia Hatch appeared before the court . . .
PCR 2:35.

The law did not define lascivious "carriages" . . .
Chapin comments that "sexual 'carriage' might be light, lewd, lascivious, naughty, unclean, filthy, wicked, sinful, wanton, unchast, or whorish" (Chapin, pp. 129–130).

In June 1655, Hugh and Mary Cole of Plymouth . . .
PCR 3:82. Details concerning their marriage date and births of their children are to be found in *PCR* 8: 72, 74.

153 **They had to monitor the behavior of everyone in their respective towns . . .**
PCR 11:90.

. . . the case of Marcy Tubbs and Joseph Rogers, as well as Jonathan Hatch and Frances Crippen.
PCR 4:42, 116–117.

. . . in seventeenth-century England there were cases of men "selling" their wives' sexual favors, and even selling the wife herself.
Keith Thomas cites the case of Thomas Heath who in 1696 was presented by the churchwardens of Thame, Oxfordshire, "for cohabiting unlawfully with the wife of George Fuller, 'having bought her of her husband at 2¼ d the pound'" ("The Double Standard," *Journal of Ideas*, 1959, vol. 20, p. 213).

154 **On October 2, 1650, the wife of Hugh Norman . . .**
PCR 2:137, 163.

Domestic Violence
This section draws on a thorough extraction and analysis of references to domestic violence in the *Plymouth Colony Records (PCR)* prepared by Jason Jordan as a paper for Anthropology 509, Historical Ethnography, at the University of Virginia in the spring of 1998.

Violence is any physical assault upon a person or property . . .
Laurel Thatcher Ulrich discusses and defines various types of violence in her chapter on viragoes, *Good Wives* (1991), pp. 184–201.

155 **In general it appears that in New England, if there was any evidence that there was undue provocation by the wife, an abusive husband was treated leniently.**
See Ulrich, p. 269, note 7.

That concerning misdemeanors . . .
PCR 11:15, 18.

156 **The 1658 code appears to have revised the law . . .**
PCR 11:143–144.

Forasmuch as John Crocker, of Scituate . . .
PCR 1:141–142.

157 **In a similar case, when Robert Ransom appealed to the court . . .**
PCR 3:63–64.

Likewise, when John Hall of Yarmouth complained . . .
PCR 3:83, 88.

The case of Joseph Gray . . .
PCR 3:46, 132, 134.

158 **At this Court, Mary, the wife of Jonathan Morey . . .**
PCR 5:16.

159 **Murder *was* the issue, however, in the case of Alice Bishop.**
PCR 2:132–133, 134.

160 **Anna Bessey, for her cruel and unnatural practice towards her father in law, George Barlow, in chopping of him in the back . . .**
PCR 4:10.

161 **. . . what Ulrich describes as "a violent underside in the village culture of New England."**
Ulrich, p. 183.

Edward Bumpas, for striking and abusing his parents . . .
PCR 6:20.

162 **One of the strangest cases in the court records is that of Ralph Earles . . .**
PCR 4:47.

In August 1665, John Dunham was brought to court . . .
PCR 4:103–104.

Some years later, Richard Marshall was sentenced to sit in the stocks . . .
PCR 6:51.

It has been said that "a man's honor depended on the reliability of his spoken word; a woman's honor on her reputation for chastity."
Lawrence Stone, *The Family, Sex, and Marriage in England: 1500–1800* (Routledge and Kegan Paul, 1977), p. 317.

John Williams Jr. was guilty of "disorderly living with his wife . . ."
PCR 4:93, *PCR* 4:125–126.

163 **In March 1655, Joan Miller was brought to court for "beating and reviling her husband . . ."**
PCR 3:75.

The case of Jane and Samuel Halloway . . .
PCR 5:29, 32; 8:65.

164 **. . . the case in 1685 in which Betty, an Indian woman, killed her husband . . .**
PCR 6:153–154.

Death
This section draws on an extraction and analysis of references to coroner's inquests in the *Plymouth Colony Records (PCR)* prepared by Jeffrey Norcross as a paper for Anthropology 509, Historical Ethnography, at the University of Virginia in the spring of 1998.

All the colonial court records reflect a similar pattern.
Homicide is discussed by Bradley Chapin in *Criminal Justice in Colonial America, 1606–1660* (1983), where he includes a table, "Homicides Brought to Trial before 1660," pp. 111–117.

165 **In February 1651 a nine-year-old boy, John Slocum . . .**
PCR 2:174–175.

166 **Accidental deaths other than by drowning varied . . .**
References for the cases cited from the court records are as follows: *PCR* 4:170; 3:223; 4:130; 5:29.

We, whose names are underwritten . . .
PCR 5:88.

167 **And, as to Persons, it is generally thought . . .**
Nathaniel Saltonstall, in "A New and Further Narrative of the State of New England," 1676, reprinted in Charles H. Lincoln (Ed.), *Narratives of the Indian Wars 1675–1699* (Scribner, 1913), p. 98. Lincoln, in a footnote, comments that another estimate from 1676 "gives the loss as 444 killed and 55 taken prisoners for the colonists, and 910 for the Indians."

Eugene Stratton notes that . . .
Stratton, p. 120.

168 **. . . Peter Benes . . . sees as evidence of the emergence of a regional cultural identity in the colony . . .**
Peter Benes, *The Masks of Orthodoxy: Folk Gravestone Carving in Plymouth County, Massachusetts, 1689–1805* (University of Massachusetts Press, 1977).

. . . Christopher Wadsworth's request in 1677 . . .
Christopher Wadsworth's will is contained in vol. 4, pt 1, pp. 19–70 of an unpublished volume of Plymouth Colony wills and inventories, cited as *Plymouth W&I*. The four original volumes of wills and inventories, 1633–1685, and a set of transcripts are held in the office of the Plymouth County court commissioner. Volumes 1 and 2 have been published as *Plymouth Colony Wills and Inventories, Vol. 1: 1633–1669*, edited by C. H. Simmons (Picton Press, 1996). Plimoth Plantation holds two sets of transcripts of the unpublished Plymouth Colony wills and inventories, volumes 3 and 4, 1670 through 1685.

. . . the will of Elizabeth Hoare of Yarmouth . . .
Plymouth W&I, vol. 4, pt 1, p. 85.

. . . the inventory of Samuel Chandler of Duxbury . . .
Ibid., p. 145.

. . . the estate of *Mayflower* passenger George Soule . . .
Ibid., p. 51.

. . . the 1684 estate of Daniel Combe of Scituate . . .
Ibid., pt 2, p. 69.

William Wetherell, a pastor in Scituate . . .
Ibid., p. 133.

. . . Peter Collamer, also of Scituate . . .
Ibid., p. 144.

169 **. . . which is consonant with the findings of Peter Benes . . .**
Benes, p. 36.

Having viewed the dead body of the said Titus Waymouth . . .
PCR 3:108–110. The inventory of his possessions, which includes burial charges, is listed on page 109, of the *Records*.

170 **Benes shows that after 1700 "generous quantities of rum were consumed by the participants and celebrants" at funerals . . .**
Benes, p. 35

. . . "more money seems to have been spent [in Plymouth Colony] on burial drinking than elsewhere.
Ibid.

Chapter 5 A Few Things Needful

172 **A knowledge of Thomas Jefferson . . .**
Henry Glassie, *Folk Housing in Middle Virginia* (University of Tennessee Press, 1975), p. 12.

. . . Darrett Rutman's *Husbandmen of Plymouth* and John Demos's *A Little Commonwealth* . . .
These two excellent, readable yet scholarly works were researched under the auspices of Plimoth Plantation. Rutman's *Husbandmen of Plymouth: Farms and Villages in the Old Colony, 1620–1692* was published for Plimoth Plantation by Beacon Press, Boston, in 1967. Oxford University Press (New York) released *A Little Commonwealth: Family Life in Plymouth Colony*, by John Demos, in 1970.

175 **. . . Harold Shurtleff in a book entitled *The Log Cabin Myth.***
Harvard University Press, 1939.

177 **Windows would have been few and small . . .**
The set of instructions found in a 1638 building contract are contained in a letter from Samuel Symonds of Ipswich, Massachusetts, to John Winthrop the Younger, also of Ipswich. They were given to Winthrop to pass on to the house carpenter whom he was employing on behalf of Symonds. *Collections of the Massachusetts Historical Society*, 1865, 4th ser., vol. 7, pp. 118-120. Edward Winslow's instructions to prospective colonists appear in *Mourt's Relation*, p. 86.

178 **This fire was occasioned by some of the seamen . . .**
Bradford, *Of Plymouth Plantation 1620-1647*, Samuel Eliot Morison (Ed.) (Knopf, 1952), pp. 136–137.

. . . after Emmanuel Altham's letter to his brother, in which he wrote that he counted twenty houses.
Sydney V. James Jr. (Ed.), *Three Visitors to Early Plymouth: Letters About the Pilgrim Settlement in New England During Its First Seven Years* (Applewood Books, 1963), p. 24.

Winslow's mention of the storm of February 4, 1621 . . .
Mourt's Relation: A Journal of the Pilgrims at Plymouth (Applewood Books, 1963), p. 48.

179 **Emmanuel Altham's letter of September 1623 . . .**
Three Visitors to Early Plymouth, pp. 23–35.

Those in New Netherland and especially in New England . . .
Cornelius van Tienhoven, "Information Respecting Land in New Netherland," *Pennsylvania Archives*, 1876, 2nd ser., vol. 5, pp. 182–183. Address delivered by secretary van Tienhoven on March 4, 1650. The address is published as part of the "Papers Relating to the Colonies on the Delaware," pp. 179–186.

Although architectural historian Abbott Lowell Cummings correctly states that no parallel for such dwellings can be found in the records of Massachusetts Bay . . .
Cummings, in his *The Framed Houses of Massachusetts Bay, 1625–1725* (Harvard University Press, 1979), p. 19.

. . . the archaeologist Ivor Noël Hume excavated what he terms a "cave house" . . .
Hume describes the excavation and provides a section of a cellar house with a conjectural reconstruction of the building in his *Martin's Hundred* (Knopf, 1983), pp. 53–59.

180 **Although this method of house construction was known in England . . . by 1600 it became the commonest construction technique in the Chesapeake Region, where it has been convincingly argued that . . .**
Cary Carson et al., "Impermanent Architecture in the Southern American Colonies," *Winterthur Portfolio*, 1981, vol. 16, nos. 2/3, pp. 135–196.

182 **In 1637, the Reverend John Lothrop, writing in the Scituate church records . . .**
Published in *The New England Historical and Genealogical Register*, vol. 10, p. 42.

So we would have it; our purpose is to build for the present such houses as, if need be, we may with little grief set afire and run away by the light.
Robert Cushman, quoted by Bradford in *Of Plymouth Plantation*, p. 363.

183 **It was agreed upon by the whole Court . . .**
Plymouth Colony Records (PCR), vol. 11, p. 4.

. . . architectural historian Dell Upton has made a very convincing argument to the contrary.
Personal communication.

In a 1586 letter to London, Ralph Lane, governor of the ill-fated Roanoke Colony . . .
Lane, in his "Discourse on the First Colony," describes a plot by the Roanokes to eliminate the English and comments: "In the dead time of night they would have beset my house, and put fire in the reeds, that the same was covered with . . ." David B. Quinn (Ed.), *The Roanoke Voyages, 1584–1590* (London: Hakluyt Society, 1955), vol. 1, p. 282. Our appreciation to Phillip W. Evans for drawing this reference to our attention.

185 **In 1969 Richard Candee published a lengthy, three-part article on Plymouth Colony architecture in *Old-Time New England.***
The article, "A Documentary History of Plymouth Colony Architecture, 1620–1700," was published in vol. 59, no. 3, 1969, pp. 59–71; no. 4, pp. 105–111, and vol. 60, no. 2, pp. 37–53, cited as Candee 1969a, b, and c, respectively. Construction methods, including vertical plank siding or plank framing, is discussed by Candee in 1969c.

186 **Second, the technique was used in England . . .**
Richard W. Brunskill, *Illustrated Handbook of Vernacular Architecture* (Faber, 1971), pp. 64–65.

187 **The plank frame house . . .**
Cummings, p. 91.

It can thus be argued . . .
Ibid., p. 90.

Although it is not impossible in theory to construct such a framed overhang using vertical planks . . .
Candee, 1969c, p. 38.

190 **Historian David Hackett Fischer . . .**
Albion's Seed: Four British Folkways in America (Oxford University Press, 1989), p. 64.

Folklorist Henry Glassie, however, has drawn attention to an important, little known fact . . .
Personal communication.

191 **Only one lean-to was in use as a kitchen, that of Captain Nathaniel Thomas (1675).**

Described as the "outer Rome Chimney leanto and Cellar," the lean-to contained the main fireplace equipment, cooking utensils, items in pewter and tin, as well as earthenware pails, milk vessels, and foodstuffs—beef, pork, and butter. The probate inventory is contained in an unpublished volume of Plymouth Colony wills and inventories, cited as *Plymouth W&I*, vol. 3, p. 137.

192 **. . . in one instance, that of the estate of John Dicksey . . . a case containing five gross of tobacco pipes . . . is listed . . .**
Plymouth W&I, vol. 3, pp. 106–108.

Probate Inventory of Will Wright, November 6, 1633
Ibid., vol. 1, pp. 21–22.

195 **It becomes clear from this listing of clothing that gives colors . . . that . . . the people of Plymouth had quite colorful wardrobes.**
Fischer discusses dress ways in the Massachusetts Bay Colony, and draws attention to the use of "sad" colors, which have persisted to this day in academic dress in New England's older universities. He also cautions against swinging from one myth to creating another concerning Puritan dress, from the grim, somber image to the constantly gaudy. The truth, as always, lies in between. This acknowledged, it must be remembered that Plymouth was not a Puritan colony, and that its population was more secular than separatist, although government tended to remain strongly influenced by separatism. The colony passed no sumptuary laws, as did the Bay Colony, where only those whose estates were valued at over £200 were permitted to wear costly clothes. So based on inventory data, it is our opinion that dress in Plymouth Colony was indeed more colorful than not. See Fischer, pp. 139–146.

196 **Robert Anderson makes a convincing argument for Wright having been a joiner . . .**
Robert Charles Anderson, *The Great Migration Begins: Immigrants to New England 1620–1633* (New England Historic Genealogical Society, 1995), p. 2075.

. . . furniture does not appear in provision lists and bills of lading for ships arriving from England.
Information from Carolyn Travers Freeman, Plimoth Plantation, July 1999, who has compiled a set of such lists.

197 ***In The Wrought Covenant*, Robert St. George provides a list of men capable of making furniture . . . in Plymouth Colony . . .**
The Wrought Covenant: Source Material for the Study of Craftsmen and Community in Southeastern New England, 1620–1700 (Brockton

Art Center/Fuller Memorial, 1979), pp. 70–102. This meticulous, thorough, and scholarly catalog of an exhibition entitled "Craftsmen and Community: The Seventeenth-Century Furniture of Southeastern New England," is an outstanding contribution to studies of material folk culture by Robert St. George, who both directed the exhibition and prepared the catalog.

198 **Robert St. George properly cautions . . .**
Ibid., pp. 22–23.

199 **In his *Reshaping of Everyday Life, 1790–1840,* Jack Larkin states . . .**
Harper & Row, 1988, p. 110.

Probate Inventory of John Rickard, July 4, 1678
Plymouth W&I, vol. 3, pt. 2, pp. 122–123.

203 **Seventeenth-century inventories from Essex, England . . .**
Francis W. Steer (Ed.), *Farm and Cottage Inventories of Mid-Essex, 1635–1749* (Phillimore, 1969).

204 **At this Court, Hester, the wife of John Rickard, Senir, of Plymouth, appeared . . .**
PCR 5: 87.

In his "Some Domestic Vessels of Southern Britain . . ."
P. Amis, "Some Domestic Vessels of Southern Britain: A Social and Technical Analysis," *Journal of Ceramic History* (George Street Press, 1968), no. 2.

205 **A modern editor was obliged to annotate the following lines . . .**
Ibid., p. 11.

Amis also quotes John Collop . . .
Ibid., p. 13.

206 **In 1688, Randle Holme published his monumental *Academy of Armory* . . .**
The Academy of Armory, or, A Storehouse of Armory and Blazon . . . (Chester: printed for the author, 1688; Scholar Press reprint, 1972).

A chamber pot or a Bed pot . . .
Randle Holme, *The Academy of Armory . . . Second Volume*, Isaac Herbert Jeayes (Ed.) (Roxburgh Club, 1905) p.2 (book 3, chap. 14).

207 **Probate Inventory of Judith Smith, Widdow, December 15, 1650**
Plymouth W&I, vol. 1, pp.110–111.

Chapter 6 Still Standing in the Ground

213 **Our knowledge of seventeenth-century ceramics . . . was rudimentary at best.**
It was not until Ivor Noël Hume published *A Guide to Artifacts of*

Colonial America (Vintage Books, 1969) that there was access to a comprehensive treatment of the subject of ceramics, as well as a wide range of other artifacts. A few similar works had been published prior to that date, but they were of relatively limited use. The most important of these was Edward B. Jelks's brief summary of the ceramics from Jamestown, but this reflects to a degree classificatory methods more typical of those used by prehistorians. Jelks is certainly not to be faulted for this; in fact, for the time, it was a sophisticated treatment of the subject. His summary, "Ceramics from Jamestown," was published as Appendix A in John L. Cotter, *Archaeological Excavations at Jamestown, Virginia*, Archaeological Research Series no. 4 (National Park Service, 1958), pp. 201–212.

214 **Harrington published his method . . .**
J. C. Harrington, "Dating Stem Fragments of Seventeenth and Eighteenth Century Clay Tobacco Pipes," Archaeological Society of Virginia, *Quarterly Bulletin*, 1954, vol. 9, no. 1, pp. 10–14.

216 **But for the vast majority of American English colonial sites, the method seems to work very well . . .**
Not everyone accepts the method uncritically, claiming that it is most accurate from the later seventeenth century through the mid-eighteenth. One reason given for its relative lack of precision in the earlier years is the possible presence of Dutch pipes. However, pipe production in Holland did not begin until about 1610, and was initiated by English pipe makers who had emigrated to Holland for a number of reasons. The earlier Dutch pipes are indistinguishable from those made in England, and begin to show a different and distinctive bowl shape around 1700. Pipe stem bore analysis of over 5,000 fragments from a Dutch fort on the Cape coast in South Africa conforms in every way with the documented dates of the outpost and fits comfortably with the ceramic evidence. Additionally, a large collection of Dutch pipes covering a date range from the earlier seventeenth century through the turn of the nineteenth show a virtual identity in bore diameter to those made in England during the same time. Finally, pipe stems from eighteen sites at Flowerdew Hundred, a Virginia plantation, provide dates that also fit the other artifactual evidence very well. The results of the Dutch pipe analysis have been published in "The Chronology of Oudepost I, Cape, as Inferred from an Analysis of Clay Pipes," by Carmel Schrire, James Deetz, David Lubinsky, Cedric Poggenpoel, *Journal of Archaeological Science* 1990, vol. 17, pp. 269–300. The pipe stems from Flowerdew Hundred sites are discussed in depth in James Deetz, *Flowerdew Hundred: The Archaeology of a Virginia Plantation, 1619–1864* (University Press of Virginia, 1993).

In 1926, Percival Lombard excavated . . .
Percival Hall Lombard, *The Aptucxet Trading Post: The First Trading Post of the Plymouth Colony* (Bourne Historical Society, Massachusetts, 1934).

William Bradford mentions that "they built a house there . . ."
William Bradford, *Of Plymouth Plantation 1620–1647*, Samuel Eliot Morison (Ed.) (Knopf, 1952), p.193.

The building was destroyed by a hurricane in 1635 . . .
Ibid., p. 279.

. . . a recent reanalysis of the material recovered by Lombard has been conducted by Craig Chartier . . .
Craig S. Chartier, "The Aptucxet Trading Post: Fact, Fiction, and a Study in 20th Century Myth Creation." Unpublished final report for the 1995 Summer Field School of the University of Ma m ssachusetts at Boston: director Barbara Luedtke; graduate student director Craig Chartier. April 2000, 2 vols.

217 **Strickland had a title search carried out . . .**
A. Rodman Hussey Jr., "Preliminary Report on Title to Land at Rocky Nook, Kingston, Plymouth County, Massachusetts owned by John Howland and Others," 1938. Manuscript on file, Plimoth Plantation.

Strickland published a brief account of his work . . .
Sidney T. Strickland, "Excavations at Site of Home of Pilgrim John Howland, Rocky Nook," in William Howland (Comp.), *The Howlands of America* (Pilgrim John Howland Society, 1939), pp. 26–30.

218 **Three years later, in 1940 and again in 1941, two sites were excavated in a thoroughly professional manner by Henry Hornblower . . .**
Hornblower's excavations of the R. M. and Winslow sites are discussed in Mary C. Beaudry and Douglas C. George, "Old Data, New Findings: 1940s Archeology at Plymouth Reexamined," *American Archeology* 1987, vol. 6, no. 1, pp. 20–30.

221 **[Isaac Allerton] was one of the busiest and most complicated men in early New England . . .**
Robert Charles Anderson, *The Great Migration Begins: Immigrants to New England 1620–1633* (New England Historic Genealogical Society, 1995), p. 38. Anderson's biographical profile of Allerton is on pp. 35–39.

Concerning Mr. Allerton's accounts.
Bradford, pp. 241–242.

As James Sherley, one of the London merchants, wrote to Bradford . . .
Ibid., p. 242.

222 **Yea, he screwed up his poor old father-in-law's account . . .**
Ibid.

231 **Concerning the frame of the house . . .**
Samuel Symonds's letter to John Winthrop the Younger, 1638. *Collections of the Massachusetts Historical Society* 1865, 4th ser., vol. 7, pp. 118–120.

233 **. . . M. W. Barley, in his pioneering study of English farmhouses and cottages . . .**
The English Farmhouse and Cottage (Routledge and Kegan Paul, 1961).

Standish must have been familiar with this type of structure from his Lancashire home near the Welsh border.
For a discussion of Myles Standish's house and its Welsh antecedents, including plans of structures, see Robert Blair St. George, *Conversing by Signs: Poetics of Implication in Colonial New England Culture* (University of North Carolina Press, 1998), pp. 100–103.

235 **As folklorist Robert St. George says . . .**
Ibid., p. 101.

237 **William Clark, the eldest son of Thomas Clark . . .**
Brief details concerning William Clark appear in the biographical profile of Thomas Clark in Anderson, p. 377. Details concerning his appointments as a deputy and selectman for Plymouth are to be found in *PCR* 5:144, 195.

Clark's house on the Eel River was "slightly fortified," according to the testimony of Keweenam, an Indian . . .
The account of the trial, including Keweenam's testimony, is to be found in *PCR* 5:204–206.

The primary sources are in conflict as to the details of what took place on that Sunday morning . . .
Apart from the official court record referred to above, contemporary accounts include: Nathaniel Saltonstall, "A New and Further Narrative of the State of New England" (London, 1676), reprinted in *Narratives of the Indian Wars 1675–1699*, Charles H. Lincoln (Ed.) (Scribner, 1913), p. 84; and William Hubbard's *A Narrative of the Troubles with the Indians in New England* (Boston, 1677), reprinted in *The History of the Indian Wars in New England*, Samuel G. Drake (Ed.) (Franklin, 1971 reprint), vol. 1, pp. 178–179, 253–254. The Benjamin Church reference is to be found in his *Diary of King*

Philip's War, 1675–1676, with an introduction by Alan and Mary Simpson (Little Compton Historical Society, R.I., 1975), p. 107.

239 **. . . Anderson . . . gives 1632 as the date of Alden's move from Plymouth to neighboring Duxbury.**
Anderson, p. 21.

In 1960 what was believed to be the site of Alden's first house . . . was excavated by Roland Wells Robbins . . .
Robbins published a report on his excavations as *Pilgrim John Alden's Progress: Archaeological Excavations in Duxbury* (The Pilgrim Society, 1969).

240 **The deed is dated February 2, 1638 . . .**
PCR 12:41.

When John Howland died in 1672, his property passed to his widow . . . and then to their son, Joseph.
John Howland's will is contained in an unpublished volume of Plymouth Colony wills and inventories, cited as *Plymouth Colony W&I*, vol. 3, pp. 49–50.

Joseph in turn devised the property . . .
Captain Joseph Howland died at Plymouth in January 1704. His will and inventory have been published in *The Mayflower Descendant* 1904, vol. 6, no. 2, pp. 86–91.

. . . recent work carried out by the University of Virginia Field School . . .
This took place in the summers of 1998 and 1999. The first field season was under the joint direction of Andrew P. Beahrs and Derek T. Wheeler, graduate students in the Department of Anthropology, assisted by Karin Goldstein, Curator of Original Collections at Plimoth Plantation, Plymouth, and the second field season was under the direction of Derek Wheeler.

242 **. . . Structure 3, which has until now been identified as the site of Joseph Howland's house.**
A report on the excavations has been published by James Deetz, "Excavations at the Joseph Howland Site (C5), Rocky Nook, Kingston, Massachusetts, 1959: A Preliminary Report," supplement to *The Howland Quarterly* 1960, vol. 24, nos. 2 and 3, pp. [1–12], and "The Howlands at Rocky Nook: An Archaeological and Historical Study," supplement to *The Howland Quarterly* 1960, vol. 24, no. 4, pp. 1–8.

According to a biography of Joseph Howland . . .
Published in *The Howland Quarterly* 1949, vol. 13, no. 4.

244 **Elizabeth [Tilley Howland] died at the home of her daughter, Lydia Browne, in Swansea in December 1687.**
Her will has been published in *The Mayflower Descendant* 1901, vol. 3, no. 1, pp. 54–57.

245 **Edward Winslow needs no introduction here . . .**
A short biographical profile of Winslow has been published in Anderson, pp. 2023–2026, and Beaudry and George also include some biographical information in their article on the R. M. and Winslow sites, which appeared in *American Archeology* 1987, vol. 6, no. 1, p. 23.

246 **The Winslow site was the only one . . . from which a large sample of animal bone was recovered and curated.**
Faunal evidence from seventeenth-century Plymouth sites is poorly represented; in many cases no effort was made to collect animal bone as a part of the site assemblage. Stanley Olsen's analysis was very preliminary and does not provide percentages. Nevertheless, we can say the following species were represented: blue runner (a fish), northern pike, Canada goose, mallard, white-winged scoter, turkey (one), cottontail rabbit, woodchuck, otter, white-tailed deer, and domestic pig, cow, and chicken. See Stanley J. Olsen and John Penman, "Faunal Remains in Association with Early Massachusetts Colonists," *Occasional Papers in Old Colony Studies*, December 1972, no. 3, pp. 22–23.

247 **Whaler's Tavern, Wellfleet, Massachusetts**
A detailed account of the excavations at Wellfleet has been published: Eric Ekholm and James Deetz, "Wellfleet Tavern," *Natural History*, August-September 1971, vol. 80, no. 7, pp. 48–56.

We hear from towns on the Cape . . .
Cited from Ekholm and Deetz, p. 53.

248 **It would be curious indeed to a countryman . . .**
Levi Whitman, *A Topographical Description of Wellfleet*, 1793. Ibid.

249 **Except a tract of oak and pines . . .**
Cited from Ekholm and Deetz, p. 49.

253 **The night being dark . . .**
Ibid.

254 **Cary Carson et al. in a seminal paper . . .**
Cary Carson, Norman F. Barka, William M. Kelso, Garry Wheeler Stone, and Dell Upton, "Impermanent Architecture in the Southern American Colonies," *Winterthur Portfolio*, Summer/Autumn 1981, vol. 16, nos. 2/3, pp. 135–196. Cited as "Impermanent Architecture."

Carson et al. suggest . . .
Ibid., pp. 136-138.

255 **It took off the boarded roof of a house . . .**
Bradford, p. 279.

Carson et al. show that there is evidence . . .
"Impermanent Architecture," p. 138.

256 **In the East Midlands and the Southeast . . .**
Leon E. Cranmer, *Cushnoc: The History and Archaeology of Plymouth Colony Traders on the Kennebec* (The Maine Archaeological Society, Fort Western Museum, and The Maine Historic Preservation Commission, 1990), p. 54.

257 **Having procured a patent . . .**
Bradford, p. 202.

But they [the traders from Plymouth Colony . . .] having made a small frame of a house ready . . .
Ibid., p. 259.

258 **I found Sir Ferdinando's house much like your barn . . .**
Cranmer, p. 57.

Carson et al. comment . . .
"Impermanent Architecture," p. 147.

260 **. . . with the first mention of porcelain not appearing until 1735 in a probate inventory.**
Marley R. Brown III, "Ceramics from Plymouth, 1621–1800: The Documentary Record," in *Ceramics in America*, I. Quimby (Ed.) (Winterthur Conference Report, 1972), p. 48. Brown's is the second of a two-part paper, the first being by James Deetz, "Ceramics from Plymouth, 1620–1835: The Archaeological Evidence," Ibid., pp. 15–40.

261 **There is the occasional reference to vessel forms . . .**
Brown III, p. 44.

262 **Jay Anderson, in his study of the foodways of the Stuart yeoman . . .**
Jay Anderson, "A Solid Sufficiency: An Ethnography of Yeoman Foodways in Stuart England," PhD dissertation, University of Pennsylvania, 1971.

263 **Comparative evidence from contemporary Essex confirms this.**
Francis W. Steer (Ed.), *Farm and Cottage Inventories of Mid-Essex, 1635–1749* (Phillimore, 1950).

266 **A complete pied-de-biche spoon was recovered; these were produced in England from around 1663 until 1700.**
Percy E. Raymond, "Latten Spoons from the Old Colony," Massachusetts Archaeological Society *Bulletin*, 1949, vol. 11, no. 1, p. 7. Raymond includes in his article an analysis of the material from the Josiah Winslow site, the R. M. site, and the John Howland site.

On the basis of Percival Raymond's typology . . .
Ibid., p. 7.

267 **The most spectacular spoon to be recovered from the Howland site . . .**
James Deetz, "Howland Spoon," *The Howland Quarterly*, January–April 1969, vol. 33, nos. 2 and 3. This two-page article is available as a reprint from the Howland House, 33 Sandwich Street, Plymouth, MA 02360, which also retails reproductions of the Howland spoon.

Chapter 7 The Time of Their Lives

273 **At Plimoth Plantation in 1997 . . .**
James W, Baker, *Plimoth Plantation: Fifty Years of Living History* (Plimoth Plantation, 1997), pp. 5–6.

274 **In 1967 he [Harry Hornblower] wrote . . .**
Ibid., p. 7.

The articles of incorporation explicitly state . . .
Ibid., p. 8.

275 **. . . similar in form to that described by Isaack de Rasieres in his classic account of Plymouth in 1627.**
Sydney V. James Jr. (Ed.), *Three Visitors to Early Plymouth: Letters About the Pilgrim Settlement in New England During Its First Seven Years* (Applewood Books, 1963), pp. 75–77.

280 **James Baker is correct in stating that constructions such as those that had emerged at the plantation in the 1960s are partly a function of the time in which they took place.**
"Haunted by the Pilgrims," pp. 353–354. In Anne Elizabeth Yentsch and Mary C. Beaudry (Eds.), *The Art and Mystery of Historical Archaeology: Essays in Honor of James Deetz* (CRC Press, 1992), pp. 343–358. Cited as "Haunted by the Pilgrims."

290 **What has been described as a remarkable cultural performance, a type of theater . . .**
The term "cultural performance" is used by Stephen Eddy Snow in his *Performing the Pilgrims: A Study of Ethnohistorical Role-Playing at Plimoth Plantation* (University Press of Mississippi, 1993). His work focuses on the transformation of this type of theater, and offers a groundbreaking analysis of the emergence of a new genre of cultural performance in the 1627 village of Plimoth Plantation.

. . . "The actors' perfect balance between earnestness and playfulness . . ."
James Carroll, "Pilgrim's Plymouth: Revisiting 1620." *New York Times*, November 4, 1984, Travel Section.

291 **Already, Baker suggests . . . "Pilgrims as 'real people' . . ."**
James Baker, "Haunted by the Pilgrims," p. 356.

Index